ULTRA-CAPACITORS IN POWER CONVERSION SYSTEMS

ULTRA-CAPACITORS IN POWER CONVERSION SYSTEMS

APPLICATIONS, ANALYSIS AND DESIGN FROM THEORY TO PRACTICE

Petar J. Grbović

Huawei Technologies Düsseldorf GmbH
Energy Competence Center Europe, Germany

IEEE PRESS

WILEY

This edition first published 2014

© 2014 John Wiley & Sons Ltd

Registered office

John Wiley & Sons Ltd, The Atrium, Southern Gate, Chichester, West Sussex, PO19 8SQ, United Kingdom

For details of our global editorial offices, for customer services and for information about how to apply for permission to reuse the copyright material in this book please see our website at www.wiley.com.

Library of Congress Cataloging-in-Publication Data

Grbović Petar J.
  Ultra-capacitors in power conversion systems : applications, analysis, and design from theory to practice / Petar J. Grbović.
    pages cm
  Includes bibliographical references and index.
  ISBN 978-1-118-35626-5 (hardback)
  1. Electric current converters–Equipment and supplies. 2. Supercapacitors. 3. Electric machinery–Equipment and supplies. I. Title.
  TK7872.C8G695 2014
  621.31′5–dc23

                                                                                        2013018944

A catalogue record for this book is available from the British Library.

ISBN: 9781118356265

Typeset in 10/12 Times by Laserwords Private Limited, Chennai, India.
Printed and bound in Malaysia by Vivar Printing Sdn Bhd

1 2014

Contents

Preface

What this Book is About

This book is about ultra-capacitors and their application in power conversion systems. It is particularly focused on the analysis, modeling, and design of ultra-capacitor modules and interface dc–dc power converters.

Power conversion systems and power electronics play a significant role in our everyday life. It would be difficult to imagine a power conversion application, such as industrial controlled electric drives, renewable sources, power generation and transmission devices, home appliances, mobile diesel electric gen-sets, earth moving machines and equipment, transportation, and so on, without power electronics and static power converters. In most of these applications, we are facing growing demands for a device that is able to store and re-store certain amounts of energy during a short period. Controlled electric drives may require energy storage to save energy during braking or provide energy in case of power supply interruption. Wind renewable "generators" may need energy storage to smooth power fluctuations caused by wind fluctuation. Power transmission devices such as static synchronous compensators (STATCOMs) need energy storage to support the power system with active power during faults and unstable operation. Mobile diesel electric gen-sets need energy storage to reduce fuel consumption and CO_2 pollution. There is a strong requirement for energy storage in transportation systems in order to improve the system's efficiency and reliability.

The energy storage device should be able to quickly store and re-store energy at very high power rates. The charge and discharge time should be a few seconds up to a few tens of seconds, while charging specific power is in the order of 5–10 kW/kg. Today, two energy storage technologies fit such requirements well: (i) flywheel energy storage and (ii) electrochemical double-layer capacitorss (EDLCs), best known as ultra-capacitors. In this book, ultra-capacitors are addressed alone.

What is Inside the Book

This book starts from a background of energy storage technologies and devices. Then, the detailed theory of ultra-capacitors follows. The fundamentals of power conversion systems and applications are also addressed. An important part of the book is the process of selection and design of ultra-capacitor modules. Finally, the book ends with a detailed analysis of interface dc–dc converters. In total, the book has five chapters.

The fundamentals of energy storage technologies and devices are given in the first chapter. All energy storage systems are classified into two categories: direct and indirect energy storage systems. Direct energy storage devices store electrical energy directly without conversion into another type of energy. Inductors and capacitors are direct energy storage devices. Particular devices with high energy density are super magnet energy storage devices (SMES) and ultra-capacitors. Indirect energy storage systems and devices convert electrical energy into another type of energy that is easier to handle and store. Typical systems are electromechanical energy storage systems, such as fly-wheel, hydro pumped, and compressed air energy storage systems. Electrochemical energy storage systems, such as electrochemical batteries and hydrogen fuel cells, are other well known energy storage systems.

The background theory of ultra-capacitors is presented in the second chapter. The ultra-capacitor model is given with particular attention to the application oriented model. The ultra-capacitor's energy and power are then defined and discussed. Different charging/discharging methods, such as voltage-resistance, current, and power control modes are analyzed. The ultra-capacitor's voltage and current characteristics are derived for different charge/discharge methods. Analysis and calculation of the ultra-capacitor's current stress and power losses under different conditions are discussed. An explanation is given of how ultra-capacitor losses depend on the charge/discharge frequency and how such losses are determined when the charge/discharge current frequency is in the range of mega-hertz (very low frequency) as well as in the range of a couple of hertz (low frequency). Some application examples, such as variable speed drives with braking and ride through capability, are given.

The fundamentals of power conversion are presented in the first part of the third chapter. Requirements for the use of a short-term energy storage device in power conversion systems are addressed and discussed. The structure of a typical power conversion system with ultra-capacitor energy storage is presented. The process of selection of an energy storage device for a particular application requirement is briefly described. Two main energy storage devices are compared: electrochemical batteries and ultra-capacitors. In the second part of the chapter we discuss different power conversion applications, such as controlled electric drives, renewable energy sources (wind, PV, and marine currents for example), autonomous diesel and natural liquid gas (NLG) gen-sets, STATCOM with short-term active power capability, UPS, and traction.

The selection of an ultra-capacitor module is intensively discussed in the fourth chapter. Design of an ultra-capacitor module is based on three main parameters, namely the module voltage, capacitance, and internal resistance. The module voltage is in fact a set of different operating voltages and the module rated voltage. The operating and rated voltages, the module capacitance and internal resistance are defined according to application requirements, such as energy storage capability, operation life span, efficiency, and so on. Ultra-capacitor losses and efficiency versus size and cost are discussed in the second part of the chapter. Some aspects of ultra-capacitor module design are presented. Series connection of elementary ultra-capacitor cells and voltage balancing issues are also discussed and the module's thermal design is considered too. The theoretical analysis is supported by several examples from some real power conversion applications.

Interface dc–dc converters are discussed in the fifth chapter. First, the background of bi-directional dc–dc power converters is given. The converters are classified in different categories, such as full power versus fractional power rated converters, isolated versus non-isolated converters, two-level versus multi-level and single-cell versus multi-cell interleaved converters. State-of-the-art topologies are compared according to the application's requirements. A detailed analysis of a multi-cell interleaved bi-directional dc–dc converter is given in the second part of the chapter where design guidelines are given too. The theoretical analysis is supported by a set of numerical examples from real applications, such as high power UPS and controlled electric drive applications.

Who Should Read this Book (and Why)

This book is mainly aimed at power electronics engineers and professionals who want to improve their knowledge and understanding of advanced ultra-capacitor energy storage devices and their application in power conversion, in the present as well as in the near future. The book could also be background material for graduate and PhD students who want to learn more about ultra-capacitors and power conversion application in general. The reader is expected to have basic knowledge in math, theory of electric circuits and systems, electromagnetics, and power electronics.

Acknowledgments

I started this story about ultra-capacitors some years ago, when I was with Schneider Electric, R&D of Schneider Toshiba Inverter (STI), in Pacy sur Eure, France. I would like to express my thanks to Dr. Philippe Baudesson and Dr. Fabrice Jadot for the support I received at that time when I first started thinking about the application of ultra-capacitors in controlled electric drives.

I would like to express my deep gratitude to Professor Philippe Delarue and Professor Philippe Le Moigne from Laboratoire d'Electrotechnique et d'Electronique de Puissance (L2EP), Lille, France, for all the creative and fruitful discussions we had and all his comments and suggestions.

I would like to express my sincere thanks to Peter Mitchell, publisher; Richard Davies, project editor; Laura Bell, assistant editor; Genna Manaog, senior production editor; Radhika Sivalingam, project manager; and Caroline McPherson, copy editor. It has been real pleasure to work with all of them.

Last but not least, I offer my deepest gratitude to my family, my wife Jelena, son Pavle, and mother Stojka, for their love and support and for their confidence in me.

Finally, let me express my deepest gratitude to God for His blessing.

Dr Petar J. Grbović
Ismaning, Germany

1

Energy Storage Technologies and Devices

1.1 Introduction

1.1.1 Energy

By definition, energy is that property of a body by virtue of which work can be done. Energy cannot be created nor destroyed; it can only be transformed from one form into another. Energy can exist in many forms, such as electromagnetic field, gravity, chemical energy, nuclear energy, and so on [1, 2]. One form of energy that we use in everyday life is so-called electrical energy. In this chapter we will discuss electrical energy storage technologies and devices.

1.1.2 Electrical Energy and its Role in Everyday Life

Electrical energy can be defined as the ability to do work by means of electric devices. Electrical energy has been used in segments of everyday life since end of 1800s, the age of Tesla and Edison. Today, electrical energy is the dominant form of energy. Approximately 60% of primary energy is converted into electrical energy and then used in diverse applications such as industry, transportation, lighting, home appliances, telecommunication, computing, entertainment, and so on.

Figure 1.1 shows a simplified block diagram of electrical energy production–transmission–consumption flow. Electrical energy is usually "generated" by electro-mechanical generators. The generators are driven by steam turbines, NLG (natural liquid gas) turbines, hydro turbines, wind-turbines, and internal combustion diesel engines. Additionally, electrical energy can be "produced" by static generators, such as photovoltaic panels and hydrogen fuel cells.

Transmission of electrical energy from the "production" to the "consumption" location is also convenient. The "production" point can be a centralized, dislocated power station

Ultra-Capacitors in Power Conversion Systems: Applications, Analysis and Design from Theory to Practice, First Edition. Petar J. Grbović.
© 2014 John Wiley & Sons, Ltd. Published 2014 by John Wiley & Sons, Ltd.

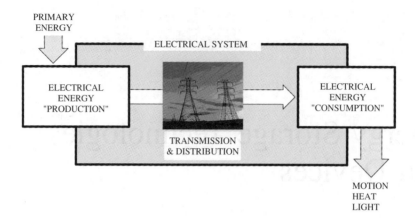

Figure 1.1   Electrical energy production–transmission–consumption process

far away from the "consumption" point, for example, a big city. The electrical energy is transferred and distributed via a high voltage transmission line and a medium/low voltage distribution network. Electrical energy is "consumed" by the end customer. In fact, electrical energy is converted to another form of energy, such as heat, light, chemical energy, linear or rotational movement, and so on.

In small-scale systems, such as diesel electric locomotives, hybrid tracks, earth moving equipment (excavators), and RTG (rubber tyred gantry) cranes, for example, electrical energy is produced by an on-board diesel–electric generator and transmitted to the on-board dislocate loads (electric motors).

It is very convenient to "produce" electrical energy from another form of energy such as mechanical or chemical. However, electrical energy cannot be easily stored. Hence, electrical energy must be "consumed" at same time that it is "produced." An imbalance between total production and consumption leads to problems of power quality, instability, and collapse of the electrical system. This makes it difficult to use electrical energy in systems with dynamic, fluctuating "production" and/or "consumption." An energy storage device is required to store or restore electrical energy and make the dynamic balance between "production" and "consumption." In this chapter we will briefly describe the major types of electrical energy storage technologies and devices.

1.1.3   Energy Storage

An energy storage device is a multi-physic device with the ability to store energy in different forms. Energy in electrical systems, so-called "electrical energy," can be stored directly or indirectly, depending on the means of the storage medium. Figure 1.2 illustrates direct and indirect energy storage processes and devices.

Devices that store the electrical energy, without conversion from electrical to another form, are called direct electrical energy storage devices. The energy storage medium is the electromagnetic field. The storage devices are electric capacitors and inductors. Devices that convert and store the electrical energy in another form of energy are called indirect

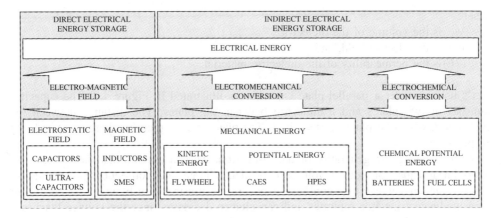

Figure 1.2   Energy storage technologies and devices

electrical energy storage devices. There are several forms of energy that can be converted from/to electrical energy. Some of the most appropriate forms of energy are mechanical and chemical. Mechanical energy can exist in two forms: energy of position, known best as potential energy and energy of motion, known as kinetic energy. The storage devices are flywheels, compressed air energy storage (CAES), and hydro pumped energy storage (HPES). Devices that use chemical energy as the form of energy to be stored are electrochemical batteries and fuel cells.

1.2   Direct Electrical Energy Storage Devices

Direct electrical energy storage devices store electrical energy directly, without conversion from electrical to another form of energy. Energy is stored in the form of an electromagnetic field in a defined Volume V. The field could be predominantly electrostatic (electric) field E and magnetic field H. Devices that predominantly use the electric field as the storage medium are known as electric capacitors. Devices that use predominantly magnetic fields to store energy are magnetic devices such as inductors. The energy storage capability of conventional capacitors and inductors is insufficient for most power conversion applications. To overcome this disadvantage, ultra-capacitor energy storage (UCES) [3–6] and super-conducting magnetic energy storage (SMES) [1, 2, 6] have been developed.

1.2.1   An Electric Capacitor as Energy Storage

Let's consider an electrostatic system composed of two metallic bodies and a dielectric in volume V between the bodies. Charging the bodies of the electrical system illustrated above, electrical energy is directly stored in the form of an electric field. Energy stored in such a system is

$$W_E = \frac{1}{2}\iiint_V \epsilon(E) E^2 dv, \tag{1.1}$$

where

    V   is the volume of the dielectric,

    E   is the electric field, and

    $\epsilon(E)$ is the permeability of the dielectric material.

Let's now consider a parallel-plate capacitor as illustrated in Figure 1.3. The capacitor consists of two plates and a dielectric between the plates. The distance between the plates is d, while the plates surface is A. The plates are charged by charge $+Q$ and $-Q$ respectively.

Without losing the generality of the discussion, we can assume that the capacitor is a nonlinear capacitor with a voltage dependent charge and consequently capacitance. Charge and capacitance of a nonlinear capacitor are illustrated in Figure 1.4. The charge of a nonlinear capacitor saturates and capacitance decreases once the voltage reaches a certain level. However, there are some examples when capacitance increases with the voltage applied. As an example, electrochemical ultra-capacitors will be discussed in the following section.

The energy of a nonlinear capacitor charged to voltage U_0 is

$$W_E = \int_0^{Q_0} u(q)dq = \int_0^{U_0} \left(C(u) + \frac{\partial C(u)}{\partial u} u \right) udu, \tag{1.2}$$

where

    $C(u)$ is the voltage dependent capacitance.

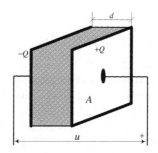

Figure 1.3   A parallel plate capacitor

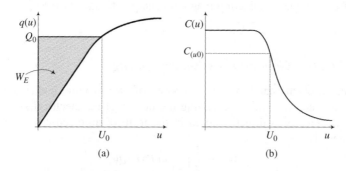

Figure 1.4   (a) The capacitor charge characteristic and (b) capacitance versus voltage

If the dielectric is linear, energy is computed from Equation 1.2 as

$$W_E = \frac{1}{2} Q_0 U_0 = \frac{1}{2} C_0 U_0^2,$$ (1.3)

where
C_0 is the capacitance of the capacitor and
Q_0 is the charge of the capacitor.

From Equations 1.3 and 1.2 we can see that the energy storage capability of an electric capacitor strongly depends on the capacitor voltage and the capacitance. It is obvious that the voltage rating and the capacitance have to be as high as possible in order to increase the capacitor's energy capability. The voltage rating and capacitance depend on capacitor technology. The most commonly used power capacitor technologies are electrolytic and polypropylene film capacitors.

1.2.1.1 Ultra-Capacitor Energy Storage

Ultra-capacitor energy storage (UCES) devices store electrical energy in the form of an electric field between two conducting plates [3, 4]. The energy storage system is composed of an ultra-capacitor and an interface power converter, as shown in Figure 1.5. The interface power converter is traditionally used for reasons of better controllability and flexibility of the UCES system. Depending on the application and nature of the electrical system, the interface power converter can be an ac–dc or a dc–dc bi-directional power converter. In some applications, the interface power converter is a cascade connected ac–dc and dc–dc power converter.

The ultra-capacitor is an electrochemical capacitor, which is composed of two porous conducting electrodes. To prevent direct contact between the electrodes, a separator is inserted between them. The electrodes are attached to metallic current collectors which are the capacitor terminals. A simplified cross-section is depicted in Figure 1.6. The electrodes and the separator are immersed in electrolyte. Each electrode forms a capacitor

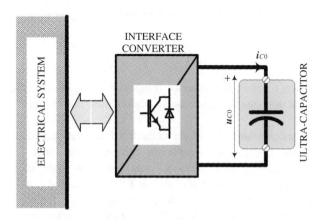

Figure 1.5 Ultra-capacitor energy storage system connected to an electrical system

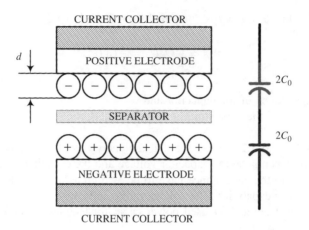

Figure 1.6 Simplified cross-section of an ultra-capacitor cell

with a layer of the electrolyte's ions. The capacitance depends on the size of the ions and the surface of the conducting electrode. Since the ion's diameter is in the range of angstroms, while the surface is in range of thousands of square meters, the capacitance is in the range of thousands of farads, which is significantly higher than that of standard electrolytic capacitors.

The ultra-capacitor is a nonlinear capacitor. The capacitance is voltage controlled capacitance defined as

$$C(u) = C_0 + k_C u, \tag{1.4}$$

where

C_0 is the initial capacitance, which represents electrostatic capacitance of the capacitor and

k_C is a coefficient, which represents the effects of the diffused layer of the ultra-capacitor.

Let the ultra-capacitor be charged on voltage U_0. The energy of the ultra-capacitor is

$$W_E = \frac{1}{2}\left(C_0 + \frac{4}{3}k_C U_0\right) U_0^2. \tag{1.5}$$

Since the capacitance C_0 is in the order of thousands of farads (F), the energy capability of the ultra-capacitor can be significantly higher than the capability of a "standard" electrolytic capacitor.

Ultra-capacitors are available as single cells from various manufacturers [5,6]. The typical capacitance of available ultra-capacitor cells is in the range 100–6000 F, while the voltage rating is 2.5–2.8 V. Figures 1.7 and 1.8 show some of the commercially available ultra-capacitor cells from manufacturer LS Mtron [5].

Ultra-capacitors are used as short-term energy storage, mainly for applications requiring high power rather than high energy. A detailed discussion on possible application fields is given in Chapter 3 of the book.

Figure 1.7 LS Mtron ultra-capacitors. Copyright LS Mtron, with permission

1.2.1.2 Ultra-Capacitors versus Electrolytic and Film Capacitors

Performances of electrostatic, electrolytic, and ultra-capacitors are summarized and compared in Table 1.1.

Electrostatic capacitors have a high voltage rating, in the range of a hundred volts up to a thousand volts or more. Specific capacitance is below $1000\,\mu F/dm^3$, while energy density is in range of 270–$350\,J/dm^3$.

(a)   (b)

Figure 1.8   LS Mtron ultra-capacitor module. Copyright LS Mtron, with permission

Table 1.1   Properties of high power capacitors and ultra-capacitors

	Voltage rating (V)	Capacitance	Energy
Electrostatic film capacitors	880	$700–900\,\mu F/dm^3$	$270–350\,J/dm^3$
Electrolytic capacitors	450	$5000–7500\,\mu F/dm^3$	$500–750\,J/dm^3$
Ultra-capacitors	2.8	$5000–7500\,F/dm^3$	$19–30\,kJ/dm^3$

In contrast to this, electrolytic capacitors have a lower voltage rating, usually up to 550 V, which is roughly half of that of the electrostatic capacitors. The specific capacitance of electrolytic capacitors is in the range $5000–7500\,\mu F/dm^3$. The energy density is in range of $500–750\,J/dm^3$, which is double that of electrostatic capacitors.

As can be seen from Table 1.1, the parameters of ultra-capacitors are different from the parameters of electrostatic and electrolytic capacitors in order of magnitude. The voltage rating is in the range 2.5–2.8 V, which is more than 2 orders of magnitude lower than that of electrolytic capacitors. However, the capacitance density is in the range of 5000–7500 F/dm^3, which is 6 orders of magnitude more than that of electrolytic capacitors. Therefore, the energy density is 25–60 times higher than that of electrolytic capacitors.

1.2.2   An Inductor as Energy Storage

In the previous section we saw how the electric field can be used as a medium to store energy. In a similar way, we can use the magnetic field to store energy. In this section, we will briefly present a magnetic device, the so-called an inductor, as an energy storage device.

Let's consider a volume V. Let the flux density and strength of magnetic field be B and H. In this instance we do not consider the source of the magnetic field. It could be pre-magnetized material, a magnetic field in the vicinity of a wire carrying a current, or a combination of the two. Energy localized in the volume V is

$$W_M = \frac{1}{2}\iiint_V \vec{H}\,\vec{B}\,dv.\qquad(1.6)$$

Let's now consider an inductor, such as that shown in Figure 1.9a. The inductor consists of a ring-core and a winding with N turns wound around the core. The winding is carrying a current I_0. The magnetic field flux density and magnetic field in the core are B and H. A general $B–H$ characteristic of a core with nonlinear characteristics is depicted in Figure 1.9b.

Energy localized in the core can be computed from the total flux and the inductor current,

$$W_E = \int_0^{\Psi_0} i(\psi)\,d\psi,\qquad(1.7)$$

where

$i(\psi)$ is the inductor current versus the total flux ψ (Figure 1.10).

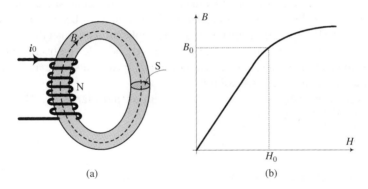

(a)   (b)

Figure 1.9   (a) A ring-core inductor and (b) $H–B$ characteristic of the inductor core

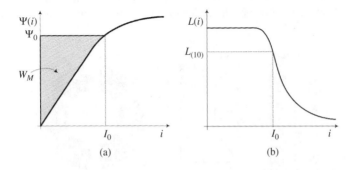

(a)   (b)

Figure 1.10   (a) Total flux versus the current and (b) inductance versus the current

The flux is defined by

$$\psi = N \iint_S \vec{B} \, d\vec{s},$$ (1.8)

where
  S is the core cross-section and
  N is the number of turns.

The energy of a nonlinear inductor carrying (being charged with) current I_0 is

$$W_E = \int_0^{I_0} \left(L(i) + \frac{\partial L(i)}{\partial i} i \right) i \, di,$$ (1.9)

where
  $L(i)$ is the current dependent inductance.

1.2.2.1 Super-Conducting Magnetic Energy Storage

An SMES device stores energy in the form of magnetic field that is created by a current in a super-conducting coil that is cryogenically cooled [1, 2, 10]. The SMES is composed of a super-conducting coil and a bi-directional interface converter, as illustrated in Figure 1.11. The interface power converter is traditionally used for reasons of better controllability and flexibility of the SMES system. Depending on the application and nature of the electrical system, the interface power converter can be an ac–dc or a dc–dc bi-directional power converter. In some applications, cascade connection of a voltage source ac–dc and dc–dc converter is used for the sake of flexibility.

A super conduction coil is a linear inductor with an inductance L_0. The energy of a SMES charged with the current I_0 is

$$W_{SMES} = \frac{1}{2} L_0 I_0^2.$$ (1.10)

From Equation 1.10, it is obvious that high energy requires large inductance and high current. The resistance of a super-conducting magnet is virtually zero. Because of this property of the super-conducting magnet, an inductance in the order of tens of henry can be easily achieved, while the current I_0 can be in the order of a thousand amperes. Therefore, an energy capability in the order of tens of megajoules can be achieved.

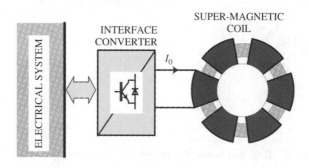

Figure 1.11 Super-conducting magnet energy storage connected to an electric system

The SMES is charged from the electric grid via an interface ac–dc or dc–dc power converter. Once the current I_0 is established in the SMES coil, the coil can be virtually disconnected from the grid. The current I_0 circulates via the SMES coil and the output terminal of the interface power converter. When the energy is required, the SMES coil current, and therefore the energy, is transferred to the grid via the interface power converter.

SMES are used in high power short-term applications. Their power level is in range of hundreds of megawatts while their charge/discharge time is less than a second. The main field of application is large-scale utility power quality restorers.

1.3 Indirect Electrical Energy Storage Technologies and Devices

Indirect electric energy storage devices are devices that use the energy conversion process to store and restore electrical energy. The energy storage device consists of an energy converter and an energy storage medium, as illustrated in Figure 1.12. Electrical energy is converted to another type of energy, such as mechanical or chemical energy. Then, the converted energy is stored using a proper storage medium. The energy conversion is performed via an energy converter, such as an electric motor/generator or an electrochemical reactor.

Mechanical energy can be stored as kinetic and potential energy. Energy storage that uses kinetic energy as storage medium is known as flywheel energy storage (FES). Energy storage devices that use potential energy as a storage medium can be divided into two groups: (i) hydro pumped energy storage (HPES) and (ii) compressed air energy storage (CAES). Energy storage devices that use chemical potential energy to store electrical energy are: (i) electrochemical batteries and (ii) hydrogen fuel cells.

1.3.1 Mechanical Energy Storage

The mechanical energy of a body can be defined by the following equation,

$$W_{MC} = \int_L \vec{F} \, d\vec{l} \, .$$ (1.11)

where
    F is the mechanical force that acts on the body and
    l is the linear distance.

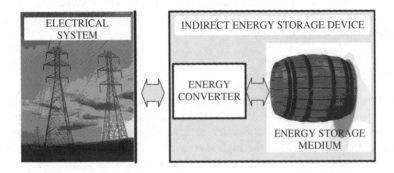

Figure 1.12   Illustration of an indirect electrical storage system

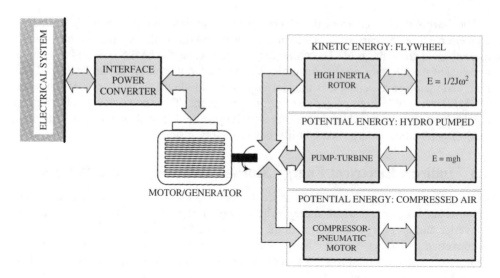

Figure 1.13   Indirect electromechanical energy storage

The force can be, for example: (i) gravity, (ii) inertia, and (iii) elastic force.

1. $F = mg$, where m is the mass of the body and $g = 9.81$ is the gravity acceleration.
2. $F = ma$, where a is the acceleration of the body in movement.
3. $F = k_c y$, where k_c is the coefficient of elasticity and y is deformation.

A structural block diagram of a mechanical energy storage system is depicted in Figure 1.13. Electrical energy is converted to mechanical energy via an electro-mechanical converter, such as a three-phase motor/generator. The mechanical energy can be stored directly as kinetic energy of a rotating mass or it can be converted and stored as the potential energy of elevated water or compressed air. The stored energy can be realized in the opposite way: kinetic or potential energy is converted to mechanical, which is further converted to electrical energy via a generator and fed back to the electric grid. For the sake of system flexibility and efficiency, the motor/generator is connected to the electrical system via an interface power converter. The converter is controlled to match the variable frequency and voltage of the motor/generator to constant frequency and voltage of the electrical system.

1.3.1.1   Flywheel Energy Storage

FES is a device that uses the kinetic energy of a rotating body as the storage medium [1, 2, 7]. As illustrated in Figure 1.14, basic FES consists of a high inertia rotor and a bi-directional electromechanical converter such as a three-phase motor/generator, which is attached to the same shaft as the high inertia rotor. The motor/generator is connected to an electrical system via an interface power converter.

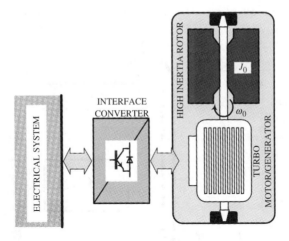

Figure 1.14 Flywheel energy storage connected to an electrical system

The kinetic energy of the flywheel is

$$W_{FW} = \frac{1}{2} J_0 \omega_0^2 \tag{1.12}$$

where

J_0 is the moment of inertia of entire rotating system that includes the
motor/generator rotor and the flywheel rotor and
ω_0 is the angular velocity of the flywheel.

The energy is being stored in the flywheel when the rotor is accelerating, and the motor/generator operates as a motor. The energy is restored when the flywheel is decelerating and the motor/generator operates as a generator. As we can see from Equation 1.12, energy depends strongly on the flywheel speed. Therefore, the voltage and frequency of the motor/generator varies significantly with the flywheel energy. It makes it both difficult and inefficient to connect the motor/generator directly to the electrical subsystem. In practice, the motor/generator is connected to the electrical subsystem via a bi-directional power converter that matches the generator voltage/frequency (which corresponds to the flywheel velocity) to the constant voltage and frequency of the electrical subsystem.

1.3.1.2 Hydro Pumped Energy Storage

HPES store energy using the potential energy of water [2]. Figure 1.15 shows a simplified block diagram of an HPES. The HPES is composed of a motor/generator, a pump/turbine, an interconnection pipe, and large upper and lower reservoirs. When electrical energy is being stored, the motor converts electrical energy into mechanical energy and runs the pump that pumps water from the lower reservoir to the upper reservoir. When the energy is being restored, the water flows from the upper reservoir to the lower reservoir via a

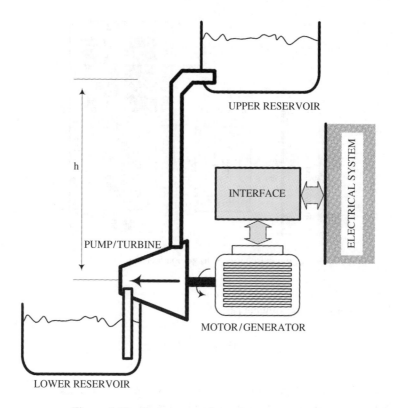

Figure 1.15   Hydro pumped energy storage system

pump that now works as a high-pressure turbine. The turbine runs the motor/generator that works as a generator and converts the mechanical energy into electrical energy. The motor/generator is directly connected to the electrical system.

The stored energy of an HPES system can be computed as

$$W_{HPES} = V_0 \rho g H,\tag{1.13}$$

where

V_0 (m^3)   is the volume of the upper reservoir,
ρ (kg/m^3) is the water density,
g (m/s^2)   is the gravity acceleration ($\cong$ 9.81), and
H (m)   is the vertical distance between the upper reservoir and the pump/turbine, the so-called hydraulic head.

HPES are used in large-scale systems that require large energy capacity and high power, such as power systems.

1.3.1.3   Compressed Air Energy Storage

CAES store energy in the form of compressed air [8]. A CAES is composed of a hermetic reservoir, a compressor, a turbine, and a motor/generator, as illustrated in Figure 1.16.

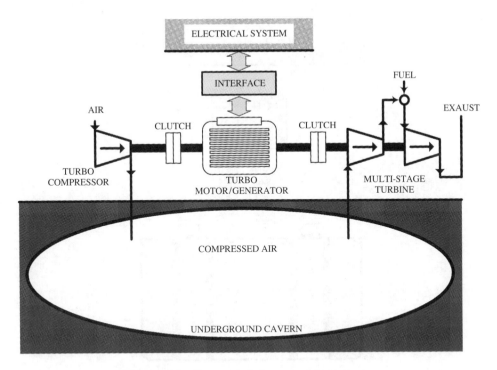

Figure 1.16 Compressed air energy storage system

The hermetic reservoir is usually an underground cavern. When the energy is being stored, the motor converts electrical energy into mechanical energy and runs the compressor that compresses air into the reservoir. When it is required, the energy can be realized by decompressing the air from the reservoir via a multi-stage gas turbine. The turbine drives the generator and converts the mechanical energy into the electrical energy that is fed back into the electrical system via an interface power converter.

The energy storage capacity depends on the deposit volume and maximum storage pressure of the compressed air. CAES are used in large-scale applications that require large energy capacity and high power, such as power systems and renewable sources.

1.3.2 Chemical Energy Storage

Chemical energy storage devices belong to the group of indirect electrical energy storage devices. Electrical energy is converted into chemical potential energy, which is further stored in a proper way. Two concepts of electrochemical energy are most often used: (i) electrochemical batteries [9] and (ii) hydrogen fuel cells [7].

1.3.2.1 Electrochemical Batteries

Battery Energy Storages (BESs), best known as electrochemical batteries, are the oldest and most established technology for storing electrical energy. Batteries are electrochemical

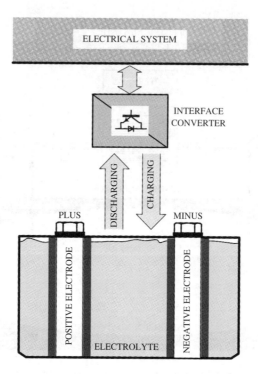

Figure 1.17 Simplified layout of an electrochemical battery cell

devices that convert electrical energy into potential chemical energy and store it during charging. When required, the stored chemical energy is realized and converted into electrical energy. An electrochemical battery as an energy storage device that is composed of one or more elementary cells connected into one unit. A battery cell consists of two electrodes, positive and negative, which are immersed in electrolyte. When the battery is being charged, an external voltage/current source is applied across the electrodes. A flow of ions is formed between the battery electrodes via the electrolyte and electrode material is transferred from one electrode to another. When required, an external load is applied between the battery electrodes. A flow of ions in the opposite direction is formed and the electrode material is transferred back from the second electrode to the first (Figure 1.17).

Depending on the electrode material and electrolyte, we can distinguish different types of electrochemical batteries. Characteristics of major state-of-the-art batteries are summarized in Table 1.2.

Lead–Acid Batteries

Lead–acid batteries are oldest and most mature battery technology. The lead-acid battery consists of a lead (Pb) negative electrode, a lead dioxide (PbO_2) positive electrode, and a separator that electrically separates the electrodes. The electrodes and separator are flooded in dilute sulfuric acid (H_2SO_4) acting as the electrolyte. Lead–acid batteries are basically

Table 1.2 Summary of state-of-the-art electrochemical batteries

	Energy density (W h/kg)	Power density (W/kg)	Life time (cycles)	Operating temperature (°C)
Lead–acid	20–35	25	200–2000	15–25
Lithium–ion	100–200	360	500–2000	−40 to 60
Lithium–polymer	200	250–1000	>1200	−40 to 60
Nickel–cadmium	40–60	140–180	500–2000	−20 to 60
Nickel–metal hydride	60–80	220	<3000	−20 to 60
Sodium–sulfur	120	120	2000	

low cost batteries that are traditionally used as car starters and UPS (uninterruptible power supply) back-up energy storage.

Lithium–Ion Batteries

Lithium–ion and lithium polymer batteries are composed of a lithium metal oxide ($LiCoO_2$, $LiMO_2$, etc.), a negative electrode, and a graphitic carbon positive electrode. The battery electrolyte consists of lithium salts dissolved in organic carbonates. The open circuit voltage is 4 V. Lithium–ion and lithium polymer batteries have been recently used in portable applications with a trend toward finding broad applications in the automotive industry.

Nickel–Cadmium Batteries

Nickel–cadmium batteries consist of nickel hydroxyl-oxide as the positive electrode and metallic cadmium as the negative electrode. The electrolyte is a concentrated solution of potassium hydroxide containing lithium hydroxide. The open circuit cell voltage is 1.3 V. The advantages of nickel–cadmium batteries are their high specific power and energy, long cycle life, rapid charge capability, wide operating temperature range, and good long-term storage capability. Major disadvantages are high initial cost, low cell voltage, and the environmental hazard of cadmium.

Nickel–Metal Hydride Batteries

Nickel–metal hydride batteries have similar characteristics to nickel–cadmium batteries. The major difference is the negative electrode. Instead of cadmium, the negative electrode is made of hydrogen absorbed in a metal hydride. When the battery is discharged, the metal hydride in the negative electrode is oxidized to form a metal alloy. The nickel hydroxyl-oxide in the positive electrode is reduced to nickel-hydroxide. When the battery is charged, the process is reversed. The open circuit cell voltage is 1.2 V.

Nickel–metal hydride batteries have similar advantages as nickel–cadmium batteries. In addition, these batteries are environmentally friendly since no toxic material such as cadmium is used. The major disadvantages are high initial cost and memory effect.

Sodium–Sulfur Batteries

Sodium–sulfur (NaS) batteries consist of a liquid sulfur positive electrode and a liquid sodium negative electrode. The electrodes are separated by a solid beta-alumina ceramic electrolyte. The battery operates at a relatively high temperature of about 300 °C. NaS batteries are used in transportation applications, but they can also be used in power smoothing applications.

1.3.2.2 Flow Electrochemical Batteries

Flow electrochemical batteries operate in a similar way as previously described for flooded batteries. The difference is that chemical energy is stored in two different electrolytes. A simplified block diagram of a flow battery is depicted in Figure 1.18. The flow battery

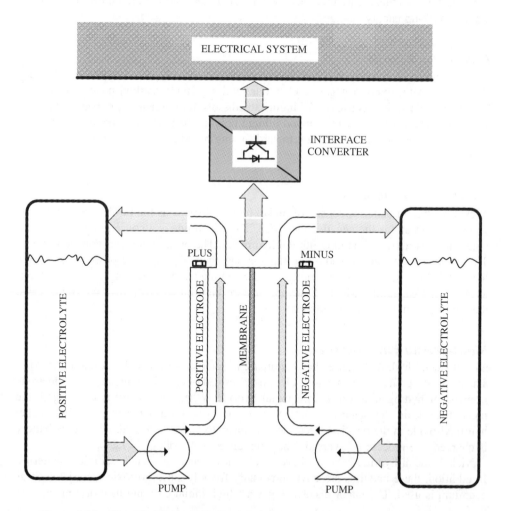

Figure 1.18 Simplified layout of a flow electrochemical battery energy storage system

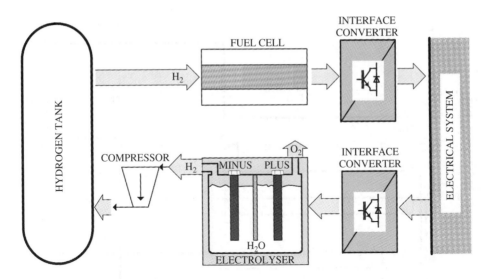

Figure 1.19   Simplified layout of a hydrogen–electric energy storage system

consists of a reactor, two tanks with positive and negative electrolytes, and two pumps. The reactor consists of two electrodes separated by an ion-selective membrane. The electrolytes are pumped through the reactor.

1.3.2.3   Hydrogen Energy Storage

Hydrogen and particularly hydrogen–electric energy storage systems have received attention recently. The energy storage process involves three steps: (i) conversion from electrical energy to hydrogen, (ii) storage of hydrogen as a gas or liquid, and (iii) conversion from hydrogen to electrical energy. A simplified block diagram of a hydrogen–electric energy storage system is depicted in Figure 1.19. The system consists of an electrolyzer, a compressor, a hydrogen tank, and a fuel cell. The electrolyzer converts electrical energy into chemical energy by converting water into hydrogen and oxygen. The compressor compresses the hydrogen and stores it in a tank. When needed, the hydrogen stored in the tank is realized through a fuel cell that produces electrical energy and water. The electrolyzer and fuel cell are connected to the electrical system via interface power converters.

1.4   Applications and Comparison

In this chapter, we have briefly described major electrical energy storage technologies and devices. Each technology and device has a specific area of application. The application area depends on requirements such as conversion power, energy storage capability, response time, and so on. Figure 1.20 shows a comparison of power capability and charge/discharge time of different energy storage technologies.

Based on the power and energy capability illustrated in Figure 1.20, energy storage technologies and application areas can be categorized and summarized as in Table 1.3.

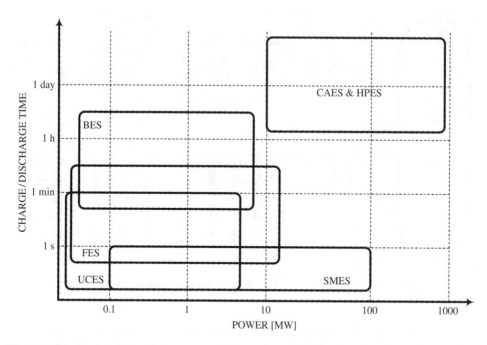

Figure 1.20 Comparison of power capability and storage time of different energy storage technologies

Table 1.3 Comparison and summary of applications and energy storage technologies

Very short term	Short term	Medium term	Long term
<1 s	Seconds to minutes	Minutes to hours	Hours to days
Dynamic voltage control	Ride through and power bridging	Power quality	Large scale utility applications
Fault management and grid stability control	Traction regenerative breaking and peak power shaving	Medium term load shifting	
Ride through and power bridging	VSD regenerative breaking and peak power shaving	Long run traction	
	Electric grid frequency control		
	Energy storage technology		
SMES	SMES	FES	HPES
UCES	UCES	BES	CAES
	FES		
	BES		

Note: VSD, variable speed drive

References

1. Ribeiro, P. F., B. K. Johnson, M. L. Crow, A. Arsoy, and Y. Liu, (2001) Energy storage systems for advanced power applications, *IEEE Proceedings*, **89**, 12, 1744–1756.
2. Huggins, R. A., (2010) *Energy Storage*, New York, Springer.
3. Conway, B. E., (1999) *Electrochemical Supercapacitors, Scientific Fundamentals and Technological Applications*, New York, Springer.
4. Halper, M.S. and J.C. Ellenbogen (2006) Supercapacitor: A Brief Overview, http://www.mitre.org (accessed 19 April 2013).
5. LS Mtron Ultra-Capacitors http://ultracapacitor.co.kr (accessed 19 April 2013).
6. Maxwell Technologies Ultra-Capacitors http://www.maxwell.com/ultracapacitors (accessed 19 April 2013).
7. Ehasani, M., Y. Gao, S. E. Gay, and A. Emadi,(2004) *Modern Electric, Hybrid, and Fuel Cell Vehicles: Fundamentals, Theory, and Design*, Boca Raton, FL, CRC Press.
8. Lemofuet-Gatsi, S. (2006) Investigation and optimization of hybrid electricity storage systems based on compressed air and supercapacitors. PhD dissertation. EPFL, Lausanne.
9. Linden, D. and T. B. Reddy, (2002) *Handbook of Batteries*, New York, McGraw-Hill.
10. Hasan Ali, M., B. Wu, and R.A. Dougal, (2010) An overview of SMES applications in power and energy systems, *IEEE Transaction on Sustainable Energy*, **1**, 1, 38–47.

2

Ultra-Capacitor Energy Storage Devices

2.1 Background of Ultra-Capacitors

An electric capacitor is a passive dynamic one-terminal electric device. In this context, dynamic means the device terminal voltage to current ratio is not constant and linear. The voltage and current are linked via a differential equation, which is generally a non-linear equation. As such, the electric capacitor has the capability to store energy as an electric charge, more precisely as an electric field between the capacitor plates. There are three different types of capacitors, namely electrostatic, electrolytic, and electrochemical capacitors. In this book, the electrochemical capacitors, so-called ultra-capacitors, are addressed only.

Ultra-capacitors are different from other types of capacitor mainly because their capacitance density (F/dm^3) and energy density (kJ/dm^3) are several orders of magnitude larger than that of electrolytic capacitors. In comparison to electrochemical batteries, the specific energy and energy density are lower, while the specific power and power density are larger, than that of the conventional batteries.

Cycling capability is also significantly better compared to batteries. Table 2.1 compares the most important properties of the ultra-capacitor versus batteries and other types of capacitor.

2.1.1 Overview of Ultra-Capacitor Technologies

Figure 2.1 shows a taxonomy of the existing types of ultra-capacitors. Here we have to highlight that the ultra-capacitor term is related to a capacitive energy storage device in general, not just so-called Electric Double Layer Capacitors (EDLCs). In this sense, the whole family of ultra-capacitors can be divided into three groups: EDLCs [1–5, 29, 30], pseudo-capacitors [1], and high voltage ceramic capacitors [6].

Ultra-Capacitors in Power Conversion Systems: Applications, Analysis and Design from Theory to Practice,
First Edition. Petar J. Grbović.
© 2014 John Wiley & Sons, Ltd. Published 2014 by John Wiley & Sons, Ltd.

Table 2.1 Properties of existing energy storage devices

	Capacitors	Ultra-capacitors	Electrochemical batteries
Energy density (W h/kg)	~ 0.1	$1-10^a$	~ 100
Peak power density (kW/kg)	10^4	$2-20$	$0.1-0.5$
Number of cycles	10^{10}	10^6	$\sim 10^3$
Life time (years)	~ 10	~ 15	~ 5

[a]Based on the technology trends and market needs, the specific energy and energy density will in near future be increased by a factor of 10 or more.

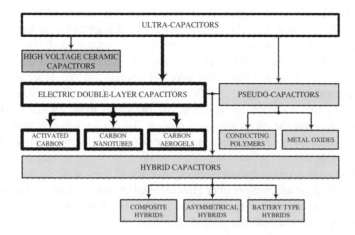

Figure 2.1 Taxonomy of the ultra-capacitors in general terminology

The EDLC is an electrochemical capacitor that is composed of two porous conducting electrodes immersed in an electrolyte and separated by a separator. Each electrode forms a capacitor with a layer of the electrolyte's ions. The energy storage process is a pure electrical process with no electrochemical reaction. In further discussion in this book, we will use the term the ultra-capacitor for EDLC only.

The pseudo-capacitor is an energy storage device that relies on an electron charge transfer reaction at the electrode–electrolyte surface to store energy. Basically, the pseudo-capacitor is very similar to rechargeable electrochemical batteries. The pseudo-capacitors are better than the EDLCs in terms of specific energy, but worse in terms of specific power.

A combination of an EDLC and a pseudo-capacitor is a so-called hybrid capacitor. Combining these two energy storage technologies into one device, it is possible to achieve moderate specific energy and power.

High voltage capacitors for energy storage have been well known for years. However, because of their low specific energy they have not been used very often in demanding energy storage applications. The multi-layer ceramic high voltage capacitor was proposed in 2006 [6] to replace the existing electrochemical batteries and EDLCs.

2.2 Electric Double-Layer Capacitors – EDLC

2.2.1 A Short History of the EDLC

The double-layer capacitor effect was discovered and described by Helmholtz in 1879 [1]. Almost a century later, a first ultra-capacitor was patented by the Standard Oil Company in 1966. A decade later, in 1978, NEC developed and commercialized this device. The first high power ultra-capacitor was developed for military applications by the Pinnacle Research Institute in 1982. Ten years later, in 1992, the Maxwell Laboratory started development of DoE (design of experiment) ultra-capacitors for hybrid electric vehicles. These days, ultra-capacitors are commercially available as cells and fully integrated modules from a number of manufacturers [7–9].

Today, ultra-capacitors are composed of two electrodes separated by a porous membrane, the so-called separator. The separator and electrodes are impregnated with a solvent electrolyte. The electrodes are made of a porous material such as activated carbon or carbon nano-tubes. The typical specific surface area of the electrode is about $2000 \, m^2/g$. Such a large surface area and very thin layer of charges, in the order of nanometers, gives specific capacitance of up to $250 \, F/g$. The rated voltage of the ultra-capacitor cell is determined by the decomposition voltage of the electrolyte. Typical cell voltage is $1-2.8 \, V$, depending on the electrolyte technology. To obtain a higher working voltage, which is determined by the application, a number of cells must be series-connected into one capacitor module.

Ultra-capacitors as energy storage devices have found very wide application in power conversion due to their advantages over conventional capacitors and electrochemical batteries: high specific energy and specific power, high efficiency, high cycling capability, wide range of operating temperatures, and long life.

2.2.2 The Ultra-Capacitor's Structure

In order to increase the capacitance of an ultra-capacitor, it is obvious that is necessary to maximize the contact surface area. To achieve this without increasing the capacitor volume, one must use a special material for the electrode. This material must have a porous structure and consequently a very high specific surface. The most frequently used material is activated carbon or carbon nano-tubes. In both cases, the specific surface may be as high as $1000-3000 \, m^2/g$. The simplified structure of an ultra-capacitor cell is depicted in Figure 2.2.

The elementary capacitor cell consists of a positive and a negative current collector, a positive and negative porous electrode made of activated carbon. The electrodes are attached to the current collectors and immersed in the electrolyte. Between the electrodes there is a separator. The separator is made of a material that is transparent to ions but is an insulator for direct contact between the porous electrodes.

2.2.3 The Ultra-Capacitor's Physical Model

Since the first development of double-layer capacitors, there have been several iterations and models of the basic structure. The very first work on double-layer capacitors was

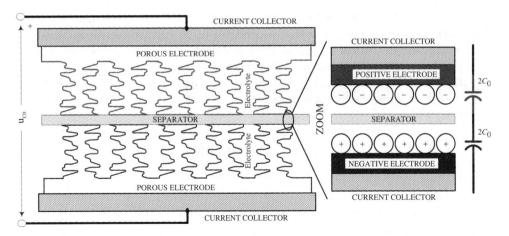

Figure 2.2 Construction of the electrochemical double layer ultra-capacitor with porous electrodes (activated carbon)

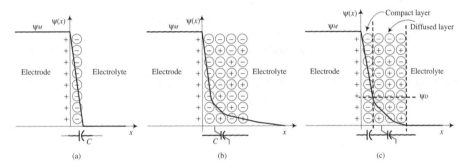

Figure 2.3 (a) Helmholtz's model of a EDLC (1857), (b) Gouy and Chapman's model (1910 and 1913), and (c) Stern's model (1924)

carried out by Helmholtz in 1853. He proposed that the layer in an electrolyte is a single layer of electrolyte molecules attached to the solid electrode, Figure 2.3a.

The specific capacitance of such a structure is

$$c' = \frac{\epsilon}{d}, \tag{2.1}$$

where

ϵ is the solvent electrolyte permittivity, and

d is the thickness of the layer, which equals the molecule diameter.

The specific capacitance computed by Equation 2.1 is overestimated compared to the experimentally obtained value. For an aqueous electrolyte with $\epsilon_R = 78$ and $d = 0.2$ nm, Equation 2.1 gives $340\,\mu\text{F/cm}^2$, which is much greater than the measured value $10–30\,\mu\text{F/cm}^2$.

Also, the model (2.1) does not take into account the fact that the capacitance is voltage dependent. In order to describe voltage dependence of the capacitance, Gouy introduced a theory of random thermal motion in 1910, and considered a space distribution of the charge in the electrolyte in proximity of the boundary between the electrolyte and electrode (Figure 2.3b). A few years later, Chapman defined the charge distribution in the electrolyte as a function of linear distance and properties of the electrolyte. The specific capacitance is estimated as

$$c' = z \sqrt{\frac{2q^2 n_0 \epsilon}{kT}} ch \left(\frac{z \Psi_M q}{2kT} \right),$$
(2.2)

where
   q is the elementary charge,
   n_0 is the concentration of anions and cations,
   z is the valence electrolyte ions,
   ϵ is the electrolyte permittivity,
   k is the Boltzmann constant, and
   T is the temperature.

In the model (Figure 2.2), the charge is considered a point charge (charge density is a Dirac function of space). Thus, the specific capacitance is overevaluated.

In 1924, Stern proposed a new model that improved Gouy and Chapman's models. He introduced the real dimension of solvent molecules and then divided the space charge into two layers: the compact layer and diffused layer (Figure 2.3c).

Total specific capacitance is estimated as

$$c' = \frac{c'_C c'_D}{c'_C + c'_D},$$
(2.3)

where
   c'_C is the compact layer capacitance, and
   c'_D is the diffused layer capacitance defined as

$$c'_D = z \sqrt{\frac{2q^2 n_0 \epsilon}{kT}} ch \left(\frac{z \Psi_D q}{2kT} \right).$$
(2.4)

The previous analysis assumed uniform distribution of the electrode pores. This means all the pores are of same size (diameter) and therefore the ions uniformly penetrate the pores. However, this is just an assumption. In reality, the size of the pores varies within large range. Hence, ions cannot penetrate all the pores. Small pores are not populated by ions at all. As a consequence, the effective surface area of the electrode is much smaller than theoretically calculated. The smaller the effective surface area, the lower the capacitance of the ultra-capacitor is. Figure 2.4a shows detail of an electrode, while Figure 2.4 b illustrates the non-uniform distribution of the electrode pores.

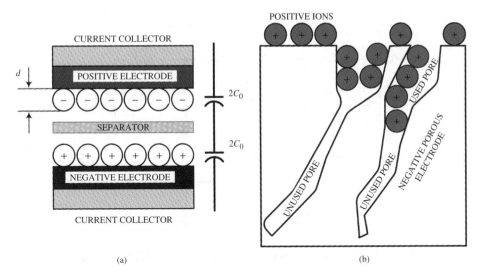

Figure 2.4 (a) Detail of double layer ultra-capacitor cell and (b) detail of the electrode with non-uniform distribution of pores

2.3 The Ultra-Capacitor Macro (Electric Circuit) Model

In this section, the ultra-capacitor macro model is analyzed and discussed in detail. The ultra-capacitor macro model is used for conversion system control analysis and design, as well as for evaluation of ultra-capacitor losses and temperature in different operating modes.

2.3.1 Full Theoretical Model

The traditional model consists of an ideal linear capacitor and equivalent series resistance (ESR). This simple model cannot be used in an ultra-capacitor model because of two phenomena: (i) the capacitance is voltage dependent and (ii) the time/space redistribution of the charge due to the porosity of the activated carbon electrodes. The porous electrode structure behaves as a nonlinear transmission line [10–15]. It is known from the theory of electric circuits that an electrically short transmission line can be approximated with an Nth order $RLCG$ ladder network. At low frequency, below 100 Hz, distributed serial inductance L can be neglected. Distributed conductance G can be neglected too, unless long-term steady state analysis is needed. Thus, an approximated model of an ultra-capacitor having porous electrodes is a serial connection of two RC leader networks of Nth order, the separator resistance R_{SP}, and the current collector resistances R_{CP} and R_{CN}. A schematic diagram of the approximated model is given in Figure 2.5.

The resistors $R_{P1} \ldots R_{P \ldots N}$ and $R_{N1} \ldots R_{N \ldots N}$ are the resistances of the positive and negative porous electrodes respectively. For more accurate modeling of the ultra-capacitor, the fact that these resistances are nonlinear and depend on the capacitor voltage must be taken into account. Nonlinear capacitances $C_{P1} \ldots C_{P \ldots N}$ and $C_{N1} \ldots C_{N \ldots N}$ are the positive and negative porous electrode capacitances.

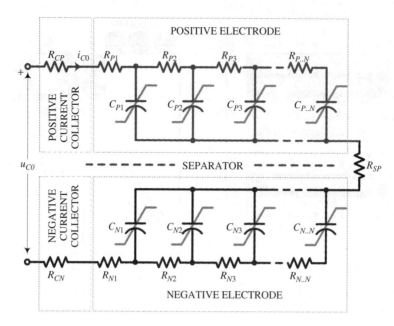

Figure 2.5 An approximated model of the electrochemical double layer capacitor taking into account the porosity of the electrodes

The voltage dependent capacitances $C_{P1} \ldots C_{P\ldots N}$ and $C_{N1} \ldots C_{N\ldots N}$ can be approximated by first-order functions of the voltage across each cell,

$$\begin{bmatrix} C_{P1} \\ C_{P2} \\ \vdots \\ C_{P..N} \end{bmatrix} = \begin{bmatrix} C_{0P1} \\ C_{0P2} \\ \vdots \\ C_{0P..N} \end{bmatrix} + \begin{bmatrix} K_{CP1} & 0 & 0 & 0 \\ 0 & K_{CP2} & 0 & 0 \\ 0 & 0 & 0 & \\ 0 & 0 & 0 & K_{CP..N} \end{bmatrix} \begin{bmatrix} u_{CP1} \\ u_{CP2} \\ \vdots \\ u_{CP..N} \end{bmatrix}, \tag{2.5}$$

$$\begin{bmatrix} C_{N1} \\ C_{N2} \\ \vdots \\ C_{N..N} \end{bmatrix} = \begin{bmatrix} C_{0N1} \\ C_{0N2} \\ \vdots \\ C_{0N..N} \end{bmatrix} + \begin{bmatrix} K_{CN1} & 0 & 0 & 0 \\ 0 & K_{CN2} & 0 & 0 \\ 0 & 0 & 0 & \\ 0 & 0 & 0 & K_{CN..N} \end{bmatrix} \begin{bmatrix} u_{CN1} \\ u_{CN2} \\ \vdots \\ u_{CN..N} \end{bmatrix}. \tag{2.6}$$

The coefficients $K_{CN1} \ldots K_{CP\ldots N}$ model the voltage dependency of the capacitance due to the diffused layer. Voltages $u_{CN1} \ldots u_{CP\ldots N}$ are the voltages across each elementary capacitor cell.

Considering that the positive and negative electrodes are symmetric, the circuit in Figure 2.5 can be reduced to a simple N^{th} order RC ladder network, depicted in Figure 2.6.

Resistances of the equivalent circuit in Figure 2.6 are

$$\begin{bmatrix} R_1 \\ R_2 \\ \vdots \\ R_{.N} \end{bmatrix} = \begin{bmatrix} R_{CP} + R_{SP} + R_{CN} \\ 0 \\ \vdots \\ 0 \end{bmatrix} + \begin{bmatrix} R_{P1} \\ R_{P2} \\ \vdots \\ R_{.P..N} \end{bmatrix} + \begin{bmatrix} R_{N1} \\ R_{N2} \\ \vdots \\ R_{.N.N} \end{bmatrix}. \tag{2.7}$$

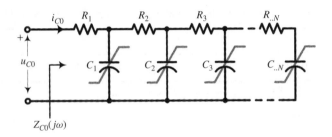

Figure 2.6   Nth order equivalent model of an electrochemical double layer capacitor

For simplicity of notation, we will use electrical elastance as the inverse variable of capacitance to define the capacitances of the equivalent model,

$$
\begin{bmatrix} \frac{1}{C_1} \\ \frac{1}{C_2} \\ \vdots \\ \frac{1}{C_{.N}} \end{bmatrix} = \begin{bmatrix} \frac{1}{C_{P1}} \\ \frac{1}{C_{P2}} \\ \vdots \\ \frac{1}{C_{P..N}} \end{bmatrix} + \begin{bmatrix} \frac{1}{C_{N1}} \\ \frac{1}{C_{N2}} \\ \vdots \\ \frac{1}{C_{.N.N}} \end{bmatrix}.
\tag{2.8}
$$

The circuit in Figure 2.6 is a nonlinear circuit because the capacitances depend on the voltage. To develop a model in the frequency domain we have to linearize the nonlinear circuit. Expanding Equations 2.5 and 2.6 into a Taylor series, taking just the zero order members and substituting them into Equation 2.8, yields

$$
\begin{bmatrix} \frac{1}{C_1} \\ \frac{1}{C_2} \\ \vdots \\ \frac{1}{C_{.N}} \end{bmatrix} \cong \begin{bmatrix} \frac{1}{C_{0P1}+K_{CP1}\cdot U_{C01}} \\ \frac{1}{C_{0P2}+K_{CP2}\cdot U_{C02}} \\ \vdots \\ \frac{1}{C_{0P..N}+K_{CP.N}\cdot U_{C0..N}} \end{bmatrix} + \begin{bmatrix} \frac{1}{C_{0N1}+K_{CN1}\cdot U_{C01}} \\ \frac{1}{C_{0N2}+K_{CN2}\cdot U_{C02}} \\ \vdots \\ \frac{1}{C_{0N.N}+K_{CN.N}\cdot U_{C0..N}} \end{bmatrix} = \begin{bmatrix} \frac{1}{C_{01}+K_{C1}\cdot U_{C01}} \\ \frac{1}{C_{02}+K_{C2}\cdot U_{C02}} \\ \vdots \\ \frac{1}{C_{0..N}+K_{C.N}\cdot U_{C0..N}} \end{bmatrix}.
\tag{2.9}
$$

Now, having a linearized model of the ultra-capacitor ladder network one can develop the ultra-capacitor input impedance $Z_{C0}(\omega)$. Since the capacitances are voltage dependent, the developed impedance is a small signal impedance, which is valid only in proximity of the capacitor voltage operating point U_{C0}.

$$
Z_{C0}(j\omega)|_{u_C=U_{C0}} = R_{C0}(\omega)|_{u_C=U_{C0}} + \frac{1}{j\omega C_C(\omega)|_{u_C=U_{C0}}}
\tag{2.10}
$$

As one can see from Equation 2.10, the ultra-capacitor ESR R_{C0} and equivalent capacitance C_C are frequency dependent properties. The resistance and capacitance are defined for zero frequency (DC operational mode) and high frequency as

$$
\lim_{\omega \to 0} C_C(\omega)|_{u_C=U_{C0}} = \sum_{i=1}^{N} C_i,
$$

$$
\lim_{\omega \to \infty} C_C(\omega)|_{u_C=U_{C0}} = C_1,
\tag{2.11}
$$

Table 2.2 Simulated parameters of a large ultra-capacitor cell 3000 F/2.5 V at $U_{C0} = 2.5$ V

R_1	C_1	R_2	C_2	R_3	C_3	R_4	C_4	R_5	C_5
0.324	124	0.000324	275	0.055	300	0.088	888	0.4	1489

The capacitance C is in (F) and the resistance R is in (mΩ).

$$\lim_{\omega \to 0} R_{C0}(\omega)|_{u_C=U_{C0}} = \sum_{i=1}^{N} R_i,$$

$$\lim_{\omega \to \infty} R_{ESR}(\omega)|_{u_C=U_{C0}} = R_1. \qquad (2.12)$$

To illustrate these properties, a large ultra-capacitor cell with the parameters $C_C = 3000$ F and $R_{C0} = 0.86$ mΩ at 2.5 V has been modeled as a fifth-order RC ladder network. The network parameters are given in Table 2.2.

The magnitude and phase of the capacitor input impedance are plotted in Figure 2.7a,b. From this plot one can see that the ultra-capacitor behaves as a pure capacitor in the very low frequency range, up to 20 mHz. In the high frequency range, let's say above 10 Hz, the ultra-capacitor bank behaves as a pure resistor. In the mid-frequency range, it behaves as an RC element.

2.3.1.1 The ESR R_{co} versus Frequency

As already mentioned, the ultra-capacitor is not an ideal loss-free device. Whenever current flows through the capacitor, regardless of the conversion process and power flow direction, an amount of energy is wasted as Joule's energy. This is due to the resistance of the ultra-capacitor collectors, porous electrodes, separator, and electrolyte. The quantity of energy wasted generally depends on the resistance and current. In power applications, the capacitor current is a varying quantity, depending on the charge/discharge cycle. In addition, the capacitor current contains a high frequency component due to the switch mode operation of the power conversion unit that charges/discharges the ultra-capacitor. The capacitor current spectra fall in the range from, let's say, millihertz up to kilohertz or tens of kilohertz. As the equivalent resistance is frequency dependent, the contribution of each spectral component to the loss is different. This is discussed in detail in Section 2.6.

The equivalent serial resistance versus frequency was calculated for the example given in Table 2.2. The resistance versus frequency is plotted in Figure 2.8a. The resistance is high and constant at low frequencies up to 200 mHz, and then decreases to a minimum value at frequencies above 10 Hz. Ratio max/min resistance is approximately 1.6.

2.3.1.2 The Capacitance versus Frequency

The capacitance varies more significantly with frequency than the resistance. The factor C_{MAX}/C_{MIN} could be up to 10 or more. What are the typical implications of this in real

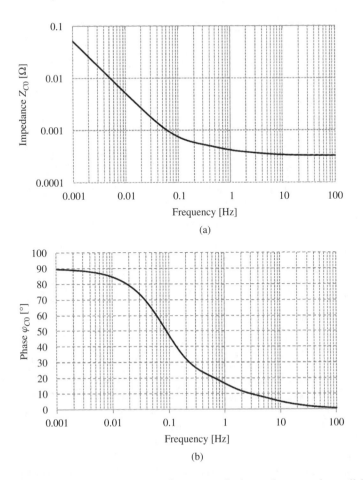

Figure 2.7   (a,b) The input impedance versus frequency of a large ultra-capacitor cell 3000 F/2.5 V

power applications? The capacitance variation with frequency means that one needs a certain time to store the required energy in the capacitor. If one charges the capacitor with high power, close to maximum, the capacitor voltage will increase quickly, and reach the maximum voltage before the capacitor is fully charged. Thus, the full energy capability of the capacitor is not used. In contrast, if one charges the capacitor with low power, the voltage increases slowly and the charge is distributed over the entire capacitor. Once the voltage reaches maximum voltage, the capacitor is fully charged and the energy capability is maximized. From this short discussion one can conclude that the total capacitance is available only at very low frequencies, and consequently light loads. Figure 2.8b shows the input capacitance versus frequency. The equivalent capacitance varies significantly in the frequency range from 0.1 to 10 Hz. Maximum to minimum capacitance ratio is approximately 10.

Figure 2.9a shows the dynamic specific energy versus pulse width. The ultra-capacitor voltage charge/discharge variation is 10% of the rated voltage. Notice that the specific

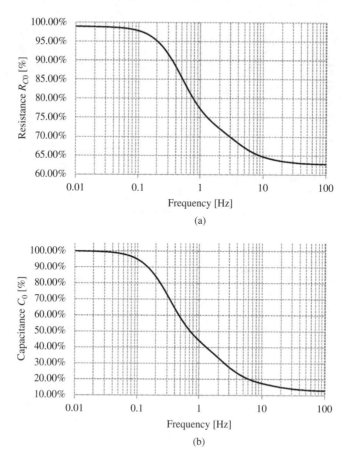

Figure 2.8   (a,b) The equivalent series resistance and the capacitance versus frequency of a large ultra-capacitor cell 3000 F/2.5 V

energy (energy capability) decreases as the pulse width decreases. The variation of the energy capability is due to the frequency dependent capacitance, which is caused by the relaxation phenomenon in the porous electrodes [1]. There is another factor that limits the ultra-capacitor energy capability: the voltage drop of the electrode and separator resistance, and porous electrode resistance close to the input. Figure 2.9b illustrates two different cases. The dark plot illustrates the charging process with a relatively high charging current, while the lighter waveforms illustrate the charging process with a current that is 10% of the previous. This issue could be solved by the control of charge/discharge process, wherein the real capacitor voltage is estimated from the capacitor model and measured input voltage and current.

The charge criterion is the voltage across the capacitor terminals, which should not be higher than U_{C0max}. The difference between the steady state voltages U_{C01}, U_{C02} represents the difference in the stored energy in the ultra-capacitor for different charging speeds.

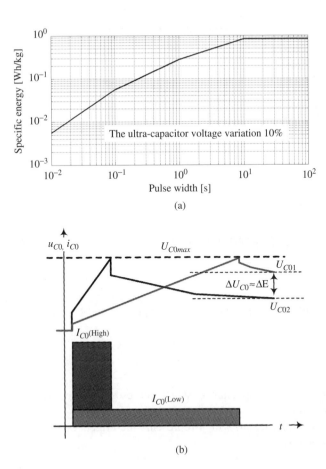

Figure 2.9 Illustration of the effect of the ultra-capacitor frequency-dependent capacitance on energy storage capability. (a) Effective specific energy versus charging/discharging pulse width and (b) time diagrams for two different cases, short pulse and long pulse

Considering a linear capacitor with frequency dependent capacitance, one can estimate the difference in energy stored as ΔE.

$$\Delta E = \frac{1}{2} C_0 \Delta U_{C0}^2 \left(1 + \frac{2U_{C01}}{\Delta U_{C0}}\right), \tag{2.13}$$

where ΔU_{C0} strongly depends on the time profile of the charging current.

2.3.1.3 The Voltage-Dependent Capacitance and ESR

The total capacitance of the ultra-capacitor is the voltage controlled capacitance. Figure 2.10 shows the voltage to capacitance characteristic of a large ultra-capacitor cell 3000 F/2.5 V. The solid line is an approximation, while the squares are measurement results.

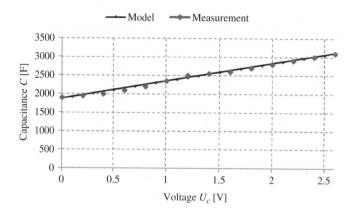

Figure 2.10   The voltage-to-capacitance characteristic of a large 3000 F/2.5 V ultra-capacitor cell

The voltage controlled capacitance can be approximated by a first-order function

$$C(u_C) = C_0 + k_C \cdot u_C,$$
(2.14)

where

C_0 is the initial linear capacitance, which represents electrostatic capacitance of the
capacitor and

k_C is a positive coefficient, which represents the effects of the diffused layer of the
ultra-capacitor.

In the case of a hybrid ultra-capacitor, the coefficient models the Faradic effect and
electrochemical processes on one side of the capacitor [1].

Using the Matlab "Polyfit" function or a similar least-squares tool, we can find the
coefficients C_0 and k_C (Equation 2.14). In the example of Figure 2.10 we find $C_0 = 1857\,\text{F}$
and $k_C = 473\,\text{F/V}$.

The capacitor current is defined as

$$i_C = \frac{\partial Q}{\partial t} = \left(C\left(u_C\right) + u_C \frac{dC\left(u_C\right)}{du_C} \right) \frac{du_C}{dt} = C_I(u_C) \frac{du_C}{dt}.$$
(2.15)

The capacitance denoted as $C_I(u_C)$ is a virtual capacitance, the so-called the current
capacitance or small signal (dynamic) capacitance. Substituting Equation 2.14 into
Equation 2.15 yields

$$C_I(u_C) = C_0 + 2k_C \cdot u_C \quad \text{and} \quad i_C = (C_0 + 2k_C \cdot u_C) \frac{du_C}{dt}.$$
(2.16)

2.3.1.4 The ESR and Capacitance versus Temperature

ESR versus Temperature

The analytical determination of the ultra-capacitor ESR and capacitance as a function of the cell temperature is difficult. As it is beyond the scope of this book, we will not discuss it further. An empirical approach will be addressed and discussed.

The ultra-capacitor ESR is a nonlinear function ψ of the cell temperature. The function ψ is a regular and continuous function of a real argument θ. Thus, it can be expanded in a Taylor series

$$R_{C0} = \Psi(\theta) = \sum_{j=0}^{\infty} \frac{1}{j!} \frac{\partial \Psi^j(\theta)}{\partial \theta^j} \theta^j \tag{2.17}$$

According to experimental results, [16], Equation 2.17 can approximated with a third-order polynomial function

$$R_{C0} \cong R_0 + k_1\theta + k_2\theta^2 + k_3\theta^3 \tag{2.18}$$

where

R_0 is the *ESR* at ambient temperature and

k_1 to k_3 are the coefficients determined experimentally.

As reported in [16], a large ultra-capacitor cell has been tested and the ESR measured at different temperatures. The parameters identified are $R_0 = 100\%$, $k_1 = -17\%$, $k_2 = 0.058\%$, and $k_3 = -0.833e - 3\%$.

Polynomial approximation (Equation 2.19) properly models the electrode resistance, but not the ionic resistance of the electrolyte. As a consequence, Equation 2.19 gives a good match to the experimental results in the temperature range from -40 to $+40\,^\circ$C. If the cell temperature is higher, the estimated ESR is significantly below the one measured.

To properly model the entire resistance, including the electrode and electrolyte, exponential approximation is proposed

$$R_{C0} \cong R_0[0.55(1 + \alpha(\theta - 25)) + 0.45\exp(-\beta(\theta - 25))], \tag{2.19}$$

where

α and β are coefficients and

R_0 is the ESR at room temperature.

Figure 2.11 shows R_{C0} versus the cell temperature θ. The function $R_{C0}(\theta)$ is computed from Equation 2.19. The parameters $\alpha = 0.007$ (1/K) and $\beta = 0.0225$ (1/K) are given for a large 3000 F/2.5 V ultra-capacitor cell [17].

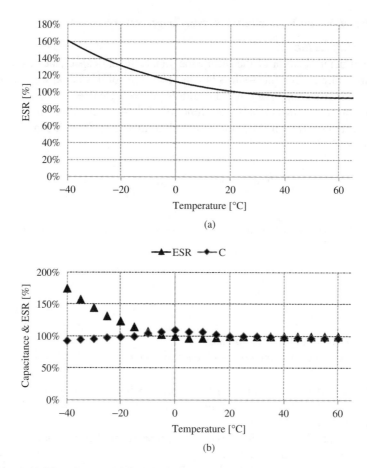

Figure 2.11 (a,b) The ultra-capacitor low frequency (dc) ESR and capacitance versus the cell temperature. Experimental results of a large cell 3000 F/2.5 V

The Capacitance versus Temperature

The capacitance is also dependent on the temperature, but is not as significant as the ESR. At low frequency ranges, below the cut-off frequency, the capacitance variation with the temperature is minor. However, the capacitance may vary significantly with temperatures in the frequency range above the cut-off frequency.

As reported in [18], the ESR varies with temperature in the low frequency range, while the variation is minor if the frequency is above few hertz. Experimental test results of a large 3000 F/2.5 V cell are presented in Figure 2.12.

2.3.2 A Simplified Model

For simplicity in the following analysis, a first order nonlinear model of an ultra-capacitor is used. The model takes into account the linear (voltage independent) internal resistance R_{C0} and total capacitance as a function of the capacitor voltage. Effects of the transmission

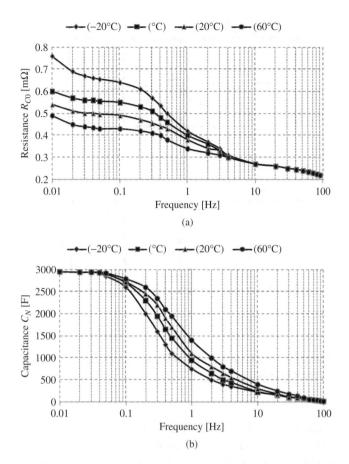

Figure 2.12 (a,b) The ESR and capacitance versus the cell temperature and frequency of a large ultra-capacitor cell (3000 F/2.5 V)

line are neglected. The internal equivalent resistance R_{C0} is modeled as a constant and frequency independent resistance. Figure 2.13 depicts a simplified model of an ultra-capacitor used in the analysis. The equivalent capacitor consists of a linear capacitor C_0 and parallel connected voltage dependent capacitor $C(u_C)$. The leakage current is modeled either by a shunt resistor R_P, Figure 2.13a or a current sink i_P, Figure 2.13b.

2.3.2.1 Leakage Current

The ultra-capacitor leakage current is defined as the charging current that is required to maintain the ultra-capacitor voltage constant at the specified value. The leakage current reported in datasheets is the value of the charging current required to maintain rated voltage after holding the ultra-capacitor at rated voltage for 72 hours at room temperature. The leakage current is measured immediately at the end of the 72-hour period. The leakage

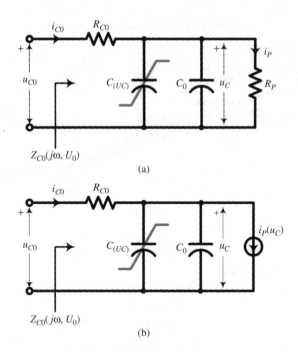

Figure 2.13 Simple model of the ultra-capacitor including leakage current model. (a) Shunt resistance R_P and (b) voltage-dependent shunt current i_P

current is not constant over time. It is also influenced by the temperature, operating voltage, the loading history, and the aging conditions.

For modeling purposes, two methods can be used. The first is a constant shunt resistor R_P, while the second is a voltage-dependent shunt current i_P. A brief discussion of both methods follows below.

Constant Shunt Resistance Model

The ultra-capacitor leakage current is usually modeled with a shunt resistor R_P. There are two ways to compute the resistance for given capacitance as a parameter.

1. The first way to determine the resistance R_P is from the criterion that the ultra-capacitor voltage decays at 4% of U_{C0max} after $T = 72$ hours. From the $R_P C_0$ circuit we have

$$R_{P1} = -\frac{T}{C_0 \ln(0.96)} \cong 24.5 \frac{T}{C_0} = \frac{6.35 \cdot 10^6}{C_0}. \tag{2.20}$$

2. Another way is to define the resistance R_P from an electrode metric of $2\,\mu\text{A/F}$. The resistance is

$$R_{P2} = \frac{U_{C0\,max}}{C_0 2 \cdot 10^{-6}}. \tag{2.21}$$

Please note that the two methods give different results. The resistance ratio is

$$\frac{R_{P1}}{R_{P2}} = \frac{12.7}{U_{C0\max}}, \tag{2.22}$$

which is roughly 5 for 2.5 V ultra-capacitors.

The first method models the leakage current over a long time (three days) after the ultra-capacitor is exposed to the full voltage. The second method gives a much lower shunt resistance and is the case of an ultra-capacitor cell that is exposed to the full voltage for the first time. For more accurate modeling, the shunt resistance has to be voltage and temperature dependent rather than constant as predicted by the previous model.

Voltage-Dependent Shunt Current Model

The natural way to model the ultra-capacitor leakage current is a voltage-controlled time and temperature variant current drain i_P, Figure 2.13b. The leakage current can be measured using two methods.

1. Charge the ultra-capacitor to a predefined level and maintain the voltage at constant by an external high precession controlled voltage source. The current measured is in fact the leakage current. Test time is 72 hours. The test can be done for different voltage levels and the ultra-capacitor temperature.
2. Charge the ultra-capacitor to a predefined level and keep it open circuited. Measure the voltage and calculate the current from measured data and Equation 2.16. Test time is 72 hours. The measurement can be done for different initial voltages and the ultra-capacitor temperature.

Once the leakage current is measured, we can determine a function that describes the current versus operating voltage, temperature, and time,

$$i_P = i_P(U_{C0\max}, \theta, t), \tag{2.23}$$

where θ is the ultra-capacitor temperature.

2.3.3 A Simulation/Control Model

In previous sections we have addressed and discussed different aspects of the ultra-capacitor models. The remaining aspect to be addressed is a model for simulation and control purposes. The model discussed below can be used for numerical simulation of the ultra-capacitor circuits. Moreover, the model is used for control system analysis and controller synthesis [19].

2.3.3.1 Large Signal Model

Let the ultra-capacitor be described by a voltage dependent capacitance $C(u_C)$ (Equation 2.14) and a constant resistance R_{C0}. The state variable equation can be derived from

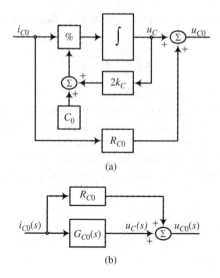

(a)

(b)

Figure 2.14 Model of a nonlinear ultra-capacitor $C = C_0 + k_C u_C$. (a) Large signal (nonlinear) model and (b) small signal (linear) model

Equation 2.16 as

$$\frac{du_C}{dt} = i_C \frac{1}{(C_0 + 2k_C \cdot u_C)},$$

$$u_{C0} = u_C + R_{C0} i_{C0} \tag{2.24}$$

where

u_C is the ultra-capacitor internal voltage as the state variable,

i_{C0} is the current as the control variable, and u_{C0} is the terminal voltage as the output variable.

Figure 2.14a shows a block diagram of the model (2.24).

2.3.3.2 Small Signal Model

The large signal model (2.24) is a nonlinear model, and as such it cannot be directly used in the control system analysis and controller synthesis using standard analysis and control techniques. It is necessary to develop a linear (small signal) model that describes the ultra-capacitor in the vicinity of the operating point. Using linearization techniques, such as the perturbation model or Taylor series expansion, we can develop a small signal model (2.25). The variables are determined as $x = X_0 + \hat{x}$, where X_0 is the steady state (operating point) and $\hat{x}$ is the small signal variation in the vicinity of the steady state.

$$\frac{d\hat{u}_C}{dt} = \frac{1}{(C_0 + 2k_C U_C)} \hat{i}_{C0} - \frac{2k_C I_{C0}}{(C_0 + 2k_C U_C)^2} \hat{u}_C$$

$$\hat{u}_{C0} = \hat{u}_C + R_{C0} \hat{i}_{C0} \tag{2.25}$$

Applying Laplace formal transformation yields the ultra-capacitor transfer function

$$u_C(s) = \frac{(C_0 + 2k_C U_C)}{s(C_0 + 2k_C U_C)_2 + 2k_C I_{C0}} i_{C0}(s) = G_{C0}(s) i_{C0}(s).$$

$$u_{C0}(s) = u_C(s) + R_{C0} i_{C0}(s) \tag{2.26}$$

Figure 2.14b shows a block diagram of the linear (linearized) model (2.26).

Remark: The above models have been developed under an assumption that the capacitance and resistance are frequency independent parameters. As discussed and explained in Sections 2.3.1.1 and 2.3.1.2, this is the case if the frequency is below the lower band and above the higher band. If this is not the case, the capacitance and resistance varies widely with the frequency. In such a case, the above developed model is not valid any more. A higher order model is required.

2.3.4 Exercises

Exercise 2.1

A large ultra-capacitor cell that can be modeled with a fifth-order RC network, Figure 2.6, has the parameters given in Table 2.2. Calculate the capacitance and internal resistance if the ultra-capacitor current frequency is 100 Hz.

Solution

The current frequency of 100 Hz is a very high frequency for an ultra-capacitor. For calculation purposes we can assume $f \rightarrow \infty$. From Equations 2.11 and 2.12 we see that the capacitance and resistance are basically reduced to the first RC element of the Nth order model, Figure 2.6. From the parameters given in Table 2.2 we can calculate the capacitance $C = 124\,F$ and resistance $R_{C0} = 0.324\,m\Omega$. Please note that the high to low frequency capacitance ratio is very low, in this case approximately 0.043. This basically means that only 4.3% of the rated capacitance is used at such high frequency.

Exercise 2.2

A large ultra-capacitor cell with the voltage-to-capacitance characteristic illustrated in Figure 2.10 is given. Calculate the dynamic capacitance at the operating point $U_{C0} = 2\,V$.

Solution

From the voltage-to-capacitance characteristic given in Figure 2.10, we can define the coefficients $C_0 = 1857\,F$ and $k_C = 473\,F/V$. From the definition of dynamic capacitance, Section 2.3.1.3, Equation 2.16, and operating voltage $U_{C0} = 2\,V$, we can calculate

$C_d = 3714\,\text{F}$. Please note that the dynamic capacitance is 35% higher than the static one defined by Equation 2.14.

2.4 The Ultra-Capacitor's Energy and Power

2.4.1 The Ultra-Capacitor's Energy and Specific Energy

Energy stored in the ultra-capacitor charged to voltage u_C is

$$E_C(u_C) = \frac{1}{2}\left(C_0 + \frac{4}{3}k_C u_C\right)u_C^2 = \frac{1}{2}C_E(u_C)u_C^2. \tag{2.27}$$

The capacitance denoted as $C_E(u_C)$ is the so-called "energetic" capacitance.

The energy available from the ultra-capacitor, discharged from the initial voltage U_{Cmax} to the final voltage U_{Cmin}, is

$$\Delta E_C = \frac{C_0}{2}(U_{C\,max}^2 - U_{C\,min}^2) + \frac{2}{3}k_C(U_{C\,max}^3 - U_{C\,min}^3). \tag{2.28}$$

Equation 2.28 defines the energy realized from the ultra-capacitor when discharged from U_{Cmax} to U_{Cmin}. However, the energy realized to the load is lower. The difference in energy is dissipated on the ultra-capacitor internal resistance R_{C0}. The energy losses depend on the resistance and the load profile.

The energy realized to the load is

$$E_{C0} = \Delta E_C - E_{LOSSES}$$
$$= \frac{C_0}{2}(U_{C\,max}^2 - U_{C\,min}^2) + \frac{2}{3}k_C(U_{C\,max}^3 - U_{C\,min}^3) - \int_0^{T_{DCH}} R_{C0}(t)i_{C0}^2(t)dt, \tag{2.29}$$

where
   $R_{C0}(t)$ is the time dependent ESR of the capacitor and
   $i_{C0}(t)$ is the load current profile.

Time dependent resistance $R_{C0}(t)$ is computed from the frequency dependent ESR $R_{C0}(\omega)$ using inverse Fourier transformation,

$$R_{C0}(t) = \frac{1}{2\pi}\int_{-\infty}^{+\infty} R_{C0}(j\omega)e^{j\omega t}d\omega. \tag{2.30}$$

More details of the ultra-capacitor loss calculation are given in Section 2.6. The ultra-capacitor design procedure is explored in detail in Chapter 4.

The ultra-capacitor's specific energy is energy stored per unit of mass of the ultra-capacitor.

$$SE_C(u_C) = \frac{\left(C_0 + \frac{4}{3}k_C u_C\right)u_C^2}{2M} \tag{2.31}$$

where M is the ultra-capacitor weight.

The specific energy unit is kJ/kg or W h/kg.

2.4.2 The Ultra-Capacitor's Energy Efficiency

Whenever the ultra-capacitor is charged or discharged, a certain amount of energy is realized on the internal resistance R_{C0}. Hence, the conversion efficiency is directly influenced by the internal resistance losses and charge/discharge profile. In this section we will define charging, discharging, and round trip efficiency, respectively.

Let's consider Figure 2.15 that illustrates a partially regenerative power conversion process. The power processor realizes the energy denoted E_{IN}. A part of that energy, denoted as charge energy E_{CH}, is stored in an energy storage device. Energy realized from the energy storage device is denoted as the discharge energy E_{DCH}. A part of the discharge energy denoted as E_{OUT} is realized to the power processor. Also, a part of the stored energy is realized as stand-by losses E_0.

Charging and discharging efficiency are defined as

$$\eta_{CH} = \frac{E_{IN} - E_{\varsigma(CH)}}{E_{IN}} = \frac{E_{CH}}{E_{CH} + E_{\varsigma(CH)}}$$

$$\eta_{DCH} = \frac{E_{OUT}}{E_{OUT} + E_{\varsigma(DCH)}} = \frac{E_{DCH} - E_{\varsigma(DCH)}}{E_{DCH}}, \tag{2.32}$$

where $E_{\zeta(CH)}$ and $E_{\zeta(DCH)}$ are charging and discharging losses respectively.

The round trip efficiency is defined by the output energy to input energy ratio,

$$\eta_{RTP} = \frac{E_{OUT}}{E_{IN}} = \frac{E_{DCH} - E_{\varsigma(DCH)}}{E_{CH} + E_{\varsigma(CH)}} = \frac{E_{CH} - E_0 - E_{\varsigma(DCH)}}{E_{CH} + E_{\varsigma(CH)}}. \tag{2.33}$$

The relation between the charging and discharging energy is

$$E_{DCH} = E_{CH} - E_0, \tag{2.34}$$

where E_0 stands for the energy storage stand-by losses.

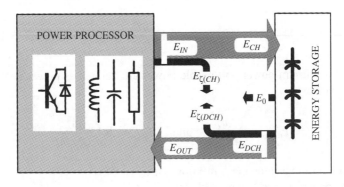

Figure 2.15 An illustration of the round trip efficiency of a power conversion system

In most fast dynamic conversion systems, the stand-by losses can be neglected. Additionally, the charging and discharging losses can be assumed as equal.

$$E_{\varsigma(DCH)} = E_{\varsigma(CH)} = E_{\varsigma}, \qquad (2.35)$$

Therefore, the round trip efficiency is

$$\eta_{RTP} \cong \frac{E_{DCH} - E_{\varsigma}}{E_{CH} + E_{\varsigma}}, \qquad (2.36)$$

Remark: The round trip efficiency can be computed directly from Equation 2.32 as

$$\eta_{RTP} = \eta_{CH}\eta_{DCH} = \frac{E_{CH}}{E_{CH} + E_{\varsigma(CH)}} \frac{E_{DCH} - E_{\varsigma(DCH)}}{E_{DCH}}, \qquad (2.37)$$

only if the stand-by losses E_0 can be neglected. If this is not the case, definition (Equation 2.33) must be used.

2.4.3 The Ultra-Capacitor's Specific Power

The ultra-capacitor power capability is defined by the so-called matched power. The matched power is the maximum power that can be extracted from a source having an internal resistance R_{int}, when the load resistance is equal to the source resistance. According to this definition, the maximum power capability of an ultra-capacitor having an internal resistance R_{C0} is

$$P_{0MAX} = \frac{u_C^2}{4R_{C0}}. \qquad (2.38)$$

Usable power of an ultra-capacitor is defined according to IEC 62391–2, [20] as

$$P_{0(D)} = 0.12\frac{u_C^2}{R_{C0}}. \qquad (2.39)$$

The specific maximum power and usable power are defined as power per unit of mass of the ultra-capacitor,

$$SP_{0\,max} = \frac{u_C^2}{4R_{C0}M} \quad \text{and} \quad SP_{0(D)} = 0.12\frac{u_C^2}{R_{C0}M} \qquad (2.40)$$

where M is the ultra-capacitor mass in kilograms.

The specific power unit is W/kg or kW/kg.

2.4.4   The Electrode Carbon Loading Limitation

Theoretically, the ultra-capacitor current is limited by the ESR and voltage. However, the manufacturers of ultra-capacitor cells and modules often define maximum peak current during one second as

$$I_{max} = \frac{1}{2} \frac{u_C C_0}{(C_0 R_{C0} + 1)}. \tag{2.41}$$

Another metric of the ultra-capacitor current limitation is the so-called carbon loading metric. Using this definition, the maximum current can be computed as

$$I_{max(CL)} = C_0 k_{CL}, \tag{2.42}$$

where k_{CL} (mA/F) is the carbon loading coefficient.

The typical carbon loading coefficient in power conversion ultra-capacitor applications is $k_{CL} = 70$ mA/F. However, it may go higher, usually 100–200 mA/F. If the carbon loading factor is as high as $k_{CL} = 500$ mA/F or higher, we say that the ultra-capacitor is extremely loaded.

2.4.5   Exercises

Exercise 2.3

A power conversion system is equipped with short-term energy storage. The energy storage requires an ultra-capacitor with the following parameters: Total capacitance $C_{TOTAL} = 3000$ F, full charge voltage 2.5 V, and charge/discharge frequency 1 Hz. Select the ultra-capacitor cell using data from Table 2.2 and Figure 2.8. Calculate the equivalent resistance of the selected ultra-capacitor module.

Solution

This is an example of an application with a relatively high charge/discharge frequency. To select the capacitor cell, the frequency dependent capacitance and resistance have to be taken into account, Figure 2.8.

From the capacitance frequency characteristic, Figure 2.8b, we find $C = \mathbf{1350\,F}$ at $f_{C0} = 1$ Hz and $U_{C0} = 2.5$ V. From these data and the required capacitance, we find that three capacitor cells have to be connected in parallel to achieve $C_0 > 3000$ F at $f_{C0} = 1$ Hz and $U_{C0} = 2.5$ V.

From the resistance frequency characteristic, Figure 2.8a, we have $R_{C0} = \mathbf{0.667\,m\Omega}$ at 1 Hz. The equivalent resistance of three parallel connected cells is $R_{C0} = \mathbf{0.222\,m\Omega}$ at 1 Hz.

Exercise 2.4

An ultra-capacitor cell with the following parameters, $C_0 = 2400$ F and $k_C = 320$ F/V, is being discharged with a constant current $I_{C0} = 50$ A. Calculate the discharge time if the discharge minimum voltage is $U_{C0min} = 1.25$ V and the initial voltage is $U_{C0} = 2.5$ V.

Solution

From Equation 2.16 we have the equation

$$I_{C0}T_{DCH} = C_0(U_{C0} - U_{C0\,min}) + k_C(U_{C0}^2 - U_{C0\,min}^2). \qquad (2.43)$$

Substituting the capacitor parameters into Equation 2.43 yields a discharge time of **$T_{DCH} = 90$ seconds**.

Exercise 2.5

Calculate the discharge time from the previous example assuming that the ultra-capacitor is a linear capacitor with $C_0 = 3200\,F$ and $k_C = 0\,F/V$.

Solution

Substituting the capacitor parameters into Equation 2.43 yields a discharge time of **$T_{DCH} = 80$ seconds**. Please note that the discharge time of a linear capacitor is slightly shorter than the discharge time of a nonlinear capacitor. The capacitance of both capacitors is the same at the maximum voltage (3200 F at 2.5 V).

Exercise 2.6

Calculate the energy and specific energy of the above examples. The ultra-capacitor cell has mass of $M = 500\,g$.

Solution

From Equation 2.27 we can compute the energy stored at the full voltage of 2.5 V. The nonlinear capacitor storage capability and specific energy are **$E_{C0} = 10.83\,kJ$** and **$SE_{C0} = 21.66\,kJ/kg$**. The linear capacitor has a storage capability and specific energy of **$E_{C0} = 10\,kJ$** and **$SE_{C0} = 20\,kJ/kg$**.

Please note that the energy capabilities of both capacitors are almost same. The difference is 8%.

Exercise 2.7

A large ultra-capacitor cell has the following parameters: $C_0 = 3000\,F$, $R_{C0} = 0.3\,m\Omega$, $U_{C0} = 2.5\,V$, and $M = 550\,g$.

1. Calculate the maximum and usable specific power of the ultra-capacitor cell.
2. Calculate the "one second" maximum current.
3. Calculate the carbon loading factor if the current is a "one second" current.

Solution

1. Substituting the ultra-capacitor cell data into Equation 2.40, yields $SP_{0max} = 10.41\,kW$ /kg and $SP_{0(D)} = 5\,kW/kg$.

2. Substituting the ultra-capacitor parameters into Equation 2.41 yields the "one second" maximum current $I_{max} = \mathbf{1973.7\,A}$.
3. From the previous results and Equation 2.42, we can find the carbon loading factor of $k_{CL} = \mathbf{657.9\,mA/F}$. According to the criterion, this is extremely high loading of the ultra-capacitor.

2.5   The Ultra-Capacitor's Charge/Discharge Methods

Figure 2.16 illustrates the interaction between the ultra-capacitor and a power conversion system denoted a "power processor." The ultra-capacitor and power processor are connected via an interface, where the interface has general meaning. The power processor in interaction with the interface charges and discharges the ultra-capacitor.

The nature of the charging/discharging depends on the interface and power processor nature. Theoretically, we can distinguish four different charging/discharging methods: (i) voltage mode, (ii) resistance mode, (iii) current mode, and (iv) power mode. The first is not applicable because the ultra-capacitor is some kind of voltage source with an internal resistance. The most important characteristics of the other three conversion methods are briefly discussed in the following section.

2.5.1   Constant Resistive Loading

Constant resistive load is the simplest charge/discharge method. The capacitor is charged from voltage source V_{BUS} via a charge resistor R_0 and discharged in the load resistor R_0 (Figure 2.17). However, because of low conversion efficiency, this method is rarely used in power applications and therefore will not be discussed.

2.5.2   Constant Current Charging and Loading

The ultra-capacitor can be charged/discharged with a constant current load/source. A constant current load/source is often found in regulated power converters, such as regulated chargers and constant torque-driven electric motors (Figure 2.18).

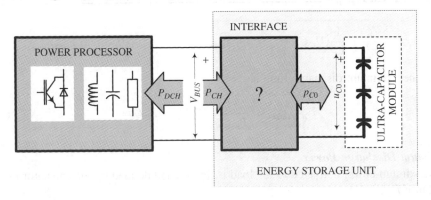

Figure 2.16   The ultra-capacitor connected to a power conversion system ("the power processor") via a charge/discharge interface

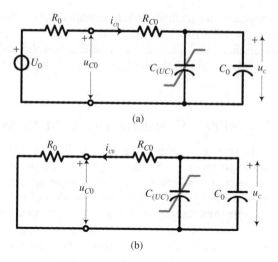

(a)

(b)

Figure 2.17   The ultra-capacitor resistive power conversion. (a) Charging and (b) discharging

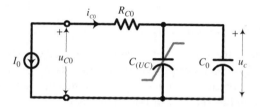

Figure 2.18   Charging/discharging the ultra-capacitor with a constant current source

2.5.2.1   Discharging

Let the ultra-capacitor initial voltage be U_C and the ultra-capacitor be discharged by constant current $I_{0(DCH)}$. The capacitor internal voltage u_C declines as

$$u_C(t) = \sqrt{\frac{C_0^2}{4k_C^2} + \frac{1}{k_C}(U_C C_0 + U_C^2 k_C - I_{0(DCH)}t)} - \frac{C_0}{2k_C} \cdots 0 < t < T_{DCH\,(max)}, \quad (2.44)$$

where the maximum discharge time is

$$T_{DCH\,(max)} = \frac{1}{I_{0(DCH)}}(C_0 U_C + k_C U_C^2). \quad (2.45)$$

Maximum Discharge Power

The maximum power delivered to the load is limited and defined by the capacitor internal resistance R_{C0},

$$P_{0MAX} = \frac{u_C^2}{4R_{C0}}. \quad (2.46)$$

The current I_0 is limited and depends on the capacitor voltage u_C and the internal resistance R_{C0}.

$$0 \leq I_{I_0(DCH)} \leq I_{0MAX} \quad \text{where} \quad I_{0MAX} = \frac{u_C}{R_{C0}}. \tag{2.47}$$

If the current exceeds the limit, the capacitor terminal voltage u_{C0} becomes negative and the load turns to be a source. However, the capacitor is still being discharged. All energy realized from the ultra-capacitor and energy delivered from the current source is dissipated in the capacitor internal series resistance R_{C0}.

Discharge Losses and Efficiency

Let the ultra-capacitor be discharged from the maximum voltage U_{Cmax} to the minimum voltage U_{Cmin} with constant current $I_{0(DCH)}$. Discharge losses are

$$E_{\varsigma(DCH)} = \int_0^{T_{DCH}} R_{C0}(t) i_{C0}^2(t) dt = R_{C0} I_{0(DCH)}^2 T_{DCH}, \tag{2.48}$$

where discharge time T_{DCH} is

$$T_{DCH} = \frac{1}{I_{0(DCH)}} (k_C(U_{C\,max}^2 - U_{C\,min}^2) + C_0(U_{C\,max} - U_{C\,min})). \tag{2.49}$$

The discharge efficiency is defined as

$$\eta_{DCH} = \frac{\Delta E_C - E_{\varsigma(DCH)}}{\Delta E_C}, \tag{2.50}$$

where E_C is the energy realized from the ultra-capacitor (Equation 2.28).

Substituting Equations 2.28, 2.48, and 2.49 into Equation 2.50 yields the discharge efficiency

$$\eta_{DCH} = 1 - R_{C0} I_{I_0(DCH)} 6 \frac{(k_C(U_{C\,max}^2 - U_{C\,min}^2) + C_0(U_{C\,max} - U_{C\,min}))}{4k_C(U_{C\,max}^3 - U_{C\,min}^3) + 3C_0(U_{C\,max}^2 - U_{C\,min}^2)}. \tag{2.51}$$

We can distinguish two cases:

1. The load profile is predefined.

   The ultra-capacitor minimum voltage is not constant. It is a function of the load profile and the capacitance C_N. The minimum voltage is

$$U_{C\,min} = \sqrt{\frac{C_0^2}{4k_C^2} + \frac{1}{k_C}(U_{C\,max} C_0 + U_{C\,max}^2 k_C - I_{0(DCH)} T_{DCH})} - \frac{C_0}{2k_C}. \tag{2.52}$$

2. The ultra-capacitor minimum voltage is defined and constant.

In this case, the minimum voltage is a constant parameter while the discharge current and discharge time are variables.

2.5.2.2 Charging

Let the ultra-capacitor be charged from an initial voltage U_{C0} with a constant current $I_{0(CH)}$. The ultra-capacitor internal voltage u_C increases according to equation

$$u_C(t) = \sqrt{\frac{C_0^2}{4k_C^2} + \frac{1}{k_C}(U_C C_0 + U_C^2 k_C + I_{0(CH)}t)} - \frac{C_0}{2k_C} \cdots 0 < t < T_{CH(max)}, \quad (2.53)$$

where $T_{CH(max)}$ maximum charging time is

$$T_{CH} = \frac{1}{I_{0(CH)}}(k_C((U_{C0\,max} - R_{C0}I_0)^2 - U_{C\,min}^2) + C_0((U_{C0\,max} - R_{C0}I_0) - U_{C\,min})). \quad (2.54)$$

The maximum charge time is limited by the ultra-capacitor terminal voltage U_{C0max}. This limit should be respected and the ultra-capacitor should not be charged beyond this limit.

Maximum Charging Power

In charging mode, the current $I_{0(CH)}$ is negative. The maximum current that can be injected into the ultra-capacitor is limited by the capacitor terminal voltage U_{0MAX},

$$I_{0\,max} = \frac{U_{0\,max} - u_C}{R_{C0}}. \quad (2.55)$$

From Equation 2.55 one can define maximum charging power as a function of the capacitor resistance and the capacitor voltage,

$$P_{0\,max} = \frac{U_{0\,max}(U_{0\,max} - u_C)}{R_{C0}}. \quad (2.56)$$

Charge Losses and Efficiency

Let the ultra-capacitor be charged from the minimum voltage U_{Cmin} to maximum voltage U_{Cmax} with constant current $I_{0(CH)}$. Charge losses are

$$E_{\varsigma(CH)} = \int_0^{T_{CH}} R_{C0}(t)i_{C0}^2(t)dt = R_{C0}I_{0(CH)}^2 T_{CH}, \quad (2.57)$$

where charge time T_{CH} is in fact the maximum charge time given by Equation 2.54

The charge efficiency is defined as

$$\eta_{CH} = \frac{\Delta E_C}{\Delta E_C + E_{\varsigma(CH)}}, \quad (2.58)$$

where E_C is the energy realized from the ultra-capacitor (Equation 2.28).

Substituting Equations 2.28, 2.54, and 2.57 into Equation 2.58 yields the charge efficiency

$$\eta_{CH} = \frac{1}{1 + R_{C0}I_{0(CH)}6\dfrac{k_C((U_{C0\,max} - R_{C0}I_{0(CH)})^2 - U_{C\,min}^2) + C_0((U_{C0\,max} - R_{C0}I_{0(CH)}) - U_{C\,min})}{4k_C((U_{C0\,max} - R_{C0}I_{0(CH)})^3 - U_{C\,min}^3) + 3C_0((U_{C0\,max} - R_{C0}I_{0(CH)})^2 - U_{C\,min}^2)}}. \quad (2.59)$$

Please note that the maximum internal voltage is

$$U_{Cmax} = (U_{C0\,max} - R_{C0}I_{0(CH)}).$$ (2.60)

As we explained in the previous section, two cases can be distinguished:

1. The charge profile is defined.
   The ultra-capacitor minimum voltage is not constant, while the maximum voltage is defined and constant. Of course, the opposite situation is possible: the maximum voltage is a variable and the minimum voltage is a constant. We will consider the case where the maximum voltage is a constant while the minimum voltage is a function of the load profile and the capacitance C_N. The minimum voltage versus the charge profile can be computed from Equation 2.54 as

$$U_{C\,min} =$$

$$\sqrt{\frac{C_0^2}{4k_C^2} + \frac{1}{k_C}((U_{C0\,max} - R_{C0}I_{0(CH)})C_0 + (U_{C0\,max} - R_{C0}I_{0(CH)})^2 k_C - I_{0(CH)}T_{CH})} - \frac{C_0}{2k_C}.$$ (2.61)

2. The ultra-capacitor's minimum and maximum voltage is defined and constant.
   In this case, the minimum and maximum voltages are constant parameters while the charge current and charge time are variables.

2.5.2.3 Round Trip Efficiency

Round trip efficiency is defined as

$$\eta_{RTP} = \eta_{DCH}\,\eta_{CH}.$$ (2.62)

Substituting Equations 2.51 and 2.59 into Equation 2.62 yields

$$\eta_{RTP} = \frac{1 - R_{C0}I_{0(DCH)}6\dfrac{(k_C(U_{C\,max}^2 - U_{C\,min}^2) + C_0(U_{C\,max} - U_{C\,min}))}{4k_C(U_{C\,max}^3 - U_{C\,min}^3) + 3C_0(U_{C\,max}^2 - U_{C\,min}^2)}}{1 + R_{C0}I_{0(CH)}6\dfrac{k_C((U_{C0\,max} - R_{C0}I_{0(CH)})^2 - U_{C\,min}^2)}{+C_0((U_{C0\,max} - R_{C0}I_{0(CH)}) - U_{C\,min})}}.$$ (2.63)

The discussion about the load profile and minimum/maximum voltages as the variables in the efficiency equations (Equations 2.51 and 2.59) also applies for the round trip efficiency (Equation 2.63).

2.5.3 *Constant Power Charging and Loading*

In most power conversion applications, load and source behave as constant power. Typical examples of constant power load are power converters with regulated output voltage, such

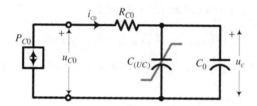

Figure 2.19 Charging/discharging the ultra-capacitor with a constant power source/sink

as pulse width modulated (PWM) variable speed drives and dc–dc converters. According to the convention of Figure 2.19, the power of the load is defined as

$$P_{C0} = -u_{C0} i_{C0},\qquad(2.64)$$

where power is positive in the drain (discharging) mode and negative in the source (charging) mode.

The circuit in Figure 2.19 is described by the following differential equation,

$$\frac{di_{C0}}{dt}(2k_C R_0^2 i_{C0}^4 + C_0 R_{C0} i_{C0}^3 + C_0 P_{C0} i_{C0} - 2k_C P_{C0}^2) - i_{C0}^4 = 0.\qquad(2.65)$$

Without losing the generality of the analysis, we can assume that the ultra-capacitor is a liner capacitor and the power P_{C0} is much lower than the matched maximum power, $P_{C0} << P_{max} = u_C^2/4R_{C0}$. Therefore, the ultra-capacitor resistance R_{C0} and the coefficient k_C can be neglected in the analysis. From these two approximations we have a simplified differential equation

$$\frac{di_{C0}}{dt} C_0 P_{C0} - i_{C0}^3 = 0.\qquad(2.66)$$

In general, the power source/drain is a time variant source. A solution of Equation 2.66 is

$$i_{C0} \cong -\frac{P_{C0(t=0)}}{U_{C(t=0)}}\sqrt{\frac{C_0 (U_{C(t=0)})^2}{C_0 (U_{C(t=0)})^2 - 2(P_{C0(t=0)})^2 \int_0^t \frac{1}{P_{C0}(\tau)} d\tau}},\qquad(2.67)$$

where
  $U_{C(t=0)}$ and $P_{C0(t=0)}$ are the initial voltage and power at $t=0$ and
  $P_{C0}(\tau)$ is the power-time profile.

In most applications, the charge/discharge power is constant in time or constant in segments of the time frame.

2.5.3.1 Discharging

The ultra-capacitor discharging current is

$$i_{C0} \cong -\frac{P_{C0(DCH)}}{U_C} \sqrt{\frac{C_0 U_C^2}{C_0 U_C^2 - 2P_{C0(DCH)}t}}, \quad P_{C0(DCH)} > 0 \qquad (2.68)$$

where the voltage U_C is the ultra-capacitor's initial voltage. The maximum discharge time $T_{DCH(max)}$ is determined from Equation 2.68 as

$$T_{DCH\,(max)} \leq \frac{C_0 U_C^2}{2P_{C0(DCH)}}. \qquad (2.69)$$

However, the discharge time (Equation 2.69) is theoretical and impossible to achieve. The system will become unstable before the discharge time (Equation 2.69) is achieved. The reason is the ultra-capacitance resistance R_{C0}, which is not considered in Equations 2.68 and 2.69. If we take R_{C0} into account, we have to compute the maximum discharge time from the differential equation (Equation 2.65).

The ultra-capacitor internal voltage is limited to the minimum

$$u_{C\,min} \geq \sqrt{P_{C0(DCH)}4R_{C0}}. \qquad (2.70)$$

The minimum discharge voltage (Equation 2.70) is determined by the system stability limit.

Figure 2.20 shows some simulation results. Discharge of an ultra-capacitor cell 3000 F/2.5 V has been simulated. The simulation parameters are: Discharge power $P_{C0(DCH)} = 1000\,\mathrm{W}$, initial voltage $U_{C0} = 2.5\,\mathrm{V}$. Figure 2.20a shows the internal voltage u_C and terminal voltage u_{C0}, while Figure 2.20b shows the terminal current i_{C0}.

Maximum Discharge Power
Just as in the case of current or resistive discharge, capacitor maximum power is limited by internal series resistance R_{C0}. The maximum power that can be delivered from an ultra-capacitor having the internal voltage u_C to the load having constant power characteristic is

$$P_{C0(DCH)\,max} = \frac{u_C^2}{4R_{C0}}. \qquad (2.71)$$

Please note that the maximum power is defined by the capacitor voltage and R_{C0}, exactly as in the two previous two cases (constant resistance and constant current load). There is, however, an essential difference. If the load is higher than the maximum power at the given ultra-capacitor voltage, the system becomes unstable and the voltage collapses.

Discharging Losses and Efficiency
Let the ultra-capacitor be discharged from the maximum voltage U_{Cmax} to the minimum voltage U_{Cmin} with constant power P_{C0}. Discharge losses are

$$E_{\varsigma(DCH)} = \int_0^{T_{DCH}} R_{C0}(t)i_{C0}^2(t)dt, \qquad (2.72)$$

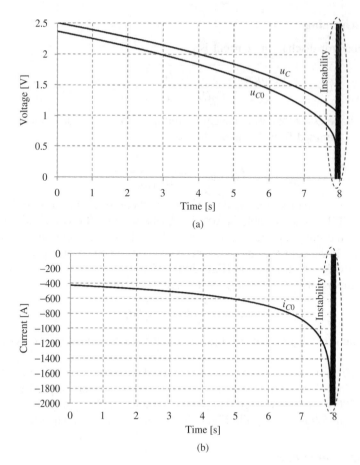

Figure 2.20 Simulation results of discharge of a large cell: $3000\,\text{F}/2.5\,\text{V}$, $P_{C0} = 1000\,\text{W}$, $U_{C0} = 2.5\,\text{V}$. (a) u_{C0}: Terminal voltage, u_C: internal voltage, and (b) i_{C0}: terminal current

where the discharge time T_{DCH} has to be computed from Equation 2.65. However, since Equation 2.65 is a first-order fourth-degree differential equation, it will not be easy and practical to use Equation 2.65 to calculate the discharge time. It can be done in several iterations using simplified model.

The First Iteration

Assume that the ultra-capacitor is an ideal capacitor without internal resistance. Under this assumption, we can compute the discharge time T_{DCH} as

$$T_{DCH(1)} = \frac{C_0(U_{C\,\text{max}}^2 - U_{C\,\text{min}}^2)}{2P_{C0(DCH)}}. \tag{2.73}$$

Now we compute the first iteration's discharge energy losses by substituting Equations 2.68 and 2.73 into Equation 2.72.

$$E_{\varsigma(DCH)(1)} \cong R_{C0} P_{C0(DCH)}^2 \int_0^{T_{DCH(1)}} \frac{C_0}{C_0 U_{C0\max}^2 - 2P_{C0(DCH)}t} dt$$

$$= \frac{R_{C0} P_{C0(DCH)} C_0}{2} \ln \left(\frac{C_0 U_{C0\max}^2}{C_0 U_{C0\max}^2 - 2P_{C0(DCH)} T_{DCH(1)}} \right). \tag{2.74}$$

The Second Iteration

Now we have a rough idea of the discharge energy losses. The next step is to compute the corrected discharge time as

$$T_{DCH(2)} = \frac{1}{P_{C0(DCH)}} \left[\frac{C_0 \left(U_{C\max}^2 - U_{C\min}^2 \right)}{2} - E_{\varsigma(DCH)(1)} \right]. \tag{2.75}$$

The second iteration's discharge energy losses are computed by substituting Equation 2.75 into Equation 2.72. Usually the second iteration gives sufficient accuracy of result and the discharge energy losses (Equation 2.76) are the final result.

$$E_{\varsigma(DCH)} \cong \frac{R_{C0} P_{C0(DCH)} C_0}{2} \ln \left(\frac{C_0 U_{C0\max}^2}{C_0 U_{C0\max}^2 - 2P_{C0(DCH)} T_{DCH(2)}} \right). \tag{2.76}$$

The discharge efficiency is defined as

$$\eta_{DCH} = \frac{\Delta E_C - E_{\varsigma(DCH)}}{\Delta E_C} = 1 - \frac{E_{\varsigma(DCH)}}{\Delta E_C}, \tag{2.77}$$

where E_C is the energy realized from the ultra-capacitor (Equation 2.28).

Substituting Equations 2.28 and 2.76 into Equation 2.77 yields the discharge efficiency

$$\eta_{DCH} = 1 - \frac{3R_{C0} P_{C0(DCH)} C_0 \ln \left(\frac{C_0 U_{C0\max}^2}{C_0 U_{C0\max}^2 - 2P_{C0(DCH)} T_{DCH(2)}} \right)}{3C_0 (U_{C\max}^2 - U_{C\min}^2) + 4k_C (U_{C\max}^3 - U_{C\min}^3)}. \tag{2.78}$$

2.5.3.2 Charging

In the charging mode of an ultra-capacitor, the power of the power source P_{C0} is negative according to the notation in Figure 2.19. Using the same method we used before, but simply applying negative power, we obtain the ultra-capacitor charging current

$$i_{C0} \cong -\frac{P_{C0(CH)}}{U_{C\min}} \sqrt{\frac{C_0 U_{C\min}^2}{C_0 U_{C\min}^2 - 2P_{C0(CH)}t}}, \quad P_{C0(CH)} < 0 \tag{2.79}$$

where the initial ultra-capacitor voltage is $U_{C\min}$.

The maximum charge time T_{CH} is determined by the ultra-capacitor maximum voltage $U_{C\max}$

$$T_{CH} \leq \frac{C_0 U_{C\min}^2}{2P_{C0(CH)}} \left(1 - \left(\frac{U_{C\max}}{U_{C\min}} \right)^2 \right). \tag{2.80}$$

Maximum Charging Power

In charging mode, the power P_{C0} is negative. Thus, the maximum power stability criterion is not applicable in this case. In other words, the system described by Figure 2.19 is stable in charging mode regardless of the power P_{C0}. In real applications, however, there is another limitation that defines the maximum power that can be transferred into the ultra-capacitor bank. It is the limitation of the ultra-capacitor and power source maximum voltage U_{0MAX} that must not be exceeded,

$$P_{C0(CH)\max} = \frac{U_{0\max}(U_{0\max} - u_C)}{R_{C0}}. \tag{2.81}$$

Charging Losses and Efficiency

Let the ultra-capacitor be charged from the minimum voltage U_{Cmin} to the maximum voltage U_{Cmax} with constant power $P_{C0(CH)}$. Charge losses are

$$E_{\varsigma(CH)} = \int_0^{T_{CH}} R_{C0}(t) i_{C0}^2(t) dt, \tag{2.82}$$

where charge time T_{CH} is defined by Equation 2.80.

Substituting Equation 2.79 into Equation 2.82 yields the charging losses

$$\begin{aligned} E_{\varsigma(CH)} &\cong R_{C0} P_{C0(CH)}^2 \int_0^{T_{CH}} \frac{C_0}{C_0 U_{C\min}^2 + 2P_{C0(CH)}t} dt. \\ &= \frac{R_{C0} P_{C0(CH)} C_0}{2} \ln\left(\frac{C_0 U_{C\min}^2 + 2P_{C0(CH)} T_{CH}}{C_0 U_{C\min}^2} \right) \end{aligned} \tag{2.83}$$

The charge time T_{CH} is computed using the two iteration methods used in Section 2.5.3.1.

$$T_{CH} = \frac{1}{P_{C0(CH)}} \left[\frac{C_0 \left(U_{C\max}^2 - U_{C\min}^2 \right)}{2} + E_{\varsigma(CH)(1)} \right] \tag{2.84}$$

where $E_{\varsigma(CH)(1)}$ is the charge energy loss of the first iteration defined as

$$E_{\varsigma(CH)(1)} \cong \frac{R_{C0} P_{C0(CH)} C_0}{2} \ln\left(\frac{C_0 U_{C0\max}^2}{C_0 U_{C0\max}^2 - 2P_{C0(CH)} T_{CH(1)}} \right), \tag{2.85}$$

where $T_{(CH)(1)}$ is the charge time of the first iteration defined as

$$T_{CH(1)} = \frac{C_0(U_{C\max}^2 - U_{C\min}^2)}{2P_{C0(CH)}}. \tag{2.86}$$

The charge efficiency is defined as

$$\eta_{CH} = \frac{\Delta E_C}{\Delta E_C + E_{\varsigma(CH)}} = \frac{1}{1 + \frac{E_{\varsigma(CH)}}{\Delta E_C}}, \tag{2.87}$$

where E_C is the energy realized from the ultra-capacitor (Equation 2.28).

Substituting Equation 2.28 into Equation 2.86 yields the charge efficiency

$$\eta_{CH} = \cfrac{1}{1 - \cfrac{3R_{C0}P_{C0(CH)}C_0 \left(\cfrac{C_0 U_{C\,min}^2 + 2P_{C0(CH)}T_{CH}}{C_0 U_{C\,min}^2} \right)}{3C_0(U_{C\,max}^2 - U_{C\,min}^2) + 4k_C(U_{C\,max}^3 - U_{C\,min}^3)}}. \qquad (2.88)$$

2.5.4 Exercises

Exercise 2.8

An ultra-capacitor cell with the parameters: $C_0 = 3000\,\text{F}$, $R_{C0} = 0.3\,\text{m}\Omega$, and $U_{C0max} = 2.7\,\text{V}$ is charged and discharged with constant current I_{C0}.

1. Calculate the maximum charging current and power at the end of the charge cycle taking into account the maximum cell voltage. The end of charge voltage is $U_{C0} = 2.5\,\text{V}$.
2. Calculate the maximum discharge current and power that corresponds to the maximum discharge current. What will happen if the ultra-capacitor is discharged with a current higher than the maximum?
3. Calculate the maximum discharge power and the current corresponding to maximum discharge power.

Solution

1. The maximum charging current and power are limited by the maximum cell voltage. From Equations 2.55 and 2.56 we compute the maximum current $I_{C0max} = 667\,\text{A}$ and the maximum power $P_{C0max} = 1800\,\text{W}$.
2. The maximum discharge current and power are limited by the capacitor's internal resistance. From Equation 2.47 we have the maximum discharge current $I_{C0max} = 8333\,\text{A}$ and the discharge power corresponding to the maximum discharge current is $P_{C0} = 0$. If the discharge current is greater than the maximum, the discharge power will become negative, which means the terminal voltage will become negative.
3. From Equation 2.46 we have the maximum discharge power $P_{C0max} = 5.2\,\text{kW}$. From Figure 2.18 we have a quadratic equation

$$R_{C0}I_{C0}^2 - U_{C0}I_{C0} + P_{C0\,max} = 0. \qquad (2.89)$$

The solution

$$I_{C0} = \frac{U_{C0}}{2R_{C0}} \pm \underbrace{\sqrt{\left(\frac{U_{C0}}{2R_{C0}}\right)^2 - \frac{P_{C0\,max}}{R_{C0}}}}_{=0}, \qquad (2.90)$$

gives the discharge current $I_{C0} = 4166\,\text{A}$.

Exercise 2.9

The ultra-capacitor cell from the previous example is charged and discharged with constant power.

1. Calculate the minimum discharge voltage U_{C0min} when the ultra-capacitor is discharged with constant power of $P_{C0} = 2500\,\text{W}$.
2. What will happen if the ultra-capacitor is discharged below the minimum?

Solution

1. The minimum discharge voltage is computed from Equation 2.71 as

$$U_{C0\,min} = 2\sqrt{P_{C0}R_{C0}}. \tag{2.91}$$

From the ultra-capacitor data we have $U_{C0min} = \mathbf{1.73\,V}$.

2. The ultra-capacitor discharged by constant power P_{C0}, Figure 2.19, is described by a quadratic function,

$$R_{C0}I_{C0}^2 - U_{C0\,min}I_{C0} + P_{C0} = 0, \tag{2.92}$$

with a solution

$$I_{C0} = \frac{U_{C0\,min}}{2R_{C0}} \pm \sqrt{\left(\frac{U_{C0\,min}}{2R_{C0}}\right)^2 - \frac{P_{C0}}{R_{C0}}}. \tag{2.93}$$

The solution of Equation 2.93 is real if

$$U_{C0\,min} \geq 2\sqrt{R_{C0}P_{C0}}. \tag{2.94}$$

That is basically the same condition as Equation 2.91. Physically, if the condition (2.94) is not respected and the ultra-capacitor is discharged below the limit, the system will become unstable.

Exercise 2.10

An ultra-capacitor cell with the following parameters, $C_0 = 2400\,\text{F}$, $k_C = 320\,\text{F/V}$, and $R_{C0} = 0.3\,\text{m}\Omega$, is being discharged with a constant current $I_{C0} = 50\,\text{A}$. The discharge time is $T_{DCH} = 50$ seconds and the initial voltage is $U_{C0} = 2.5\,\text{V}$.

1. Calculate the energy realized to the load and energy dissipated on the internal resistance.
2. Calculate the same if the discharge current and discharge time are $I_{C0} = 500\,\text{A}$ and $T_{DCH} = 5$ seconds.

Solution

1. From Equation 2.44 we have

$$U_{C0\,min} = \sqrt{\frac{C_0^2}{4k_C^2} + \frac{1}{k_C}(U_{C0}C_0 + U_{C0}^2 k_C - I_0 T_{DCH})} - \frac{C_0}{2k_C}, \qquad (2.95)$$

end of discharge voltage $U_{C0min} = 1.84\,\text{V}$. From Equation 2.29 we can find

$$\Delta E_C = \frac{C_0}{2}(U_C^2 - U_{C\,min}^2) + \frac{2}{3}k_C(U_C^3 - U_{C\,min}^3). \qquad (2.96)$$

energy realized from the ultra-capacitor $\Delta E_C = 5440.5\,\text{J}$ and dissipated on the resistance $\Delta E_\zeta = 37.5\,\text{J}$. Hence, the energy realized to the load is $\Delta E_C = 5403\,\text{J}$.

2. If the capacitor is discharged with a higher current but shorter period, the end of discharge voltage is the same as in the previous case $U_{C0min} = 1.84\,\text{V}$. Therefore the ultra-capacitor energy is the same $\Delta E_C = 5440.5\,\text{J}$. However, the internal resistance losses are 10 times higher, $\Delta E_\zeta = 375\,\text{J}$. The total energy realized to the load is $\Delta E_C = 5065.5\,\text{J}$.

2.6 Frequency Related Losses

As mentioned in Section 2.3.1, the ultra-capacitor is a nonlinear device, with voltage and frequency-dependent properties. In this section, the effect of the frequency-dependent resistance on conversion losses is discussed.

One can distinguish two different frequency ranges in the spectrum of the capacitor current. The first is a low frequency, which is related to the capacitor's operational mode and cycle. The second is a high frequency current due to the nature of the power converter used to charge/discharge the ultra-capacitor. Low frequency currents are normally aperiodic, while high frequency currents are periodic, where the basic period is a multiple or fraction of the switching period T_S.

The serial ESR is the frequency-dependent resistance. Based on the time to frequency transformations, one can conclude that the resistance is also time dependent.

$$R_{C0} = R_{C0}(\omega) \Rightarrow R_{C0} = R_{C0}(t). \qquad (2.97)$$

Due to the frequency dependent resistance, instantaneous voltage and current are not linked by a simple coefficient R_{C0},

$$u_{RC0}(t) \neq R_{C0} \cdot i_{C0}(t). \qquad (2.98)$$

Considering that the excitation current i_{C0} is sinusoidal function, one can write that the instantaneous voltage and current are linked by a simple coefficient R_{C0}, where R_{C0} is the resistance at a specified frequency,

$$u_{RC0}(t) = U_{RC0}\sin(\omega_0 t) = R_{RC0}(\omega)|_{\omega=\omega_0}I_{C0}\sin(\omega_0 t) = R_{RC0}(\omega)|_{\omega=\omega_0}i_{C0}(t). \qquad (2.99)$$

The instantaneous power of the resistor R_{C0} carrying a current i_0 is

$$p(t) = u_{RC0}(t) \cdot i_{C0}(t) = f(i_{C0}(t)) \cdot i_{C0}(t), \qquad (2.100)$$

where $u_{RC0} = f(i_{C0}(t))$ is the voltage across the ESR of the capacitor as a function of the current.

This nonlinearity is a kind of hidden nonlinearity, which exists due to the frequency-dependent resistance of the capacitor.

The average power calculated over a period T is

$$P_{C0(S)}(T) = \frac{1}{T} \int_t^{t+T} f(i_{C0}(\tau)) i_{C0}(\tau) d\tau. \qquad (2.101)$$

where the period T represents the fundamental period in the case of periodic function. If the current is nonperiodic, T represents the period of observation.

2.6.1 The Current as a Periodic Function

Consider the current i_{C0} as a periodic function with a period T_0, and angular frequency ω_0. This current can be expanded in a Fourier series:

$$i_{C0}(t) = \sum_{k=0}^{+\infty} I_{C0(k)} \sin(k\omega_0 t + \varphi_k). \qquad (2.102)$$

The voltage across ESR R_{co} can also be expanded in a Fourier series

$$uR_{co}(t) = \sum_{k=0}^{+\infty} R_{C0}(k\omega_0) \cdot I_{C0(k)} \sin(k\omega_0 t + \psi_k), \qquad (2.103)$$

where $R_{C0}(k\omega_0) = R_{C0}(\omega)$ is the frequency-dependent R_{C0} of the capacitor.

Since the resistor is quasi-linear, the phase displacement for each harmonic is zero as a consequence. Inserting Equations 2.102 and 2.103 into Equation 2.100 yields instantaneous power dissipated on the resistor R_{C0},

$$p(t) = \sum_{k=0}^{+\infty} I_{C0(k)} \sin(k\omega_0 t + \varphi_k) \cdot \sum_{k=0}^{+\infty} R_{C0}(k\omega_0) I_{C0(k)} \sin(k\omega_0 t + \varphi_k). \qquad (2.104)$$

Using the Lagrange identity [21],

$$f_1 \cdot f_2 = \sum_{k=0}^{+\infty} F_{1(k)} \cdot \sum_{k=0}^{+\infty} F_{2(k)} = \sum_{k=0}^{+\infty} F_{1(k)} F_{2(k)} + \sum_{n=0}^{+\infty} \sum_{\substack{m=0 \\ n \neq m}}^{+\infty} F_{1(n)} F_{1(m)}, \qquad (2.105)$$

where f_1 and f_2 are regular functions that can be expanded in potential series, yields

$$p(t) = \frac{1}{2} \sum_{k=0}^{+\infty} R_{C0}(k\omega_0) I_{0(k)}^2 (1 - \cos(2k\omega_0 t + 2\phi_k)).$$

$$+ \sum_{n=0}^{+\infty} \sum_{\substack{m=0 \\ n \neq m}}^{+\infty} I_{0(n)} \cdot R_{C0}(m\omega_0) I_{0(m)} \sin(n\omega_0 t + \phi_n) \cdot \sin(m\omega_0 t + \phi_m) \qquad (2.106)$$

The average power dissipated over a period T is

$$P_{C0(\varsigma)} = \frac{1}{T}$$

$$\times \int_{t}^{t+T} \left(\begin{array}{c} \frac{1}{2}\sum_{k=0}^{+\infty} R_{C0}\left(k\omega_0\right) I_{C0(k)}^2 (1 - \cos(2k\omega_0\tau + 2\varphi_k)) + \\ + \sum_{n=0}^{+\infty}\sum_{m=0}^{+\infty} I_{C0(n)} \cdot R_{C0}(m\omega_0) I_{C0(m)} \underset{n \neq m}{\sin(n\omega_0\tau + \varphi_n) \sin(m\omega_0\tau + \varphi_m)} \end{array} \right) d\tau.$$

$$(2.107)$$

Using the orthogonal property of the *sin* and *cos* functions,

$$\int_{t}^{t+T} \cos(n\omega_0\tau)\cos(m\omega_0\tau)d\tau\big|_{n\neq m} = 0, \quad \int_{t}^{t+T} \sin(n\omega_0\tau)\sin(m\omega_0\tau)d\tau\big|_{n\neq m} = 0$$

$$\int_{t}^{t+T} \cos(n\omega_0\tau)\cos(m\omega_0\tau)d\tau\big|_{n\neq m} = 0$$

$$(2.108)$$

yields an average power

$$P_{C0(\varsigma)} = \frac{1}{2}\sum_{k=0}^{+\infty} R_{C0}(k\omega_0) \cdot I_{0(k)}^2 = I_{RMS}^2 \sum_{k=0}^{+\infty} \left[R_{C0}\left(k\omega_0\right) \frac{I_{0(k)}^2}{I_{RMS}^2} \right] = I_{RMS}^2 R_{C0(EQ)}. \quad (2.109)$$

The resistance denoted $R_{C0(EQ)}$ is the ultra-capacitor equivalent resistance. Please note from Equation 2.109, that the total average power depends strongly on the frequency spectrum of the capacitor current. Thus, to estimate total losses one has to take account of the real time profile of the capacitor current.

2.6.1.1 Low Frequency Current

In general, the low frequency current will be either pseudo-periodic or periodic, depending on the application. The dominant time constant and fundamental frequency depends on the charge/discharge cycle and power/current level. This can be expressed as

$$i_{C0\_LF}(t) = i_{C0\_LF}(P_0, U_{C0}, R_{C0}, C). \quad (2.110)$$

Two application examples are illustrated in Figures 2.21 and 2.22. The first is a tooling machine application with intermittent load. The ultra-capacitor is used as an energy storage device to filter peak power from the mains supply. Figure 2.21a shows an experimental waveform of the ultra-capacitor current and voltage. The current amplitude spectrum is shown in Figure 2.21b. Note that the dominant frequency (first harmonic) is 0.48 Hz and the higher harmonics fall in the mid-frequency range (see example in Figure 2.7). Hence, the losses have to be computed taking all spectral components of the current and the frequency dependent resistance into account, as given in Equation 2.109.

Another typical application is a lift or hoisting application with a requirement for braking. The waveforms over one entire cycle are shown in Figure 2.22a, while the

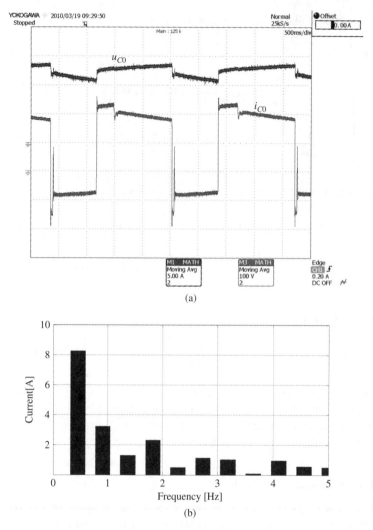

(a)

(b)

Figure 2.21 (a) Experimental waveform of the ultra-capacitor current i_{C0} (5 A/div) and voltage u_{C0} (100 V/div). (b) The current amplitude spectra. Variable speed drive with the ultra-capacitor as the energy storage device used to shave the drive input peak power

amplitude spectrum is shown in Figure 2.22b. In this example, the current fundamental frequency is below the ultra-capacitor mid-frequency range. Therefore, the losses can be computed using the total RMS current and the constant (frequency independent) equivalent resistance.

2.6.1.2 High Frequency Current

The ultra-capacitor's high frequency current is the current ripple caused by the power converter used to charge/discharge the ultra-capacitor. The current ripple depends on the

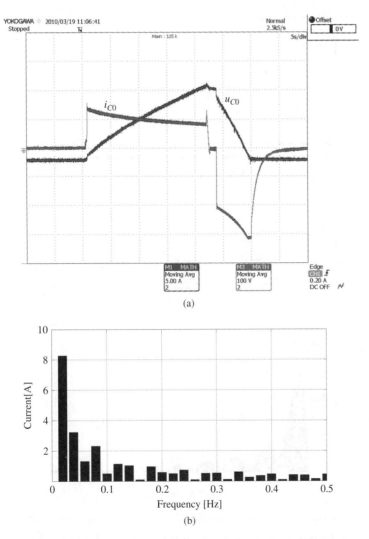

Figure 2.22 (a) Experimental waveform of the ultra-capacitor current i_{C0} (5 A/div) and voltage u_{C0} (100 V/div). (b) The current amplitude spectrum. Braking and energy recovery cycle of a variable speed drive with the ultra-capacitor as the energy storage device

power converter topology. This will be discussed in more detail in Chapter 5. For the moment, we can assume that the power converter is designed in such a way as to have the current ripple significantly smaller than the average current (selecting an appropriate topology or adding a low pass filter between the converter and the ultra-capacitor). Also, it is assumed that the fundamental frequency of the current ripple is far above the ultra-capacitor cut-off frequency (the capacitance and resistance can be assumed as constant properties).

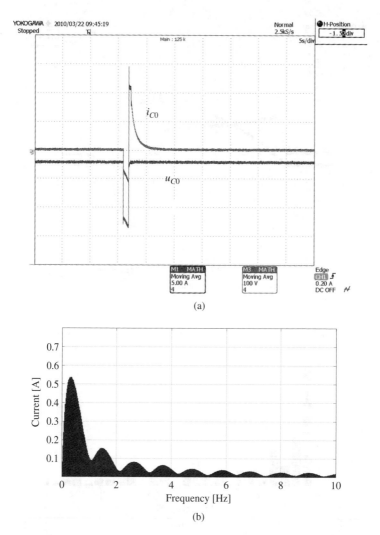

(a)

(b)

Figure 2.23 (a) Experimental waveform of the ultra-capacitor current i_{C0} (5 A/div) and voltage u_{C0} (100 V/div). (b) The current amplitude spectra. Variable speed drive with emergency supply based on the ultra-capacitor as an energy storage device

2.6.2 The Current as a Nonperiodic Function

Let's consider that the capacitor current is a nonperiodic function. A typical example is a variable speed drive with extended ride-through capability. The ultra-capacitor is employed as the energy storage to supply the drive system in case of a short power interruption. An example is illustrated in Figure 2.23. The waveforms of the ultra-capacitor current and voltage are shown in Figure 2.23a, while Figure 2.23b shows the current amplitude spectrum computed within a window $T = 50$ seconds.

The ultra-capacitor current and voltage drop on the equivalent resistor R_{C0} could be represented by Fourier integrals

$$i_0(t) = \frac{1}{2\pi} \int_{-\infty}^{+\infty} I(j\omega)e^{j\omega t}d\omega, \, u_{RC0}(t) = \frac{1}{2\pi} \int_{-\infty}^{+\infty} R_{C0}(j\omega)I(j\omega)e^{j\omega t}d\omega. \quad (2.111)$$

Then, instantaneous power dissipated on the R_{C0} is

$$p(t) = \frac{1}{4\pi^2} \int_{-\infty}^{+\infty} I(j\omega)e^{j\omega t}d\omega \cdot \int_{-\infty}^{+\infty} R_{C0}(j\omega)I(j\omega)e^{j\omega t}d\omega. \quad (2.112)$$

From the conditions $\int_{-\infty}^{+\infty} i_0(t)dt < \infty$ and $\int_{-\infty}^{+\infty} u_{RC0}(t)dt < \infty$ it follows that average power is zero

$$P_{C0(\varsigma)} = \lim_{T \to \infty} \frac{1}{T} \int_{-\frac{T}{2}}^{+\frac{T}{2}} p(t)dt = 0. \quad (2.113)$$

Please note that average power (Equation 2.113) is zero, and as such it has no practical value. However, the question is: how to evaluate the ultra-capacitor losses? Is it sufficient to compute the losses or is there some other important parameter? Two parameters are important, namely: (i) the energy dissipated on the internal resistor and (ii) the module temperature increase.

2.6.2.1 The Energy Losses

The energy dissipated on the ultra-capacitor internal resistance is

$$E_{C0(\varsigma)} = \int_{-\infty}^{\infty} p(t)dt = \frac{1}{4\pi^2} \int_{-\infty}^{\infty} \left(\int_{-\infty}^{+\infty} I(j\omega)e^{j\omega t}d\omega \cdot \int_{-\infty}^{+\infty} R_{C0}(j\omega)I(j\omega)e^{j\omega t}d\omega \right) dt. \quad (2.114)$$

2.7 The Ultra-Capacitor's Thermal Aspects

2.7.1 Heat Generation

Before we develop the ultra-capacitor cell thermal model, let's first define the heat generation mechanism. We can distinguish two kinds of heat source in the ultra-capacitor [28].

1. Irreversible heat generation due to ionic transport of charge in the electrolyte, and electronic transport of charge in the porous electrode and metallic collector. This was already discussed in Section 2.5.3.2.2, where we defined the ultra-capacitor losses due to the ultra-capacitor equivalent resistance.
2. Reversible heat generation due to the thermodynamic phenomena taking place inside the ultra-capacitor [22].

For the sake of simplicity, reversible heat generation can be neglected in the analysis. This assumption is correct if the thermal time constant is larger than the period of charge/discharge. In other words, if the dominant frequency of the ultra-capacitor current is lower than the cut-off frequency of the thermal model, the reversible heat generation can be neglected. Hence, the total power losses are basically just losses of the internal resistance.

The power losses can be computed in the frequency domain from Equation 2.112 and using convolution in the frequency domain.

$$P(j\omega) = \frac{1}{2\pi} \int_{-\infty}^{+\infty} (R_{C0}(jv)I(jv)I(j(\omega - v)))dv, \qquad (2.115)$$

where

R_{C0} is the ultra-capacitor resistance as a function of the frequency (Equation 2.10),
I is the ultra-capacitor current spectrum and ω is the angular frequency.

2.7.2 Thermal Model

A thermal model of an ultra-capacitor cell can be developed using a thermo-electrical analogy. Figure 2.24a shows a model that takes into account the losses and temperature distribution inside the ultra-capacitor cell. The entire cell is divided into j domains. The losses of each domain are denoted as $P_{C1(\zeta)}$ to $P_{Cj(\zeta)}$, while the domains' temperatures are θ_1 to θ_j. Each domain is represented by thermal capacitance C_{TH1} to C_{THj} and the domain to the cell terminal thermal resistance R_{TH1} to R_{THj}.

The heat generated in the ultra-capacitor cell is transferred to the ambient in two ways. The first is convection from the case surface to the ambient. The convection is modeled by a thermal resistor $R_{TH(conv)}$ connected between the cell case and ambient. The second is conduction from the cell terminal to the cell case $R_{TH(case)}$ and the terminals to the module $R_{TH(con)}$. In high power applications, the ultra-capacitor cells are large cells with large terminals. Since the ultra-capacitor current can be large, massive bas bars are used to connect the cells into a module. The bas bars are usually made of copper or aluminum, and therefore thermal conductivity can be significant. The heat transfer via radiation is neglected since the cell temperature and the radiation coefficient are small.

For the sake of simplicity, the ultra-capacitor cell can be modeled as a bulk source of heat having a lumped thermal capacitance C_{core} and resistance $R_{TH(core)}$, as shown in Figure 2.24b. The thermal model can be further simplified using the first order RC approximation (Figure 2.24c). The thermal resistance R_{TH} is total resistance from the cell core to the ambient. The thermal capacitance C_{TH} is total capacitance of the cell including capacitance of the electrodes and the cell case.

2.7.3 Temperature Rise

A simplified model of Figure 2.24c is used to estimate the cell temperature. For a detailed analysis of temperature distribution within the ultra-capacitor cell, the complete model of Figure 2.24a should be used. This, however, requires precise modeling of the ultra-capacitor cell. It is usually done by ultra-capacitor manufacturers using precise 3D thermal analysis [23].

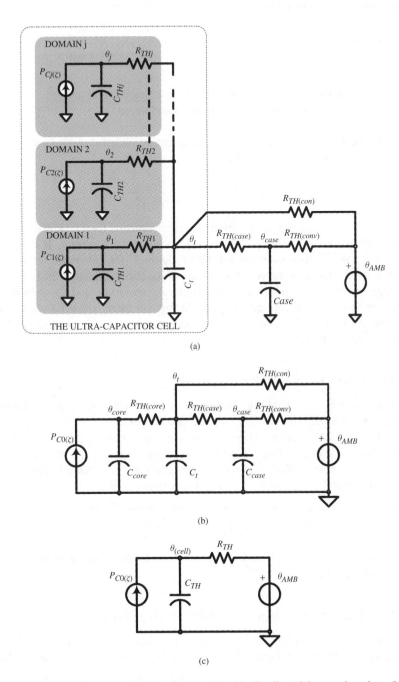

Figure 2.24 (a) Model of the ultra-capacitor cell with distributed losses domains, (b) lumped simplified thermal model, and (c) equivalent first order RC model

The ultra-capacitor temperature rise is

$$\Delta\theta(j\omega) = P_{C0(\varsigma)}(j\omega)Z_{TH}(j\omega) = P_{C0(\varsigma)}(j\omega)\frac{R_{TH}}{1+j\omega R_{TH}C_{TH}}, \qquad (2.116)$$

where

$Z_{TH}(j\omega)$ is the thermal impedance of the ultra-capacitor cell and
$P_{C0(\varsigma)}$ are the power losses in the frequency domain.

In some cases, it is easier to compute the temperature increase using Laplace transformation rather than Fourier transformation,

$$\Delta\theta(s) = P_{C0(\varsigma)}(s)\frac{R_{TH}}{1+sR_{TH}C_{TH}}. \qquad (2.117)$$

The losses $P_{C0(\varsigma)}(s)$ are computed from the time domain losses (Equation 2.109 or 2.112), depending on the ultra-capacitor current profile and spectrum. If the current is a periodic function, the losses are defined by Equation 2.109, otherwise we have to use general form losses given by inverse Fourier transformation (Equation 2.112).

Depending on the ultra-capacitor losses (current) spectrum and its relation to the ultra-capacitor thermal time constant, we can distinguish three cases: (i) very short non-repetitive pulse, (ii) very long pulse or constant power, and (iii) a pulse of moderate duration or repetitive pulses.

2.7.3.1   Very Short Non-Repetitive Pulse

Let the ultra-capacitor be loaded with a very short pulse, when the pulse width is much shorter than the thermal time constant of the ultra-capacitor, $T \ll T_{TH}$. The losses are assumed to be constant or quasi-constant. Since the pulse is short, the thermal process is mainly an adiabatic process. There is no heat exchange between the ultra-capacitor and ambient. The entire energy dissipated on the ultra-capacitor resistance is absorbed by the ultra-capacitor thermal capacitance C_{TH}. The temperature rise is

$$\Delta\theta(t) = \frac{1}{C_{TH}}\int_0^t P_{C0(\varsigma)}(t)dt = \frac{E_{C0(\varsigma)}}{C_{TH}}, \qquad (2.118)$$

where the losses $p_{C0(\varsigma)}(t)$ are given by Equation 2.112. The dissipated energy $E_{C0(\varsigma)}$ is given by Equation 2.114.

2.7.3.2   Very Long Pulse or Repetitive Pulses at High Frequency

When the ultra-capacitor load profile is such that the losses are constant or slightly varying over a long period, the thermal capacitance can be neglected. The temperature rise is

$$\Delta\theta(s) = P_{C0(\varsigma)}(s)R_{TH} \quad \text{and} \quad \Delta\theta(t) = p_{C0(\varsigma)}(t)R_{TH}. \qquad (2.119)$$

Let the ultra-capacitor load profile be a repetitive sequence with the repetition period T_{RP}, where the period T_{RP} is much shorter than the ultra-capacitor time constant,

$T_{RP} \ll T_{TH}$. In this case the ultra-capacitor thermal capacitance plays the role of a low pass filter. Thus, the average temperature is determined by the thermal resistance and the average losses. The ultra-capacitor cell temperature rise can be approximated by

$$\Delta\theta(t) \cong \left[\frac{1}{T_{RP}} \int_0^{T_{RP}} p_{C0(\varsigma)}(t)dt \right] R_{TH}, \tag{2.120}$$

where the losses $p_{C0(\varsigma)}(t)$ are given by Equation 2.112.

2.7.3.3 A Pulse of Moderate Duration or Repetitive Pulses at "the Critical" Frequency

When the ultra-capacitor load profile is such that its frequency falls in the same range as the thermal model cut-off frequency, the temperature is calculated from Equation 2.116 or 2.117.

2.7.4 Exercises

Exercise 2.11

An energy storage unit is composed of ultra-capacitor cells with the following parameters: $C_0 = 3000\,F$, $R_{C0} = 0.3\,m\Omega$, $R_{TH} = 3.2\,K/W$, $C_{TH} = 588\,J/K$. The ultra-capacitor energy storage is used in a short term UPS (uninterruptible power supply) system.

Calculate the ultra-capacitor cell temperature if the ambient temperature is $\theta_{AMB} = 40\,°C$ and the ultra-capacitor is discharged with current $I_{C0} = 500\,A$ during $T_{DCH} = 10$ seconds.

Solution

The ultra-capacitor time constant is $\tau_{TH} = 1880$ seconds. Since the load pulse is much shorter than the time constant, the first case (a short pulse) can be applied. From Equation 2.118

$$\Delta\theta = \frac{1}{C_{TH}} \int_0^t p_{C0(\varsigma)}(t)dt = \frac{R_{C0}I_{C0}^2 T_{DCH}}{C_{TH}}, \tag{2.121}$$

and the above parameters we can find the temperature rise $\Delta\theta = \mathbf{1.28\,°C}$ and the ultra-capacitor absolute temperature of $\theta = \mathbf{41.28\,°C}$.

Exercise 2.12

A large ultra-capacitor cell with the parameters given in Table 2.2 and Figure 2.8 is used in a power converter. The ultra-capacitor cell's thermal parameters are $R_{TH} = 3.2\,K/W$, $C_{TH} = 588\,J/K$. The ultra-capacitor's current is square-waveform with magnitude $I_0 = 100\,A$, frequency $f_0 = 1\,Hz$, and duty cycle $d = 0.5$. The ambient temperature is $\theta_{AMB} = 40\,°C$.

1. Calculate the cell temperature if the ultra-capacitor is loaded during $T_{LOAD} = 15$ minutes. The load repetition period is $T_P = 4$ hours.

2. Calculate the maximum load current of the ultra-capacitor cell if the cell's allowed temperature is $\theta_{max} = 65\,^{\circ}\text{C}$.

Solution

1. The ultra-capacitor current has a frequency that falls in the range where the ESR strongly depends on the frequency. Therefore, the losses have to be computed taking into account frequency-dependent ESR. The current can be expanded in Fourier series

$$i_{C0}(t) = \sum_{k=0}^{+\infty} I_{C0(k)} \sin(k\omega_0 t + \varphi_k) = I_0 \frac{4}{\pi} \sum_{k=0}^{+\infty} \frac{1}{k} \sin(k\omega_0 t). \tag{2.122}$$

Figure 2.25a shows the current harmonics up to the 19th harmonic. From the ultra-capacitor data given in Table 2.2 and Figure 2.8 we can compute the ESR at the fundamental frequency and its harmonics, Figure 2.25b.

From the losses equation (Equation 2.109) and the ESR versus frequency (Figure 2.25b), we compute the losses spectrum and total losses. The losses spectrum is depicted in Figure 2.26. The total losses are $P_{C0(\varsigma)} = 7.78\,\text{W}$.

Now, from Equation 2.117 and the ultra-capacitor losses we can compute the cell temperature versus time. First we need to calculate the initial temperature that depends on the losses and the losses profile. Since the load duty cycle $T_{LOAD}/T_P = 0.0625$ is very low, we can approximate the load profile with a single moderate duration pulse. The cell temperature θ versus time is

$$\theta(t) = \theta_{AMB} + \Delta\theta(t) = \theta_{AMB} + P_{C0(\varsigma)} R_{TH} \left(1 - \exp\left(-\frac{t}{R_{TH} C_{TH}}\right)\right). \tag{2.123}$$

Substituting data of the ultra-capacitor thermal and the load time T_{LOAD} into Equation 2.123 yields the cell temperature $\theta = 49.5\,^{\circ}\text{C}$.

2. From Equation 2.117 we can find the losses given the cell maximum temperature $\theta_{max} = 65\,^{\circ}\text{C}$.

$$P_{C01(\varsigma)} = \frac{\theta_{max} - \theta_{AMB}}{R_{TH} \left(1 - \exp\left(-\frac{T_{LOAD}}{R_{TH} C_{TH}}\right)\right)}. \tag{2.124}$$

From Equation 2.124 and the data given in the previous example we find the ultra-capacitor losses $P_{C01(\varsigma)} = 20.54\,\text{W}$.

The losses of an ultra-capacitor loaded with an arbitrary current with RMS value I_0 can be expressed in the form

$$P_{C0(\varsigma)} = I_0^2 R_{C0(EQ)}, \tag{2.125}$$

where $R_{C0(EQ)}$ is the ultra-capacitor equivalent ESR normalized to the current fundamental frequency, Equation 2.109.

From the losses ratio we can define the current I_{01}

$$I_{01} = I_0 \sqrt{\frac{P_{C0(\varsigma)}}{P_{C0(\varsigma)1}}}. \tag{2.126}$$

From losses $P_{C01(\zeta)} = 20.54\,\text{W}$, $P_{C0(\zeta)} = 7.78\,\text{W}$, and $I_0 = 100\,\text{A}$ we find the ultra-capacitor current $I_{01} = 162.5\,\text{A}$ that gives the maximum cell temperature $\theta_{max} = 65\,°\text{C}$ at the end of the load cycle $T_{LOAD} = 900\,\text{seconds}$ (15 minutes).

The above analyzed scenarios have been simulated. The ultra-capacitor cell temperature versus time is depicted in Figure 2.27.

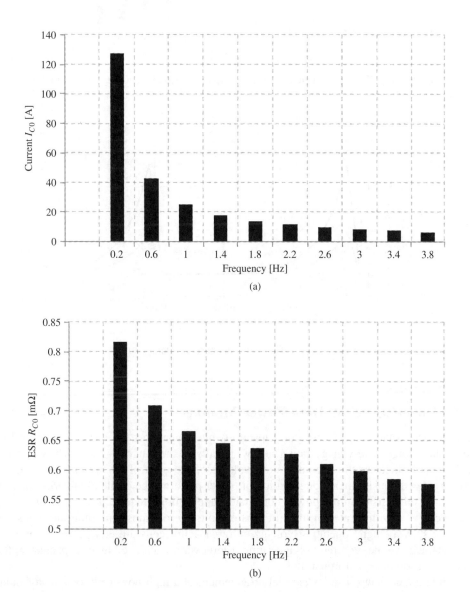

(a)

(b)

Figure 2.25 (a) The ultra-capacitor current spectrum and (b) ESR at fundamental frequency and harmonics

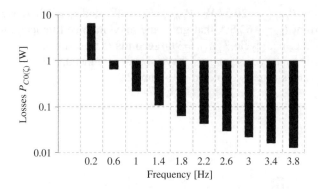

Figure 2.26   The ultra-capacitor losses at the fundamental frequency and harmonics

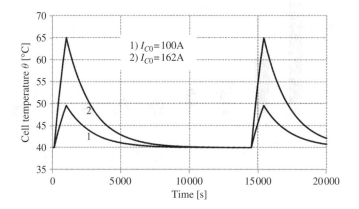

Figure 2.27   The ultra-capacitor cell temperature

2.8   Ultra-Capacitor High Power Modules

Unlike electrostatic and electrolytic capacitors, ultra-capacitors are characterized by low terminal voltage. The voltage rating is low, usually in the range 1–2.8 V. The operating voltage is limited by the decomposition voltage of the electrolyte and cannot be easily increased without significant degradation of power performance of the ultra-capacitor cell.

On the other hand, most power conversion applications require a short term energy storage device with a rated voltage of tens of volts up to almost one thousand volts. To achieve such a "high" operating voltage, ultra-capacitor cells are connected in series in order to achieve the required voltage rating. Moreover, to achieve higher capacitance, the cells are also connected in parallel.

Figure 2.28a shows a series/parallel arrangement of a high power ultra-capacitor module. In total N cells are series connected and M cells are parallel connected. The capacitance and voltage of each individual cell are $C_{0(cell)}$ and $U_{C0(cell)}$ respectively. The voltage

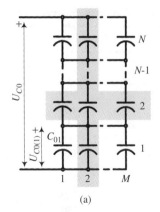

(a)

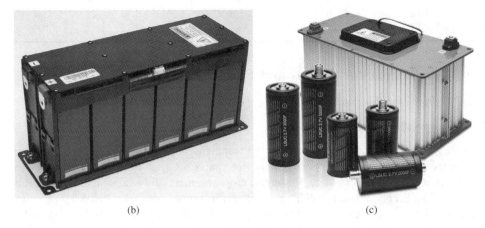

(b)   (c)

Figure 2.28 (a) Circuit diagram of an ultra-capacitor module. (b,c) LS MTRON ultra-capacitor module for UPS applications [31]. Copyright LS Mtron, with permission

rating and capacitance of the module are

$$U_{C0} = U_{C0(cell)}N \quad \text{and} \quad C_0 = C_{0(cell)}\frac{M}{N}. \tag{2.127}$$

The energy capability of the module is

$$E_{C0} = NM\,E_{C0(cell)}, \tag{2.128}$$

where $E_{C0(cell)}$ is the energy of one individual cell. The state-of-the-art modules available on the market are usually rated on the following voltages: 16, 48, 56, 125, and 200 V. Some of the available modules are summarized in Table 2.3.

If the power conversion application requires higher voltage and/or capacitance than provided by existing modules, more submodules have to be connected in series and in parallel.

Table 2.3 High power modules available on the market

Module	Voltage (V)	Capacitance (F)	SE (W h/kg)	Weight (kg)
Maxwell technologies				
BMOD0500	16	500	3.2	5.51
BMOD0083	48	83	2.6	10.3
BMOD0130	56	130	3.1	18
BMOD0094 P075	75	94	2.9	25
BMOD0063 P125	125	63	2.3	65
LS MTRON ultra-capacitors				
LS 16.2 V/500 F	16.2	500	3.5	5.1
LS 33.6 V/250 F	33.6	250	4	9.8
LS 50.4 V/167 F	50.6	166	3.43	17.2
LS 201.6 V/41 F	201.1	41	2.21	104

2.9 Ultra-Capacitor Trends and Future Development

2.9.1 The Requirements for Future Ultra-Capacitors

2.9.1.1 Specific Energy and Power

Two key features are relevant for the development of the new generation of ultra-capacitors: (i) specific energy and (ii) the internal ESR that basically determines the specific power and power density. Presently, most commercially available ultra-capacitors have specific energy around 5 W h/kg and specific power of up to 20 kW/kg. Existing ultra-capacitor specific energy is approximately 5% of the specific energy of the widely used lithium–ion battery. For most applications this is not sufficient. There is a need for specific energy of 20 W h/kg or more.

2.9.1.2 State of the Charge Characteristic versus Voltage

Another characteristic of ultra-capacitor energy storage is related to application requirements. If the ultra-capacitor is a pure linear capacitor, the terminal voltage and the energy extracted (stored) are linked by the following equation

$$\frac{U_{C0\,min}}{U_{C0\,max}} = \sqrt{1 - \frac{2\Delta E_C}{C_0 U_{C0\,max}^2}},\qquad(2.129)$$

where

U_{C0min} and U_{C0max} are the ultra-capacitor minimum and maximum voltage and
ΔE_C is the energy restored from the ultra-capacitor when discharged from maximum to minimum voltage.

If we need to restore 75% of the stored energy, the minimum voltage U_{C0min} will be 50% of the maximum voltage U_{C0max}. This is a big drawback of capacitor-based energy storage. Additional measures are necessary in order to ensure proper operation of the system when the ultra-capacitor voltage varies in such a wide range. This, as a consequence, increases the complexity and cost of the entire energy storage system.

The only way to minimize this effect is to design an ultra-capacitor as a nonlinear capacitor with a strongly voltage-dependent capacitance. In this case, the ultra-capacitor voltage is less dependent on the energy. Let the capacitance versus voltage be approximated by a first-order function (Equation 2.14). From Equation 2.28, we can find the ultra-capacitor minimum voltage as a function of the coefficient k_C, where the function has a strictly positive first derivative over k_C,

$$\left. \frac{\partial U_{C0\,min}}{\partial k_c} \right|_{\substack{U_{C0\,max} = Const \\ \Delta E_C = Const}} > 0. \qquad (2.130)$$

The higher the coefficient k_C the higher the minimum discharge voltage U_{C0min}. On the other hand, the higher the minimum voltage U_{C0min} the lower the stress and the better the utilization of the ultra-capacitor is.

2.9.2   The Technology Directions

Currently, there are four different mainstream ultra-capacitor developments: (i) carbon nano-tube technology, (ii) nano-gate technology, (iii) so-called EeStore technology, and (iv) mega farad super-capacitor technology. A summary of all four ongoing ultra-capacitor technologies is given in Table 2.4.

A team at Massachusetts Institute of Technology (MIT) led by Professor Joel Schindall has started development of a new ultra-capacitor based on carbon nano-tube technology [24, 25]. An ultra-capacitor based on such an approach could have specific energy as high as 25% or even 50% of the energy density of existing chemical batteries.

Okamura Laboratory and Power Systems announced the first significant development results of a new generation of ultra-capacitors based on very promising nano-gate technology in September 2007 [26]. The expected specific energy is 50–80 W h/kg, which is in the same order as the specific energy of the existing electrochemical batteries.

The third, quite different, ultra-capacitor technology is so-called EeStore technology, which promises to increase specific energy up to 280 W h/kg, and the operating voltage up to 3000 V [6]. These ultra-capacitors are based on high voltage multi-layer ceramic technology, and therefore the internal series resistance is expected to be very low compared to existing technologies.

Table 2.4   Existing ultra-capacitor technology and technologies under development

	Activate carbon	Nano-tube	Nano-gate	EesE	Mega farad super-capacitor
Specific energy (W h/kg)	~5	20–40	50–80	~280	~500
Operating voltage (V)	< 2.8	~3	~3.9	3000	< 2.8
Internal resistance (mΩ)	$0.66 + 1000/C^a$	As the existing	As the existing	Lower than the existing	As the existing

[a]This is an approximation where C is the cell capacitance (F).

A recently announced technology is the double-layer capacitor with a thin layer of high permittivity material on top of the activated carbon electrode [27]. The expected specific energy is 2 orders of magnitude greater than the existing technology, $\sim 500\,\mathrm{W\,h/kg}$.

2.10   Summary

In this chapter, the ultra-capacitor as an energy storage device dedicated to power conversion applications has been discussed. In comparison to state-of-the-art electrochemical batteries, ultra-capacitors have higher specific power, greater efficiency, longer lifetime, and greater cycling capability. In comparison to state-of–the-art electrolytic capacitors, the ultra-capacitors have higher specific energy. All these advantages make ultra-capacitors a good candidate for many power conversion applications with a need for short term, $0.1-15$ seconds, energy storage. The potential applications are industrial variable speed drives and systems, power transmission/distribution networks, traction, building, and IT centers.

The ultra-capacitor macro model has been discussed. Depending on the application requirements, a simplified first-order or higher-order RC model is proposed. The model can be used to estimate the ultra-capacitor losses and temperature. The first-order model is sufficient if the ultra-capacitor current frequency is well below or above the transition frequency. Otherwise, a second- or even a third-order model is necessary. The first-order model is sufficiently accurate for the interface power converter controllers' analysis and synthesis.

Activated carbon double-layer capacitors are state-of-the-art technology [2–5]. Among these, there are four different technologies under development: (i) nano-tube capacitors [24, 25], (ii) nano-gate capacitors [26], (iii) EeStore high voltage multi-layer capacitors [6], and (iv) mega Farad ultra-capacitors [27]. All technologies under development promise orders of magnitude higher specific energy in comparison to the state-of-the-art technology. Some of them, for example, technology (iii) and (iv), promise specific energy even greater than state-of-the-art electrochemical batteries.

References

1. Conway, B. E. (1999) *Electrochemical Supercapacitors, Scientific Fundamentals and Technological Applications*, New York Springer.
2. Schneuwly, A. and R. Gallay, (2000) Properties and applications of supercapacitors: from the state-of-the-art to future trends. Proceedings of PCIM.
3. Hermann, V., A. Schneuwly, and R. Gallay, (2001) High performance double-layer capacitor for power electronic applications. Proceedings of PCIM.
4. Burke, F.A. (2003) Ultracapacitors: present and future. Proceedings of the Advanced Capacitor World Summit 2003, Washington, DC.
5. Halper, M.S. and J.C. Ellenbogen, (2006) Supercapacitor: A Brief Overview, http://www.mitre.org (accessed 19 April 2013).
6. Weir, R.D. and C.W. Nelson, (2006) Electrical energy storage unit (EESU) utilizing ceramic and integrated circuit technologies for replacement of electrochemical batteries. US Patent 7 033 4060 B2, Apr. 2006.
7. Eskorad, S., Knoke, S., Majeski, J. and Smit, K. (2005) *Electrochemical Capacitors for Utility Applications*, Palo Alto, CA, EPRI, p. 1012151.
8. LS Mtron Ultra-Capacitors Data Sheet, http://www.ultracapacitor.co.kr (accessed 19 April 2013).
9. Maxwell Technologies Ultra-Capacitors http://www.maxwell.com/ultracapacitors (accessed 19 April 2013).

10. Belhachemi, F., S. Raiel, and B. Davat, (2000) A physical based model of power electric double-layer supercapacitors. *Proceedings of Industry Applications Conference*, **5**, pp. 3069–3076.
11. Buller, S., Karden, E., Kok, D. and Doncker, R.W. (2002) Modeling the dynamic behavior of supercapacitors using impedance spectroscopy. *IEEE Transactions on Industry Application*, **38**(6), 1622–1626.
12. Buller, S. (2003) Impedance-based simulation models for energy storage devices in advanced automotive power systems. PhD Dissertation. Institute for Power Electronics and Electrical Drives RWTH Aachen University.
13. Riu, D., N. Retière, and D. Linzen, (2004) Half-order modeling of supercapacitors. *Proceedings of Industry Applications Conference*, pp. 2550–2554.
14. Farande, R., M. Gallina, and D.T. Son, (2007) A new simplified model of double-layer capacitors. Proceedings of International Conference Clean Electrical Power, ICCEP '07, pp. 706–710.
15. Funaki, T. and T. Hikihara, (2008) Characterization and modeling of the voltage dependency of capacitance and impedance frequency characteristics of packed EDLCs. *IEEE Transactions on Power Electronics*, **23**, 3, 1518–1525, .
16. Gualous, H., D. Bouquain, A. Berthon, J.M. Kauffmann, (2003) Experimental study of supercapacitor serial resistance and capacitance variations with temperature, *Journal of Power Sources*, **123**, 86–93, .
17. Miller, J.M. (2011) *Ultracapacitor Applications, IET Power and Energy Series*, Vol. 59, Stevenage, The Institute of Engineering and Technology.
18. Rafik, F., H. Gualous, R. Gallay, A. Crausaz, A. Berthon, (2007) Frequency, thermal and voltage supercapacitor characterization and modeling, *Journal of Power Sources*, **165**, 928–934, .
19. Grbović, P.J. (2010) Ultra-capacitor based regenerative energy storage and power factor correction device for controlled electric drives. PhD dissertation. Laboratoired'Electrotechniqueetd'Electronique de Puissance (L2EP) Ecole Doctorale SPI 072, Ecole Centrale de Lille, Lille.
20. IEC (2006) 62391-2. *Fixed Electric Double Layer Capacitors for Use in Electronic Equipment-Part II: Sectional Specification Electric Double Layer Capacitors for Power Applications*. International Electrotechnical Commission.
21. Kurtz, M. (1991) *Handbook of Applied Mathematics for Engineers and Scientists*, McGraw-Hill, New York.
22. Shiffer, J., D. Linzen, and D. Uwe Sauer, (2006) Heat generation in double layer capacitors, *Journal of Power Sources*, **160**, 1, 765–772, .
23. Gualous, H., H. Louahlia-Gualous, R. Gallay, and A. Miraoui, (2009) Supercapacitor thermal modeling and characterization in transient state for industrial applications *IEEE Transaction on Industry Applications*, **45**, 3, 1035–1044, .
24. Signorelli, R., D. Ku, J. Kassakian, and J. Schindall, (2007) Fabrication and electrochemical testing of the first generation carbon-nanotube based ultracapacitor cell. 17th International Seminar on Supercapacitors and Hybrid Energy Storage Systems, pp. 70–79.
25. Signorelli, R., J. Schindall, D. Sadoway, and J. Kassakian, (2009) High energy and power density nanotube ultracapacitor design, modelling, testing and predicting performance. 19th International Seminar on Supercapacitors and Hybrid Energy Storage Systems.
26. Okamura, M., K. Hayashi, T. Tanikawa, and H. Ohta (2007) The nanogate-capacitor has finally been launched by our factory. Proceedings of the 17th International Seminar on Double Layer Capacitors and Hybrid Energy Storage Devices, December 10–12, 2007.
27. Ezzat, B. G. (2009) New mega-farad ultracapacitors. *IEEE Transactions on Ultrasonics, Ferroelectrics, and Frequency Control*, **56** (1), 14–21.
28. Linzen, D. (2005) Impedance-based loss calculation and thermal modeling of electrochemical energy storage devices for design considerations of automotive power systems. PhD dissertation. Institute for Power Electronics and Electrical Drives RWTH Aachen University.
29. Schindall, J. (2007) New [Old] Technologies for Energy Storage, NRC Emerging Technologies Study, October 16.
30. Miller, J.M. and U. Deshpande, (2007) Ultracapacitor technology: state of technology and application to active parallel energy storage systems. 17th International Seminar on Supercapacitors and Hybrid Energy Storage Systems, pp. 9–25.
31. Maxwell K2 Series Ultra-Capacitors Data Sheet, http://www.maxwell.com/products/ultracapacitors /products/k2-series (accessed 19 April 2013).

3

Power Conversion and Energy Storage Applications

3.1 Fundamentals of Static Power Converters

Static power converters are electric devices that's main role is to modify the presentation of one electric quantity to another, where the quantity is voltage or current. Figure 3.1 shows a block circuit diagram of a static power converter. The converter consists of a network of static devices, where a static device is one without moving parts. Static devices are passive components such as inductors, transformers, capacitors, and resistors, and power semiconductors such as diodes and controlled switches. Passive devices are connected in a subcircuit denoted as the input and output filter in Figure 3.1. Power semiconductors are connected in a subcircuit denoted as the switching matrix in Figure 3.1. Controlled power semiconductor switches are Insulated Gate Bipolar Transistor (IGBT), Thyristor and its derivatives such as Gate Turn-off Thyristors (GTO) and Integrated Gate Commutated Thyristors (IGCT), Metal Oxide Semiconductor Field Effect Transistor (MOSFET), Bipolar Junction Transistor (BJT), and Junction Field Effect Transistor (JFET) [1, 2].

3.1.1 Switching-Mode Converters

Switching-mode power converters are devices that transfer power (energy) from the input to the output periodically. Figure 3.2a illustrates a mathematical representation of a switching-mode power converter. The "converter" consists of a multiplexer MUX and a passive low pass filter (LPF). The input denoted x_{IN} is the main "power" input, while the input $s(t)$ is the control input. The control input is a so-called switching function that takes two values: 0 and 1. Details are given at the top of Figure 3.2b.

The multiplexer output denoted x_{SW} is basically a product of the power input x_{IN} and control input s,

$$x_{SW}(t) = x_{IN}(t)s(t). \tag{3.1}$$

Ultra-Capacitors in Power Conversion Systems: Applications, Analysis and Design from Theory to Practice,
First Edition. Petar J. Grbović.
© 2014 John Wiley & Sons, Ltd. Published 2014 by John Wiley & Sons, Ltd.

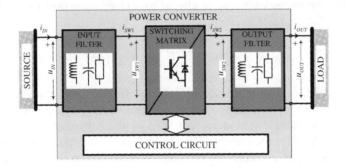

Figure 3.1 A static power converter as an interface between two electrical systems

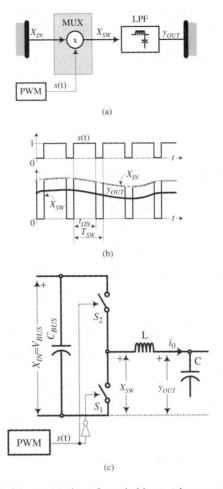

Figure 3.2 (a) Mathematical representation of a switching mode power converter, (b) waveforms, and (c) realization of a voltage source converter

The output of the multiplexer, x_{SW}, is filtered by an LPF, in order to extract y_{OUT} which is the average value of x_{SW},

$$y_{OUT}(t) = \frac{1}{T_{SW}} \int_t^{t+T_{SW}} x_{SW}(t)dt. \tag{3.2}$$

The output variable y_{OUT} is usually (in the majority of applications) voltage that is applied to a load. The input variable x_{IN} and switched variable x_{SW} can be voltage as well as current. If the variables x_{IN} and x_{SW} are voltage, the converter is the voltage source converter. If the variables are current, the converter is the current source converter.

Figure 3.2c illustrates a realization of a cell of a voltage source converter. The cell consists of two switches S_1 and S_2, a dc bus capacitor C_{BUS} and LPF LC. The dc bus voltage is the converter input variable $x_{IN} = v_{BUS}$. The output of the switching cell S_1S_2 is the variable x_{SW}. The filter output voltage is the output variable y_{OUT}. In some cases, the inductor current i_0 is defined as the system output variable.

The switch S_2 is controlled by the switching function $s(t)$, while the switch S_1 is controlled by the complementary signal of the switching function s(t). When the switching function $s(t) = 1$, the switch S_2 is conducting and the cell output voltage is $x_{SW} = x_{IN}$. When the switching function $s(t) = 0$ the switch S_1 is conducting and the cell output voltage is $x_{SW} = 0$. Hence, it is obvious that the switching cell S_1S_2 plays the role of a multiplexer from Figure 3.2a.

3.1.2 Power Converter Classification

We can distinguish four basic static power converter topologies, as illustrated in Figure 3.3:

(a) ac–dc power converters, known as rectifiers, which convert an ac quantity to a dc quantity of constant or varying magnitude.
(b) dc–dc power converters, known as choppers, which convert a dc quantity of one magnitude to a dc quantity of another constant or varying magnitude.
(c) dc–ac power converters, known as inverters. Inverters convert a dc quantity to an ac quantity at a certain magnitude, frequency, and phase which can be constant or varying.
(d) ac–ac power converters, known as cyclo-converters. These power converters convert an ac quantity with fixed or variable magnitude and frequency to another ac quantity with a different or the same magnitude, frequency, and phase.

Depending on the input/output quantity, static converters can be a state-of-the-art voltage source and current source converter [3], or the recently introduced impedance (Z) source converter [4]. In this book, we will address voltage-source power converters only.

3.1.3 Some Examples of Voltage-Source Converters

As an example of voltage-source converters, let's have a look Figure 3.4, which shows the circuit diagram of a three-phase pulse width modulated (PWM) dc–ac (ac–dc) power converter (Figure 3.4a) and a bi-directional dc–dc converter (Figure 3.4b). The dc–ac

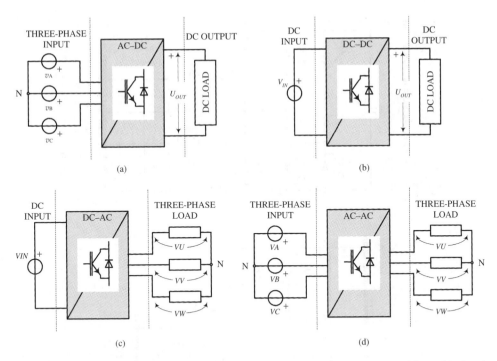

Figure 3.3 Four basic power converter topologies. (a) ac–dc converter (rectifier), (b) dc–dc converter (buck, boost, buck-boost, Ćuk), (c) dc–ac converter (inverter), and (d) direct ac–ac converter (cyclo-converter, matrix converter)

converter consists of six switches (IGBT+FWD) connected as a three-phase bridge. The dc side of the converter consists of a dc bus capacitor C_{BUS}. The dc bus voltage is V_{BUS}. The ac side of the converter consists of three inductors L_{OUT} and three capacitors C_{OUT}. The converter is bi-directional and therefore the power flow could be from the dc side to the ac and from the ac to the dc side. In the first case, we say that the converter is an inverter, while in the second case we say the converter is a rectifier. The converter switches (IGBTs) are driven by PWM signals at the switching frequency f_{SW}.

Figure 3.4b shows the circuit diagram of a high power bi-directional dc–dc power converter. Please note the similarity between the three-phase dc–ac and dc–dc power converters. Figure 3.5 shows the simulated output current and voltage of one phase of the two-level three-phase PWM inverter.

3.1.4 Indirect Static AC–AC Power Converters

Indirect ac–ac power converters are often used in certain applications, such as controlled electric drives, uninterruptible power supplies (UPS), and so on. A simplified circuit diagram is depicted in Figure 3.6. The converter is composed of two basic converters, an ac–dc (rectifier) and a dc–ac (inverter), which are connected via a common dc link. The primary power source can be the electric grid, a renewable power source such as

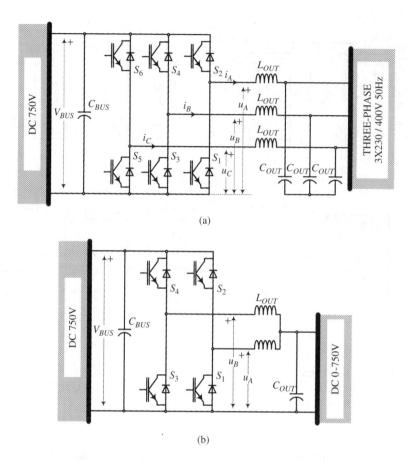

(a)

(b)

Figure 3.4 (a) Voltage source three-phase dc–ac (ac–dc) and (b) dc–dc power converter

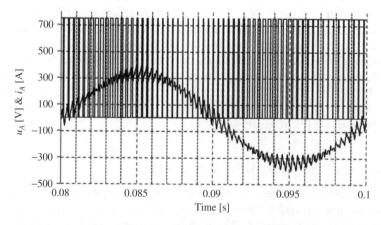

Figure 3.5 Simulated voltage and current waveforms of a three-phase dc–ac power converter. $V_{BUS} = 750\,\text{V}$, $U_{OUT} = 220\,\text{V}$, $f_{OUT} = 50\,\text{Hz}$, and $f_{SW} = 3\,\text{kHz}$

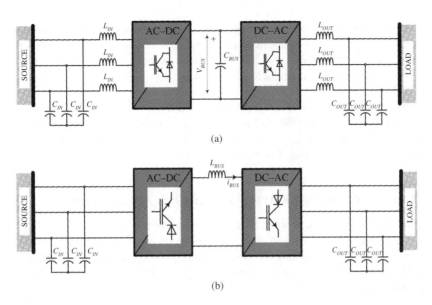

Figure 3.6 Indirect ac–ac back-to-back power conversion system (a) voltage source and (b) current source

wind or solar, or an autonomous power source such as a diesel engine or gas turbine (GT). The primary source voltage can be low voltage, in the range from 230 up to 690 V, or medium voltage, in the range from 1380 up to 6600 V.

Numerous different topologies can be used as the input rectifier. The topology selection depends on the application requirement. In most industrial low voltage/power applications, the rectifier is a three-phase diode rectifier with or without a dc bus filter inductor. In some applications, such as telecom and data center applications, uni-directional three-level boost rectifiers, known as Vienna Rectifiers are used [5]. A hybrid boost rectifier has been recently proposed [6]. In medium voltage high power applications, the rectifiers are mainly IGBT and IGCT-based PWM rectifiers or multi-phase diode/silicon controlled rectifiers (SCR)s. The input filter inductor L_{IN} and capacitor C_{IN} are part of the rectifier circuit. A modular multi-level converter has been recently proposed as a good alternative for medium voltage applications [7].

The dc link of a voltage source rectifier/inverter illustrated in Figure 3.6a is composed of a capacitor C_{BUS} which supports the dc bus voltage V_{BUS}. The dc bus capacitor plays two roles: the role of a filter capacitor and of energy storage on a very short term scale, up to 100 ms. This capacitor is traditionally an electrolytic capacitor or a high quality polypropylene film capacitor.

The dc–ac converter is a single-phase or three-phase three or four wire voltage-source inverter. The topology varies depending on the application requirement and voltage/power rating. In low voltage, low-to-medium power applications, the inverter is a two-level or three-level neutral point clamped (NPC) or flying capacitor (FC) PWM inverter [3]. In medium voltage high power applications, the inverters are multi-level topologies, such as FC, stacked H bridge converters, and modular multi-level converters (MMC) [7].

Regarding the inverter switch technology, two main Si-based switch technologies are well established. In low power applications, the IGBTs are a very well adopted solution, while in high power application, the IGCTs are used. Recently, significant effort has been put into the research and development of SiC-based high voltage switches, such as JFET and MOSFET [8].

The system load could be a three-phase induction as well as a synchronous motor, electric micro grid or public grid, and so on. In some of the applications, such as UPS or grid connected inverters, the load is connected to the inverter via an output filter inductor L_{OUT} and a capacitor C_{OUT} (see Figure 3.6).

3.2 Interest in Power Conversion with Energy Storage

In the previous section we briefly addressed static power converters. A static power converter has been defined as a device without energy storage capability, which means that the converter input and output instantaneous power are equal. However, is there any interest or need to integrate an energy storage device within the power conversion process? This will be addressed in this section.

3.2.1 Definition of the Problem

Let's consider an electric system as illustrated in Figure 3.7. The entire system consists of two subsystems, A and B. The nature of the subsystems A and B can be very different. These systems could be electric grids, variable speed drives, electric arc furnaces (EAFs), renewable energy sources, autonomous generators, and so on. The subsystems are interconnected via a device called the power processor. In this terminology, the power processor is a power converter that controls power flow between two subsystems. The power converter could be one of the basic topologies as well as a combination of them. For the sake of simplicity, we will assume that the power processor is an indirect ac–ac power converter, as described in the previous section. It was explained that the dc link voltage V_{BUS} is supported by relatively small dc bus capacitor. The dc bus capacitor is usually selected according to the current or voltage ripple criterion. The capacitor selected on such a criterion can be fully charged and discharged within a fraction of a second.

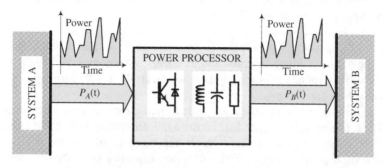

Figure 3.7 Power system A and B interconnected via a power processor without energy storage capability

Therefore, any variation in the system B power P_B instantaneously affects the system A power P_A and vice versa.

$$P_A(t) \cong P_B(t) \tag{3.3}$$

There are numerous examples of applications that show such behavior. Variable speed drives with intermittent load, such as lifts, cranes, and servo drives are typical examples [9–13]. Other examples are autonomous electric supply systems, such as rubber tyred gantry cranes (RTGC)s [14], heavy duty hybrid traction, excavators and earth moving equipment [15], and mobile gen-sets [16]. The local electric network is supplied from a diesel internal combustion engine (ICE) or GT gen-set. The network is loaded with three-phase motors with heavy intermittent characteristics.

Similar examples are high power arc furnaces that have extremely nonlinear and varying load characteristics [17–19]. In all these applications the output power (the load) is highly varying, even negative when an electric drive brakes, while the input power (grid or micro grid) is expected to be as smooth as possible.

Another example is renewable energy sources with strongly varying power production capacity, such as wind and ocean tidal turbines [20–25]. The system input power is produced by the turbine and is highly varying. The output power is expected to be as smooth as possible regardless of the variation of the input power (for example, the wind speed variation or sudden change of sun radiation).

An extreme case of variation of power supply capability is voltage interruptions and sags [17]. There are some applications that require ride-through capability, such as variable speed drives and UPS systems in applications such as critical industrial processes, hospitals, data centers, telecom base stations, military radar centers, and so on. All these applications are primarily supplied from the electric grid. However, due to unexpected interruptions caused by faults in the electric grid, supply of the critical load may be interrupted. This is unacceptable in most cases, and therefore an alternative power supply must be provided so that the load is supplied continuously. These are known as uninterruptible power supply (UPS) applications [26–28].

All the above mentioned applications are characterized by a common factor; power drawn from or injected into the electric grid has to be as smooth as possible regardless of the variation in power on the opposite side of the power conversion system. The main reasons for this requirement are:

- a fluctuating load may cause fluctuations and flickers in the grid voltage,
- additional losses in the distribution network,
- the distribution equipment, such as transformers, fuses, circuit breakers, and cables are sized on peak power,
- the peak power penalty, and
- critical loads must be supplied without interruption.

3.2.2 The Solution

What is the solution of the above described issue? To summarize, the issue is the lack of energy storage capability of the power processor that interconnects two electrical systems. This case is illustrated in Figure 3.7. Hence, the solution is the integration of energy storage

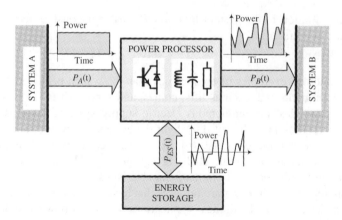

Figure 3.8 Power system A and B interconnected via a power processor with energy storage capability

into the power processor, as illustrated in Figure 3.8. With energy storage integrated within the power processor, power fluctuation of one system (A or B) fill will be filtered by the energy storage and will not be reflected to the second system (B or A). Average (smooth) power circulates between the systems A and B, while instantaneous fluctuating power circulates between the energy storage and the source of the fluctuation, not influencing the system that requires smooth power.

3.2.3 Which Energy Storage is the Right Choice?

In Chapter 1, we briefly addressed different energy storage technologies and devices used in power conversion systems. All energy storage technologies and devices can be split into two main groups: (i) large-scale and (ii) small-scale energy storage. Large-scale energy storages are hydro pumped and compressed air energy stores, which are used for long-term energy storage in power systems. Power levels are in the range of hundred MW, while charge/discharge time is in the order of couple of hours or days. Small-scale energy storages are super-conducting magnetic energy storage (SMES), ultra-capacitor energy storage (UCES), flywheel energy storage (FES), electrochemical batteries, and hydrogen fuel cells.

In the power conversion applications addressed in this book, small-scale (short-term) energy stores are the only ones considered. Due to their complexity and relatively low energy capability, SMES are used only in specific applications that are not considered here. Traditionally, three types of energy storage technologies are used: (i) flywheels, (ii) ultra-capacitors, and (iii) electrochemical batteries.

Flywheels and ultra-capacitors have similar energy density and power density characteristics. However, due to some constraints of FES, such as mechanical construction complexity and efficiency, ultra-capacitors are the preferred solution rather than flywheels. Thus, we will compare ultra-capacitors versus electrochemical batteries, while the flywheel will not be considered.

3.2.4   Electrochemical Batteries versus Ultra-Capacitors

3.2.4.1   Comparison of the Basic Properties

Before we start the discussion of which type of energy storage is an appropriate solution for the specific application requirement, let's briefly compare the basic properties of capacitors, ultra-capacitors, and electrochemical batteries. Their basic properties are summarized in Table 3.1. The energy density and power density of capacitors, ultra-capacitors, and electrochemical batteries is illustrated on the plot in Figure 3.9.

3.2.4.2   Selection and Sizing of the Energy Storage

Selection and sizing of an electrochemical battery depends on two main parameters: the storage capability and peak power. If the required energy capability is such that the autonomy time is several tens of minutes, the battery is selected on the grounds of energy capability. However, if the autonomy time is shorter than the critical $T_{CR(BATTERY)}$, let's

Table 3.1   Comparison of basic properties of ultra-capacitors and electrochemical batteries

	Ultra-capacitor	Electro-chemical battery
Specific energy (Wh/kg)	<10	50–150
Cost of energy ($/Wh)	5	—
Specific power (kW/kg)	5–10	<0.5
Cost of power ($/kW)	5–10	—
Cycling capability (cycles)	>500 000	<1000
Life time (years)	10–15	<5
Operating temperature (°C)	—	—

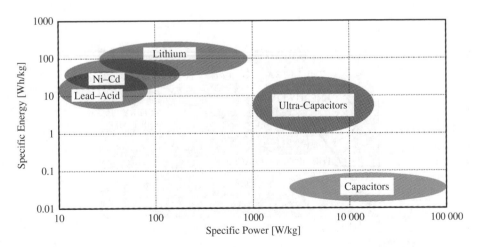

Figure 3.9   Power density versus energy density of capacitors, ultra-capacitors, and electrochemical batteries

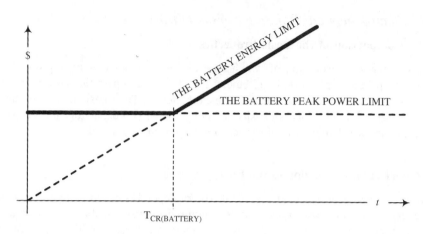

Figure 3.10   The energy storage cost versus the autonomy time. An electrochemical battery example

say couple of minutes, the battery selection and sizing must be based on peak power rather than energy capability. This is illustrated graphically in Figure 3.10.

The same sizing philosophy can be applied to the ultra-capacitor selection process. If the autonomy time is longer than the critical time $T_{CR(UC)}$, the ultra-capacitor is selected based on the energy capability, otherwise the ultra-capacitor is selected on the peak power demand. In both cases, peak power demand can be determined as the power at specific efficiency or maximum matched power that can be delivered or absorbed by the storage device.

Figure 3.11 illustrates the selection process of energy storage in a power conversion system. If the autonomy is shorter than the critical time T_{CR}, the UCES is a better and

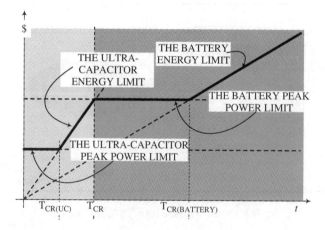

Figure 3.11   The energy storage cost versus the autonomy time. Comparison of an electrochemical battery and an ultra-capacitor

more cost-effective solution. Otherwise, battery energy storage is the better option. The selection process and a detailed discussion on the critical time calculation are given in Chapter 5.

3.2.4.3   Integration of Ultra-Capacitors in a Power Conversion System

As mentioned in Chapters 1 and 2, the ultra-capacitor is an electric capacitor, not an electrochemical battery. Therefore, the capacitor energy strongly depends on the capacitor voltage,

$$E_C(u_C) = \frac{1}{2}C_0 u_C^2, \qquad \Delta E_C = E_{\max}\left(1 - \left(\frac{U_{C\min}}{U_{C\max}}\right)^2\right). \tag{3.4}$$

where $E_{\max}$ is the maximum energy storage capability, and $U_{C\min}$ and $U_{C\max}$ are the ultra-capacitor minimum and maximum voltage respectively. For the sake of simplicity the ultra-capacitor is assumed as a linear capacitor C_0.

Please note from Equation 3.4 and Figure 3.12 that the ultra-capacitor voltage varies a lot with the energy. For example, let's assume the capacitor is fully charged on the nominal voltage. If the application requires 75% of the stored energy to be realized, the ultra-capacitor has to be discharged to 50% of the nominal voltage. However, such a large variation in ultra-capacitor voltage is very often unacceptable for the main converter. The dc link voltage is normally allowed to vary in a narrow range around the nominal voltage. To solve this issue, three solutions are theoretically possible.

The first solution is to select the ultra-capacitor such that the voltage variation for the energy stored/extracted is acceptable for the dc bus. This is, however, not an optimal solution. The ultra-capacitor will be oversized and therefore unacceptably large and expensive.

The second option is to design the main power converter in such a way as to have dc bus voltage in a broad range. This is also unacceptable because it requires a higher voltage rating of the converter switches and therefore means lower efficiency and higher cost, which is not competitive, nor practical.

The third solution is to connect the ultra-capacitor and the dc bus via a voltage matching device. The voltage matching device can be an ideal dc transformer with variable transfer ratio M, Figure 3.13. However, an ideal transformer does not exist. It is only a representation and model of real power converters. In this case, the ideal dc transformer is a model of a bi-directional dc–dc power converter, Figure 3.13b. The dc–dc converter is controlled in such a way as to control the dc bus voltage regardless of the variation of the ultra-capacitor voltage. The bus voltage can be controlled to be constant or to follow a reference, which depends on the application requirement.

The solution with the bi-directional dc–dc interface converter is standard in power conversion application with integrated UCES. The topology of the interface converter varies from case to case. It could be a non-isolated or isolated converter, two-level or multi-level, single-cell or multi-cell interleaved, and so on [29, 30]. More details of the interface dc–dc converter are given in Chapter 5.

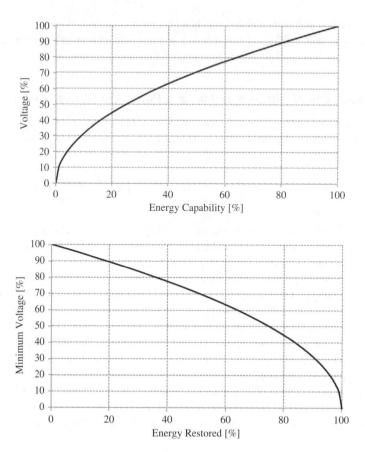

Figure 3.12 Ultra-capacitor voltage and energy relation. (a) Voltage versus energy capability. (b) Minimum voltage versus energy restored from the ultra-capacitor

3.3 Controlled Electric Drive Applications

Electric drives are electromechanical converters that convert electrical energy into mechanical energy in order to make rotational as well as linear movement. Figure 3.14 shows a block diagram of a controlled electric drive. The drive consists of an electric motor as an electromechanical converter, a mechanical load, a power converter, and a primary power source. The primary power source could be a public single-phase or three-phase grid, dc distributed grid, an electrochemical power source, such as a battery or fuel cell and diesel ICE, or a gas turbine gen-set.

3.3.1 Controlled Electric Drives from Yesterday to Today

Throughout the centuries, "production power" was the power of animals and slaves, hydro power, and wind power. In the 1800s, after James Watt's invention, it was the power of

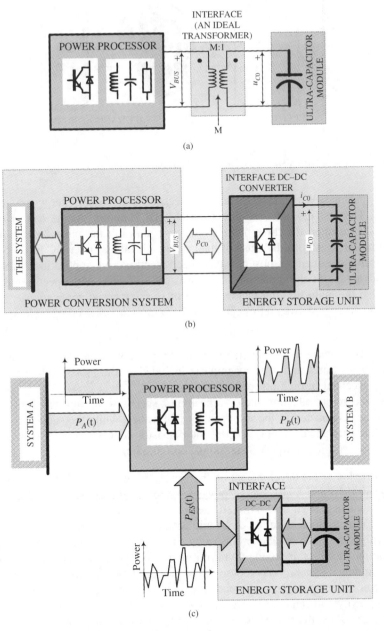

Figure 3.13 Integration of the ultra-capacitor within the power processor. (a) An ideal dc transformer, (b) a bi-directional controlled dc–dc converter as an emulation of an ideal dc transformer, and (c) an entire conversion system

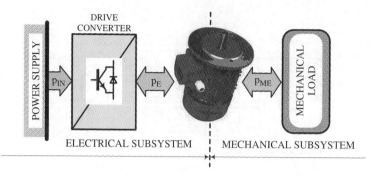

Figure 3.14   Controlled electric system as an electrical to mechanical energy conversion system

the steam machine. With the discovery of electricity, electrical energy gradually came into focus. The first motors were direct current (DC) motors. At the end of the 1800s, Nikola Tesla invented the three-phase voltage system and the most famous motor, the induction motor, was born. Because of their many advantages over DC motors, induction motors became dominant in most constant speed electric drive applications. However, the basic disadvantage of the induction motor was the difficulty with speed regulation, and this was also the main limiting factor for use in variable speed applications.

At the beginning of twentieth century, a few configurations of variable speed drives were used.

1. The Ward–Leonard motor-generator group. Since power conversion in such a system is done three times and a dc machine is included in the loop, this concept was not broadly accepted in high power applications.
2. The wound rotor induction motor. The motor speeds were adjusted and "regulated" by a circuit connected to the rotor via set of brushes, while the stator was connected to the fixed frequency supply: (i) the rotor resistor control and (ii) "constant power" Kramer or "constant torque" Scherbious configurations.

The first period of the development of "power electronic" controlled electric drives was that between 1910 and 1940. Early "power electronic" drives were based on triggered-arc power switches, such as controlled mercury-arc rectifiers, thyratrons, and ignitrons. The drive configurations were the electronic Kramer configuration using an uncontrolled rectifier bridge, the electronic Scherbius using a rectifier-inverter configuration, the Brown Boveri commutator-less drive, the thyratron motor configuration, and an early version of the load commuted synchronous motor drive. In the 1930s, the first cyclo-converter was used. None of these topologies had broad success in industrial applications, simply because of the complexity and reliability issues of the "power electronic" switches.

In the early 1960s, the first SCR was invented. This invention was a large step toward development of power electronics controlled electric drives. The Kramer and Scherbius drive configuration, load commuted synchronous motor drives, and cyclo-converter drives became dominant in most high power applications. Later on, the current sourced inverter with a variable output frequency became a very competitive scheme for induction motor applications. Voltage source drive topologies became competitive with the invention of the GTO in the 1970s and IGBTs in the 1980s.

Today, 70% of the world's electricity production is consumed by some kind of controlled electric drive; traction and transportation drives, industrial drives, home appliance drives, and so on. This indicates the importance of controlled electric drives in everyday life.

Modern controlled electric drives are exclusively based on three-phase motors, either induction or permanent magnet synchronous motors. The motor is powered from a primary power source via a power converter, the so-called drive converter. The drive converter controls the motor rotor speed and position. The converter can be a voltage source as well as the current source topology [3]. Voltage source converters are traditionally used in low voltage and low to medium power applications. Current source converters are used in high power and high voltage applications.

3.3.2 Application of Controlled Electric Drives

Applications of controlled electric drives are in a very broad range, starting from general purpose drives for pumps and fans, hoisting applications such as cranes and lifts, conveyers, servo drives, and so on. One particular group of electric drive applications are traction drives. As traction drive applications are very important in everyday life, they will be addressed in Section 3.8 as an independent application group.

The first type of application is the hoisting type application. Figure 3.15a shows a photo of a typical on-port rubber tyred gantry (RTG) crane. The hoisting drive time–power profile is sketched in Figure 3.15b. When lifting the load, the hosting drive takes energy from the primary supply, in this case a diesel engine generator. When lowering the load, the drive operates in braking mode. As a diesel engine generator is not reversible, the braking energy cannot be pumped back into the primary power source. The braking energy is dissipated as heat in the brake resistor. RTG applications are discussed in more detail in Section 3.5.

Lift applications are similar to the RTG crane applications, except for two differences. The mains, as the primary power supply source, is reversible in the case of lift applications. This means that the drive braking energy can be fed back to the mains. The lift load (cabin) is balanced by a counterweight. Thus, the drive operating mode depends on the load direction (up or down) and the ratio of load to counterweight. Figure 3.16a,b shows a picture of a typical lift and power-time load profile.

The second type of application is industrial machines with intermittent load. Such applications are characterized by a low ratio of average to peak power. The input power is highly positive when the drive accelerates, and the power is highly negative when the drive decelerates. During constant speed operation, the input power is normally low. Typical applications are industrial robots and tool carriers in automatic milling machines (Figure 3.17).

3.3.3 Definition of the Application Problems

Three issues can be identified in the application of controlled electric drives, [9]:

1. recovery of the braking energy,
2. the drive's ride-through capability, and
3. fluctuation of the drive's power.

(a)

(b)

Figure 3.15 (a) RTGC as example of a controlled electric drive application with a highly intermittent load (Copyright Liebherr Container Cranes Ltd, with permission) and (b) power–time profile

3.3.3.1 Dynamic Braking

The drive power is normally positive, where positive means the power flow from the input supply to the motor. However, in some cases the motor can turn to be generator. In this instance, power flows from the generator (motor) into the input supply. This happens when the drive decelerates or operates in the constant speed mode with negative torque. For example, this could be in hoisting applications or the deceleration of drive load with a large inertia. If the rectifier is uni-directional, as is case in most low cost

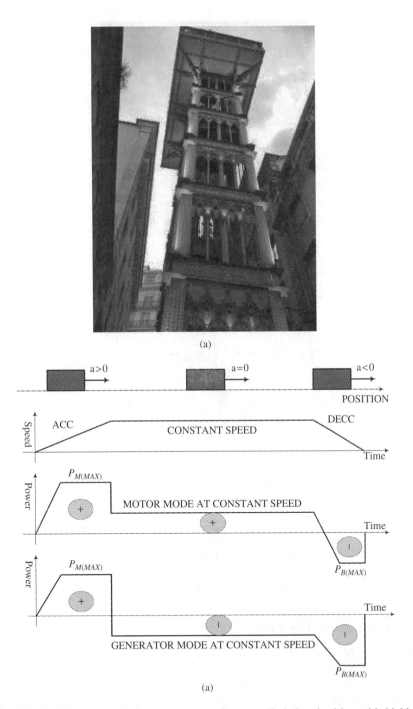

(a)

(a)

Figure 3.16 (a) Lift as an application example of a controlled electric drive with highly intermittent load (image of Elevador de Santa Justa, copyright © 2007 Soulman of XD400) and (b) power–time profile.

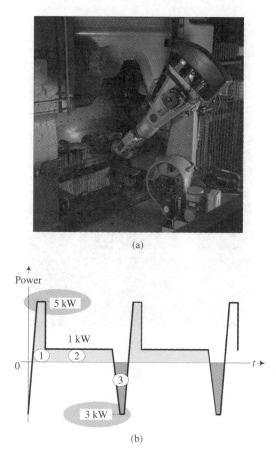

Figure 3.17   (a) An industrial robot as an example of a highly intermittent load and (b) power–time profile. Copyright © 2003 KUKA Roboter GmbH

low power applications, an additional passive circuit must be used to dissipate the energy from the load. A typical example is a lift drive with a brake resistor, [9]. As reported in the literature, 20–30% of consumed energy is burned by the braking resistor. In today's energy crisis, energy efficiency has become an issue requiring an urgent solution.

3.3.3.2   Ride-Through Capability

Ride-through capability of a device is defined as the capability of the device to survive (ride-through) short power interruption of the main power supply. A short power interruption may last from a few cycles (20–40 ms) up to a few seconds.

Modern controlled electric drives are sensitive to mains supply disturbances. The most frequent disturbances are voltage dips/sags. Voltage sag is defined as the instantaneous decrease in the RMS voltage, where the decrease is in the range of 10–90% of the nominal voltage, and the sag duration is in the order of a half cycle to up to a minute. Such a power interruption causes the dc bus voltage to drop below its lower limit, and then the entire drive system trips. In critical process industries, such as the semiconductor and glass industry, power interruptions are very costly, in the range of 10 k US$ up to 1 M US$ per single interruption.

3.3.3.3 Input Peak Power

When the drive accelerates, the input power is high and positive. During deceleration the power is high and negative. At constant speed operation, the input power is low and could be positive as well as negative. Figure 3.18 shows the power–time profile of a typical controlled 30 kW electric drive. Typical application examples are hoisting drives, industrial tolling machines with intermittent load, and public transportation traction drives.

We can distinguish three operating modes:

1. Acceleration, when the drive power is highly positive. The energy is being stored as the drive kinetic energy.
2. Constant speed, when the drive power is low and positive.
3. Deceleration, when the drive power is highly negative. The kinetic energy is being restored.

The peak to average power ratio could be greater than 10. Such strong input power fluctuation directly affects the supply grid. Power fluctuation on the grid side causes voltage fluctuation and additional losses and requires oversized installation.

3.3.4   The Solution

To solve the issue of power fluctuation, which can be the load power variation or the supply power variation, short-term energy storage can be integrated within the drive converter. The energy storage filters the load power fluctuation and provides power in case of a mains power supply interruption. The energy storage device is an ultra-capacitor with an interface dc–dc converter, as illustrated as a general case in Figure 3.13b. The ultra-capacitor is

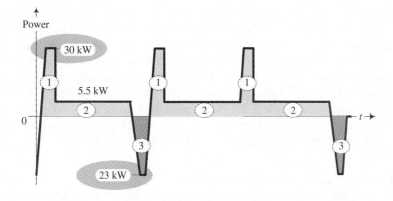

Figure 3.18   Power–time profile of a controlled electric drive in fast start/stop mode

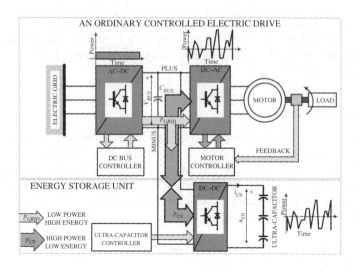

Figure 3.19   A controlled electric drive with an ultra-capacitor as an energy storage device

used as energy storage for different functions, such as the saving of braking energy and extension of ride-through capability [9–11] and smoothing of peak power [12, 13].

Figure 3.19 shows a block diagram of a controlled electric drive with integrated UCES. The drive converter consists of a PWM three-phase voltage source inverter, a dc bus capacitor C_{BUS} and an input rectifier. The rectifier is fed from the electric grid via a grid filter and an interconnection transformer (which is optional, depending upon the drive power and voltage rating). The UCES is connected to the drive converter via an interface bi-directional dc–dc converter.

3.3.4.1 The DC–DC Converter and Control

The interface dc–dc converter is usually a non-isolated bi-directional buck converter, but it could also be another topology. More details of the interface converter topology are given in Chapter 5. The primary control objective of the dc–dc converter is to asymptotically regulate the dc bus voltage to a desired reference, where the reference depends on the system-operating mode. The secondary control objective is to smooth the drive input power, regardless of the variation of the drive load. The last control objective is to control the ultra-capacitor state of the charge, including control of the ultra-capacitor current.

One example of the interface converter control structure proposed and analyzed in [9, 11–13] is depicted in Figure 3.20. Please note the two control units. The top one is the drive rectifier control, while the bottom one is the dc–dc converter control. The dc–dc converter control scheme is a cascade control with four control levels:

1. PWM generator and converter protection.
2. The ultra-capacitor current control and limitation. The current controller is denoted as $G_{iC0.}$

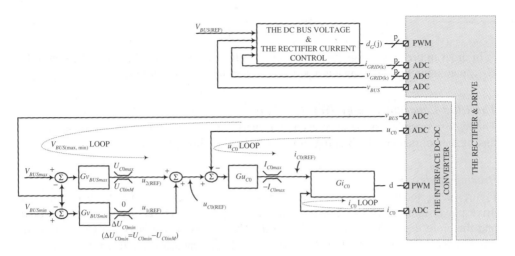

Figure 3.20 An example of the interface dc–dc converter control scheme

3. The ultra-capacitor voltage control and limitation. The voltage controller is denoted as G_{uC0}.
4. The dc bus voltage control. The dc bus voltage is controlled by two controllers, $G_{vBUSmax}$ and $G_{vBUSmin}$.

The rectifier control and the dc–dc converter controls are coupled via the common dc bus voltage. The dc bus voltage is controlled by the dc–dc converter and rectifier in parallel. The rectifier reference voltage is $V_{BUS(REF)}$, while the references for the dc–dc are V_{BUSmax} and V_{BUSmin}. To guarantee functionality of the entire system, the rectifier and dc–dc converter dc bus voltage references have to be

$$V_{BUSmin} < V_{BUS(REF)} < V_{BUSmax}. \qquad (3.5)$$

3.3.4.2   Control Modes

The control system can operate in three different modes: the rectifier mode, the ultra-capacitor energy transfer mode, and the mains peak power filtering mode.

The Rectifier Mode
The drive is supplied from the mains. The dc bus voltage is equal to the reference $V_{BUS(REF)}$ (being actively controlled by the drive rectifier). The dc bus voltage is lower than the reference V_{BUSmax} and greater than the reference V_{BUSmin}. Hence, the controller $G_{vBUSmax}$ is saturated at U_{C0inM}, while the controller $G_{vBUSmin}$ is saturated at zero. Please, note that the controllers $G_{vBUSmax}$ and $G_{vBUSmin}$ are designed in such a way as to have *out* ↓ *if error* > 0 & *out* ↑ *if error* < 0, where *out* is the controllers' output and *error* is the controllers' input (the dc bus voltage control error). The symbols ↑ and ↓ denote that the variable is increasing and decreasing respectively. The ultra-capacitor voltage reference is therefore given by Equation 3.6.

$$u_{C0(REF)} = U_{C0inM} + 0 = U_{C0inM}. \tag{3.6}$$

In order to prevent energy flow between the ultra-capacitor and the drive, the controller G_{uC0} maintains the ultra-capacitor voltage as constant at the intermediate level U_{C0inM}.

The Ultra-Capacitor Energy Transfer Mode
The system operates in the ultra-capacitor energy transfer mode. Two sub-modes can be distinguished, namely the braking mode and the ride-through mode.

Braking Mode
Figure 3.21a shows the profile of an ultra-capacitor current and a voltage profile. The drive power is negative (the braking mode). Let's assume the rectifier is uni-directional, and therefore the current cannot be reversed. The dc bus capacitor is charged and the dc bus voltage v_{BUS} increases. Once the dc bus voltage reaches the reference V_{BUSmax}, the

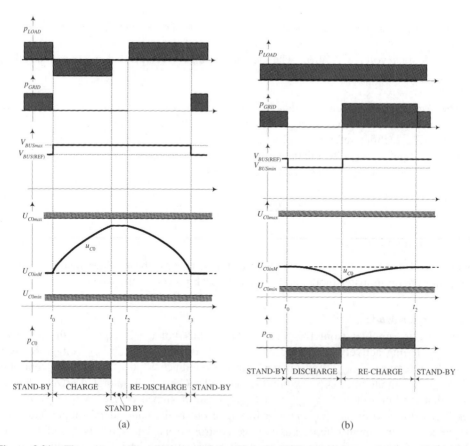

Figure 3.21 The system voltage/current time profiles of a VSD (variable speed drive). (a) Braking and energy recovery mode and (b) ride-through and the grid recovery mode

dc bus voltage controller $G_{vBUSmax}$ is de-saturated, while the controller $G_{vBUSmin}$ remains saturated at zero. Thus, the ultra-capacitor voltage reference u_{C0ref} starts increasing from U_{C0inM} toward U_{C0max},

$$U_{C0inM} < u_{C0(REF)} \uparrow \leq U_{C0\,max}. \tag{3.7}$$

The ultra-capacitor current reference is set by the controller G_{uC0} at such a level as to maintain the dc bus voltage constant $v_{BUS} = V_{BUSmax}$.

If the braking energy is greater than the ultra-capacitor capacity, the ultra-capacitor will be fully charged at the maximum voltage U_{C0max} before the braking phase is finished. The dc bus voltage controller $G_{vBUSmax}$ will be saturated at U_{C0max}, and the ultra-capacitor voltage reference will stop increasing. The ultra-capacitor voltage will stay constant and the current i_{C0} will fall to zero. Charging of the ultra-capacitor is finished. Then, the dc bus voltage will start increasing until activation of the safety braking resistor or the drive over-voltage (over-braking fault) protection.

When the drive operates in motoring mode, the ultra-capacitor has to be discharged on the intermediate value U_{C0inM} in order to keep the system ready for the next braking phase. The dc bus voltage controller $G_{vBUSmax}$ maintains the dc bus voltage at V_{BUSmax}. The output of the controller $G_{vBUSmax}$ is decreasing, and therefore the ultra-capacitor voltage reference is decreasing toward U_{C0inM}.

$$U_{C0inM} \leq u_{C0(REF)} \downarrow < U_{C0\,max}. \tag{3.8}$$

The ultra-capacitor is discharged, supplying the drive load. Once the ultra-capacitor has been discharged to the intermediate level U_{C0inM}, the dc bus voltage controller $G_{vBUSmax}$ is saturated at the reference U_{C0inM}. The ultra-capacitor voltage reference is therefore

$$u_{C0(REF)} = U_{C0inM} + 0 = U_{C0inM}. \tag{3.9}$$

The ultra-capacitor voltage is constant, $u_{C0} = U_{C0inM}$. The ultra-capacitor current falls to zero and the discharging phase is finished. The dc bus capacitor is discharged and the voltage decreases until the input rectifier starts conducting. Then, the drive is again supplied from the mains.

The Ride-Through Mode (RT)
Figure 3.21b shows the profile of the system current and the voltage profile when the drive operates in ride-through mode. The mains power supply is interrupted and the dc bus voltage starts decreasing until it reaches the minimum V_{BUSmin}. The controller $G_{vBUSmin}$ is then de-saturated and its output starts decreasing below zero toward $\Delta U_{C0min} = U_{C0min} - U_{C0inM}$. Since the controller $G_{vBUSmax}$ stays saturated at U_{C0inM}, the ultra-capacitor voltage reference starts decreasing below U_{C0inM} toward U_{C0min}.

$$(U_{C0inM} + U_{C0\,min} - U_{C0inM} = U_{C0\,min}) \geq u_{C0(REF)} \downarrow > U_{C0inM}. \tag{3.10}$$

The ultra-capacitor is discharged lower in order to regulate the dc bus voltage as constant at the minimum level $v_{BUS} = V_{BUSmin}$. Once the mains has recovered, the dc bus voltage increases to the nominal voltage (defined by the mains voltage). The controller $G_{vBUSmin}$ is saturated at zero and therefore the ultra-capacitor voltage reference rises to

U_{C0inM}. The ultra-capacitor is charged to the intermediate level U_{C0inM} in order to keep the system ready for the next power interruption. If the power interruption is too long, longer than the specified, the ultra-capacitor will be discharged to the minimum U_{C0min}. Then, the ultra-capacitor current falls to zero and the dc bus voltage starts decreasing until it reaches the under supply fault (USF) level. The drive fails in USF and the whole system is stopped.

Power Filtering Mode

Whenever the load is greater than the input power, the ultra-capacitor is discharged. The drive rectifier bandwidth is low and therefore the input power cannot follow fast changes in the dc bus voltage. The input power is smooth. The dc–dc converter (via the controller $G_{vBUSmin.}$) regulates the dc bus voltage at a lower reference V_{BUSmin}. During the complementary state, when the input power is greater than the drive power, the ultra-capacitor is charged. The dc–dc converter regulates the dc bus voltage at a lower reference V_{BUSmax}.

The input power and drive power are equal when the drive power is constant (smooth). The dc bus voltage is regulated by the drive rectifier to the reference $V_{BUS(REF)}$. The ultra-capacitor current is zero and the voltage is regulated to U_{C0inM}.

3.3.4.3 Some Experimental Examples

Figures 3.22 and 3.23 show experimental waveforms measured in a 5.5 kW general purpose controlled electric drive with UCES. The drive parameters are: the input is three-phase 400 V and 50 Hz, the dc bus voltage $V_{BUS} = 650$ V, the dc bus capacitor $C_{BUS} = 820\,\mu$F, the output nominal load $P_{LOAD} = 5500$ W.

Figure 3.22a illustrates the case of dynamic braking and energy recovery mode. Figure 3.22b illustrates the case of the mains power interruption and the mains recovery.

Figure 3.23 illustrates the capability of the drive system to reduce the input peak power. The dc bus load is cycling between 10 and 100% with a repetition period of 2.2 seconds and a duty cycle of 40%. Figure 3.23a illustrates the case when the peak power filtering function is de-activated. As one can see from the waveform, the ultra-capacitor current is zero; there is no energy exchange between the ultra-capacitor and the drive. The mains current is modulated with the load and it varies from 1 to 11 A. Figure 3.23b illustrates the case when the peak power filtering function is activated. The ultra-capacitor current commutates between −8 A (the dc bus load is 100%) and 6 A (the dc bus load is 10%). The mains current magnitude is approximately 6 A, and is continuous without significant variation.

3.4   Renewable Energy Source Applications

3.4.1   Renewable Energy Sources

3.4.1.1   Wind "Generator"

Wind energy has become one of the mainstream renewable energy sources. A windmill consists of three main parts; a wind turbine (WT), an electric generator, and a tower. The turbine converts the energy of the wind into rotation, while the electric generator

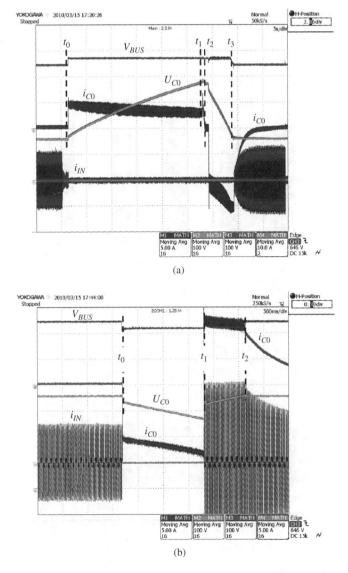

Figure 3.22 Experimental waveforms of the ultra-capacitor current i_{C0} (5A/div) and voltage u_{C0} (100 V/div), the dc bus voltage v_{BUS} (100 V/div) and the mains current i_{IN} (10A/div). $V_{BUS} = 650$ V, $V_{MAINS} = 400$ V, $C_{BUS} = 820\,\mu F$, $P_{LOAD} = 5500$ W. (a) Braking and energy recovery mode and (b) ride-through and the grid recovery mode

converts the mechanical energy of the turbine rotor into electrical energy. The power of the windmill varies from a couple of kilowatts up to a couple of megawatts. Figure 3.24 shows Siemens' farm of 2.3 MW wind power turbines at Wildorado Wind Ranch (2010).

The turbine-generator set can be a constant speed or variable speed system. Today, variable speed wind turbines are state of the art. Constant speed wind turbines are almost

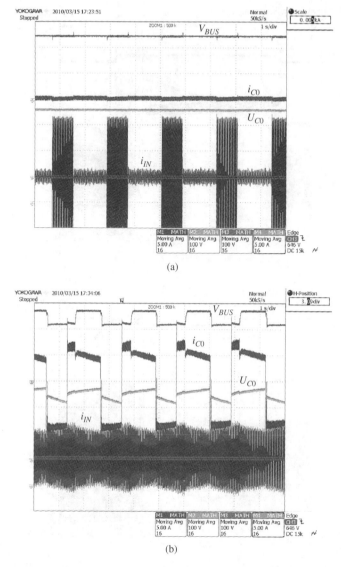

Figure 3.23 Experimental waveforms of the dc bus voltage v_{BUS} (100 V/div) and the mains current i_{IN} (5A/div) the ultra-capacitor current i_{C0} (5A/div) and voltage u_{C0} (100 V/div), when the load is cycling (10 to 100 to 10%). $V_{BUS} = 650$ V, $V_{MAINS} = 400$V, $C_{BUS} = 820\,\mu$F, $P_{LOAD} = 5500$ W. (a) The input power smoothing function is *OFF*. (b) The input power smoothing function is *ON*

ubiquitous. A solution that has received particular attention over the last couple of years is that of a full power rated interface converter.

A simplified electric circuit diagram of a typical wind turbine and double conversion full power rated power converter is given in Figure 3.25. The system consists of a variable speed turbine, a three-phase permanent magnet synchronous generator (PMSG), and a

Figure 3.24   Siemens 2.3 MW Wind Power turbines at Wildorado Wind Ranch (2010). Copyright © 2010 Billy Hathorn

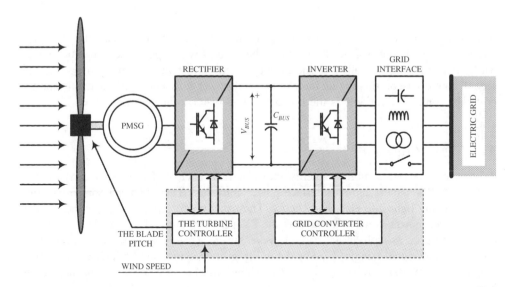

Figure 3.25   Block circuit diagram of variable speed wind turbine with full-scale power converter

full-scale power converter. The converter consists of a rectifier, a voltage dc bus, and a grid inverter. The inverter is connected to the grid via an interface unit, which basically consists of a line filter, a circuit breaker, and a step-up transformer (Figure 3.25).

The control scheme of a variable speed wind turbine can vary from concept to concept. The most common control scheme consists of two independent controllers. The first

Figure 3.26 Nellis Solar Power Plant, 14 MW powerplant installed in 2007 in Nevada, USA. Copyright © 2007 U.S. Air Force photo/Airman 1st Class Nadine Y. Barclay

controller controls the turbine via two variables, namely the turbine pitch and the generator torque. The pitch control is used to control the ratio of wind speed and generator power, while the torque is controlled in such a way as to extract maximum wind power. The second controller controls the dc bus voltage and power flow between the dc bus and the grid. The controlled variable is the line active and reactive current (power).

3.4.1.2 PV "Generator"

PV systems convert sunlight's energy into electrical energy. The system consists of a solar panel (PV panel) and an interface power converter. The PV panel converts energy from incoming light into electrical energy. The panel output is a dc voltage and current. The interface power converter connects the PV panel output and a load. In general, the load can be a dc as well an ac load. However, in most cases, the load is an ac system, such as an individual (single) load or the utility grid. In this case, the interface converter is a dc–ac converter (inverter). Figure 3.26 shows 14 MW solar power plant installed in Nevada, USA.

Figure 3.27 shows a circuit diagram of a grid connected PV system. The system consists of a PV field, a PV interface, a double conversion inverter, and a grid interface. The PV interface connects the PV field and the inverter. It is basically a set of fuses, cables, and EMI filters. The grid connected inverter can be single conversion or double conversion inverter. A double conversion inverter is illustrated in Figure 3.27. In the case of a single conversion inverter, the PV output is connected directly to the grid inverter.

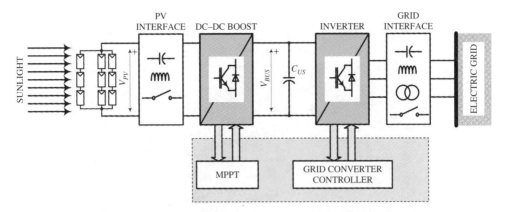

Figure 3.27   Block circuit diagram of grid connected double conversion PV system

3.4.1.3   Marine Current "Generator"

Another source of significant energy is the ocean. Ocean energy can be classified into a few different categories: (i) marine tidal current energy, (ii) wave energy, (iii) thermal energy, (iv) marine biomass energy, and (v) salinity osmotic energy [21]. Taking into account the technology status and economical aspects, marine tidal current energy is the most appropriate source of electric energy. The main reasons are its relatively simple realization and deterministic characteristic. The deterministic characteristic is very important to accurately predict the production capability.

A marine current energy recovery system is very similar to a windmill. The system consists of a marine current turbine (MCT) and an electric generator. The turbine converts the energy of moving water into mechanical energy of the turbine rotor. The mechanical energy is then converted into electrical energy via an electric generator. Figure 3.28 shows an MCT SeaFlow with its rotor raised.

A simplified circuit diagram of a marine current energy recovery system is depicted in Figure 3.29. A three-phase PMSG is connected to a boost rectifier. The rectifier generates the dc bus voltage from the generator three-phase voltage. The dc bus voltage is further converted into three-phase symmetrical voltage via the output inverter. Normally the inverter is connected to a fixed voltage/frequency grid via the grid interface (fuses, cables, and circuit breakers). The rectifier and inverter are controlled in very similar manner to that of the wind and PV system. The rectifier is controlled in such a way as to extract the maximum power from the turbine within the limitation of the dc bus voltage, while the inverter controls the output voltage or active/reactive power fed-back to the grid.

3.4.2   Definition of the Problem

There are two main issues in the operation of renewable energy sources. The first is the fluctuation of power generated by the system. The second issue is the system's behavior and performance in the case of grid faults and disturbances.

Figure 3.28 Marine current turbine SeaFlow with its rotors raised. Copyright © 2003 Fundy

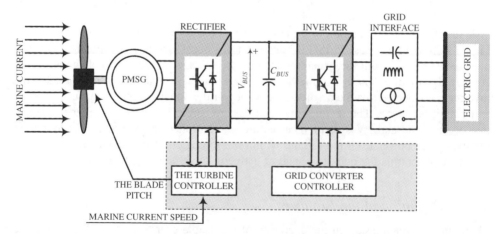

Figure 3.29 Block circuit diagram of marine current energy recovery system with a variable speed turbine and full scale interface grid side power converter

3.4.2.1 Power Variation and the Effects on the Grid

Wind Turbine (WT)

Although wind energy is considered one of the most promising energy sources, short term variation of the production capacity is one of its most critical drawbacks. The wind turbine output power strongly depends on the wind speed,

$$P_{OUT} = cv_W^3, \tag{3.11}$$

where v_W is the wind speed and c is a coefficient [20].

A slight variation in the wind speed creates significant variation in the output power. The output power variation can be controlled and reduced by control of the coefficient C_P controlling the blade pitch. However, the power variation cannot be eliminated for two reasons. The first reason is the dynamic response of the pitch control. The typical response of the blade pitch control system is around 5°/s. Basically, this means that wind variation in the frequency range of 10–100 mHz can be controlled and de-coupled from the turbine output power. However, wind speed variations in a higher frequency range, above 100 mHz, cannot be effectively controlled and the wind speed variation will be directly reflected to the turbine output power. The second limitation of the pitch control of the turbine power is the saturation of the blade angle.

Another source of power fluctuation is caused by tower shadow effects. Each time the turbine blade passes its tower, a pulsating torque is produced. The torque pulsation produces a pulsation of the turbine power, which is further transmitted to the generator, the converter, and finally into the grid. The power fluctuation could be up to 20% of the rated power, while the frequency is up to few hertz.

Marine Current Turbine (MCT)

The situation with marine current energy is similar to that of the wind turbine. The MCT output power strongly depends on the underwater marine current speed,

$$P_{OUT} = c_M v_{TIDE}^3, \tag{3.12}$$

where v_{TIDE} is the tide speed and c_M is a coefficient [21].

There are two different mechanisms that create fluctuation of the MCT speed and power. The first is the long-term variation of the water current speed within a period of 6–12 hours. This is an astronomical phenomenon and as such is predictable with high accuracy. The second mechanism is fluctuation of the MCT power due to long wavelength swells that affect the underwater marine current speed. This is a short-term disturbance with a period of a few seconds [21].

Effects on the Grid

Variation in the power being injected into the grid may seriously affect it. The first effect is the fluctuation of the point of common coupling (PCC) voltage. Figure 3.30 shows a single line simplified circuit diagram of a power system grid. Wind turbines WT_1 to WT_n are connected to the PCC. The PCC is connected to an ideal grid via coupling impedance Z_{PCC}. The variation of the power being injected into the PCC causes variation

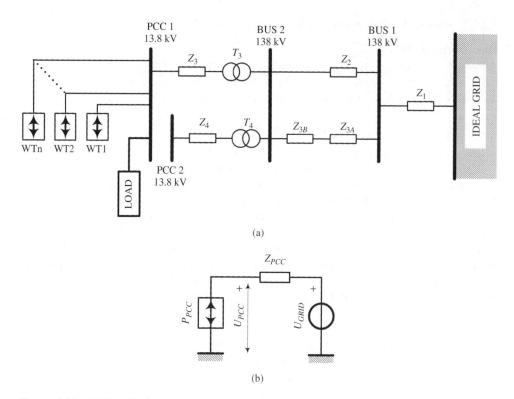

(a)

(b)

Figure 3.30 (a) Detail of a power system grid with connected renewable (wind) energy sources. (b) An equivalent circuit diagram

of the voltage on the PCC impedance Z_{PCC}, which is reflected as variation of the PCC voltage. The voltage variation known as flickers is undesirable [17]. It causes problems with lighting and sensitive equipment connected to the grid.

Power fluctuation may also affect the system's stability. If the power of the renewable source is significant in comparison to the system's power, the power fluctuation may cause system instability. In the worst case scenario, the system may collapse.

3.4.2.2   Low Voltage and Zero Voltage Ride Through

Another issue that very often arises in renewable energy applications is the system's behavior in the case of grid disturbance, such as voltage sags and interruptions. This is known as the low voltage ride-through (LVRT) and zero voltage ride-through (ZVRT) capability of a renewable energy "generator" [24].

Let's assume the WTR is running at a certain speed and power being connected to the grid. At some instant, the grid is exposed to a fault and the WT terminal voltage drops to zero. Since the terminal voltage is zero, power fed-back to the grid is also zero. Because of imbalance between the WT power and output power, the converter dc bus voltage will quickly rise over the nominal one. The rectifier control will therefore reduce the generator torque to zero to prevent destruction of the dc bus and the entire converter system. Since the turbine power is not yet reduced, the WT will start to accelerate. To prevent damage or even total destruction of the turbine, the blades pitch system will react and try to "close" the blades. However, the pitch system cannot react fast enough to prevent overcharging of the dc bus and acceleration of the turbine. A traditionally used solution is the dynamic braking resistor. The resistor is used as the bus load in the case that the turbine power cannot be fed-back to the grid, such as in a LVRT or ZVRT event. As per the grid code requirement, the wind turbine must stay connected to the grid in order to support the grid with reactive power [24].

3.4.2.3 The Blade Pitch Emergency Power Supply

Today's advanced wind turbines are variable speed turbines, usually with three blades. The blades' angle is flexible and can be adjusted depending on the turbine operating point and wind speed. Basically, the pitch angle is controlled via an electromechanical control system as a part of the turbine maximum power point tracking (MPPT) system.

In the case of turbine or power converter failure, the blades must be driven into the 90° position quickly in order to prevent the turbine over-accelerating and mechanical damage. The problem arises when the blade pitch system loses power supply from the turbine electric system. To prevent such a critical situation and ensure the blades will be "closed," independent energy storage must be installed for each blade pitch system.

Traditionally, electrochemical batteries are used as emergency energy storage for the blade pitch system. The discharge time is very short, in the range of 15–30 seconds, while the discharge power can be a couple of kilowatts. Therefore, the battery size must be based on the peak power not on the energy capability. Another issue with state-of-the-art batteries is operating temperature and life span. Since the wind turbine may operate at very low temperatures, an additional heating system must be installed in the turbine hub to keep the battery temperature above the minimum. All this makes it difficult and impractical to use electrochemical batteries as emergency energy storage for the turbine blade pitch system.

3.4.3   Virtual Inertia and Renewable Energy "Generators"

3.4.3.1   Traditional Synchronous Generators

Traditional power systems are based on centralized power generation. Generators are large synchronous three-phase generators driven by hydro, steam, or, recently, gas turbines. Such systems are characterized by their large inertia, which is basically the inertia of the generators and turbines. On top of the generator and turbine inertia, there is also inertia of the large three-phase electro-motors directly connected to the grid.

Power generation and consumption of a power system must be balanced. If there is imbalance between production and consumption, the generators and turbines will accelerate or decelerate, depending on the total unbalanced power. Variation of the generators' speed is directly reflected to the system frequency which will also change around the set point frequency. Variation of the power system frequency is very undesirable and the turbine control must react and prevent large variation. The initial variation of the frequency at the moment of power imbalance depends on the system's inertia. Systems with very large inertia will react with very small variation of the frequency around the set point. In contrast, systems with small inertia will react with large variation of the frequency. In some cases, the system may become unstable and it may collapse. Hence, it is obvious the system inertia is very important and necessary for power system stability.

3.4.3.2 Renewable Energy "Power Electronics Generators"

Today, the contribution of renewable "generators" to the total power of the power system is increasing rapidly. In the near future it is expected that 30–40% of power will be produced by renewable energy "generators." Is this an issue, and if so why?

Advanced renewable energy "generators" are connected to the power system via full-power rated power converters. The converter totally decouples the source of the energy (wind turbine, PV, MCT) from the power system. The converter is synchronized to the grid voltage via a fast phase locked loop (PLL). Hence, the converter output frequency will track the grid frequency within a few miliseconds. Coming back to the traditional large inertia synchronous generators, it is obvious that the renewable "generators" behave as no-inertia generators. The output frequency can charge with a rate of 10 Hz/s or even more.

What will happen in the case of a system power imbalance or fault? In that case, the converter will continue to track the system's "frequency." Since the converter is trying to follow the frequency, the "generator" will not help in stabilizing the system. On the contrary, the generator may create instability of the system. This is an issue in small-scale power systems, such as isolated systems on small islands. In the near future, when renewable energy contributions become significant, it will become an issue even for large-scale systems.

3.4.3.3 Virtual Inertia–Virtual Synchronous Generator

To solve the problem of no-inertia renewable "generators," the converter has to be controlled in a completely different way. Instead of tracking the system frequency, the converter has to behave like a large inertia generator [31–34]. The output frequency should not change quickly as the system frequency changes. However, the key issue is not how to control the converter, it is how to provide the inertia feature, where the inertia literally means energy. Hence, the issue is to provide the short-term energy capability of the converter. The converter should be able to absorb energy in the case of a system frequency increase, and to provide energy in the case of a system frequency decrease. In another words, the converter power corresponds to the first derivative of the system frequency.

The PV panel as a purely static generator has no inertia. Hence, the PV panel power cannot be negative and it cannot absorb (store) energy. Also, it cannot provide more power than that defined by the maximum power set point. Wind turbines and MCT have a certain inertia of the turbine rotor and mass of the blades. Thus, theoretically, it is possible to use the inertia of the turbine and the blades. However, wind turbines and blades are sensitive to strong torque pulses that can happen in the case of a sudden change of frequency. Therefore, the kinetic energy of the turbine and blades cannot be used as real inertia for system stabilization. The only solution is the integration of an external energy store into the converter. The energy storage will play the role of large inertia and the converter will behave as a large synchronous generator.

3.4.4 The Solution

To solve the problem of power fluctuation and LVRT/ZVRT, and provide the virtual inertia feature of a renewable energy "generator," energy storage can be integrated into the "generator." Basically, there is no fundamental difference between WT, PV, and MCT. Thus, we will explain energy storage integration with a WT example only.

The energy storage can be integrated at the wind turbine level or the wind farm level. Since most of the new generation of variable speed wind turbines are connected to the electric grid via a full scale power converter, it can be proven that integration of energy storage at the turbine level is the better option. The main reason for this is the fact that the interface power converter is an indirect ac–dc–ac converter with an accessible dc link. Additionally, the energy storage output is dc output, which makes connection between the main converter and the energy storage easier. If the energy storage is connected at the wind farm level, an additional dc–ac converter is required. Another advantage of the turbine level connection is the possibility of locally controlling the short-term power flow of each turbine, which makes the system control more flexible and robust.

Which type of energy storage is the most appropriate option depends on the requirements, particularly the frequency of the power fluctuation (time scale). This has already been discussed in Section 3.2.4. If the storage time is shorter than the critical time T_{CR}, the energy storage should be an ultra-capacitor, otherwise an electrochemical battery is an optimal choice.

An example of a wind turbine with UCES is illustrated in Figure 3.31. The wind converter is a back-to-back converter with a dc link voltage V_{BUS}. The ultra-capacitor is connected to the dc bus via an interface dc–dc converter. The dc–dc converter is controlled in such a way as to achieve smooth converter output power regardless of the variation in turbine power. One example of an interface converter control scheme is illustrated in Figure 3.32. The control scheme is almost the same as the control of the dc–dc converter for controlled electric drive applications [9, 11]. The only difference is that the dc–dc converter control operates in parallel to the control of the grid connected inverter.

Figure 3.33 shows the power–time profile of a WT equipped with UCES. The WT power fluctuates around an average, while the grid power is expected to be smooth. The difference between the instantaneous power of the WT and the grid is absorbed by the ultra-capacitors. The ultra-capacitor is charged whenever the WT power is higher than the grid power, and the ultra-capacitor is discharged whenever the WT power is lower than the grid power.

3.5 Autonomous Power Generators and Applications

3.5.1 Applications

Autonomous power generation systems are energy sources used to produce electrical energy from fuel, such as diesel, natural liquid gas (NLG), or coal. Most of these applications belong to three main groups: (i) autonomous hoisting applications, (ii) hybrid earth moving equipment, and (iii) autonomous mobile gen-sets.

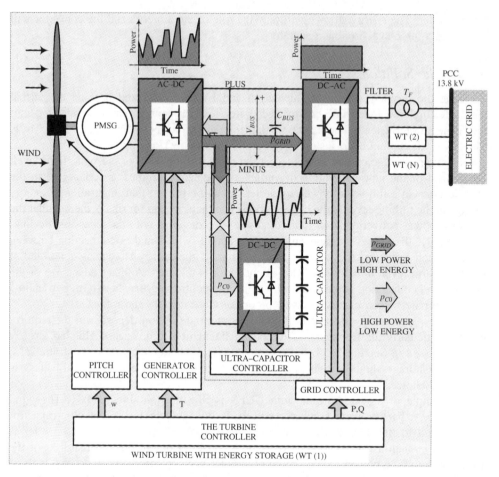

Figure 3.31 Block diagram of a variable speed wind turbine with full scale interconnection power converter and ultra-capacitor energy storage

3.5.1.1 Autonomous Hoisting Applications

Typical examples of autonomous hoisting applications are RTGCs, which are used to load/unload containers from trucks and ships in big ports, Figure 3.34, [14]. An ICE, usually a diesel engine, drives an electric generator that feeds the RTGC local electric network. A simplified circuit diagram is given in Figure 3.36. Hoisting and traction electric motors are supplied from the local dc network.

3.5.1.2 Hybrid Earth Moving Machines

The second group of these applications is diesel electric generators in hybrid dampers, Figure 3.35a, and hybrid excavators, Figure 3.35b [15]. Traditionally, these machines were pure diesel engine driven systems. Recently, the hybrid diesel–electric concept

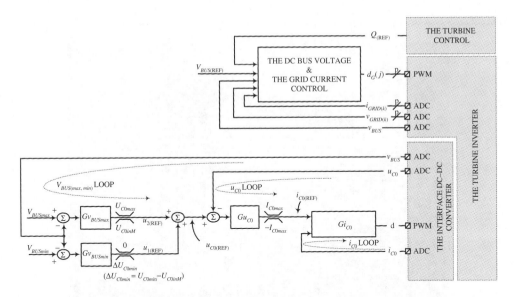

Figure 3.32 One example of the interface dc–dc converter control scheme

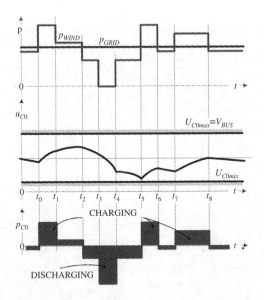

Figure 3.33 Power–time profile and voltage of the ultra-capacitor integrated in a wind turbine

has been used. A diesel engine drives an electric generator that supplies the machine's local electrical network. All actuators are electrically driven, some of them directly via an electric motor, some of them indirectly via an electric motor, hydraulic pump, or a hydraulic motor. A simplified electric circuit diagram is given in Figure 3.36.

(a)

(b)

Figure 3.34 (a) and (b) RTGC as an example of an autonomous power generation system. Copyright Liebherr Container Cranes Ltd, with permission

(a)

(b)

Figure 3.35 (a) Hybrid diesel-electric mining truck Komatsu 930E. Copyright © 2009 Cvmontuy. (b) Hybrid diesel-electric excavator Komatsu PC200LC. Copyright Komatsu

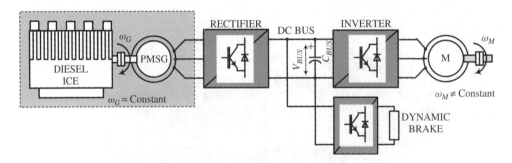

Figure 3.36   A simplified circuit diagram of a diesel-electric generation drive system

3.5.1.3   Autonomous Mobile Gen-Sets

The third group of applications covers autonomous mobile gen-sets [16]. Constant speed synchronous generators driven by diesel engines have been used over the last couple of decades. Applications are varied, starting from small home-used remote gen-sets, up to large stationary gen-sets used as long-term UPS in hospitals, military bases, critical industrial processes, and so on. A typical diesel ICE and gas turbine gen-set are shown in Figure 3.37. The gen-set consists of a prime mover, such as a diesel ICE or gas turbine. The mover runs a three-phase generator, usually a synchronous permanent magnet generator (PMG). A simplified circuit diagram is depicted Figure 3.38. The generator speed is controlled via the fuel governor of the prime mover. The speed has to be constant in order to keep the generator frequency constant.

3.5.2   Definition of the Problem

The main problem with all of the abovementioned applications is that the ICE (diesel engine for instance) feeds the load directly, without a significant energy buffer. Any fluctuation of the load is reflected directly to the engine. Typically, the load in such systems is highly intermittent, with a very high peak to average power ratio. This has three main effects on the system cost and fuel consumption.

- The diesel engine has to be sized on the peak power demand. This makes the engine and generator very oversized.
- Most of time, the engine operates at a non-optimal operating point with poor efficiency and high fuel consumption.
- The third problem, which is in fact an extreme case of the previously described problem of power fluctuation, is the dynamic braking of the system's electric drive. It could be, for example, lowering the container of RTGC or decelerating of a mining damper. Since a diesel engine is uni-directional, power cannot flow from the system to the source. A commonly used solution in this case is a brake chopper and brake resistor. This is a very simple solution, but it is inefficient. The braking energy, which in some applications represents a significant portion of the energy consumption, is dissipated on the brake resistor.

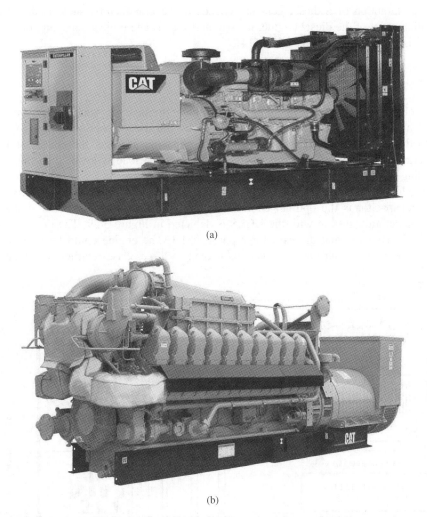

(a)

(b)

Figure 3.37   (a) Diesel ICE gen-set and (b) gas turbine gen-set for small local network. Reprinted courtesy of Caterpillar Inc

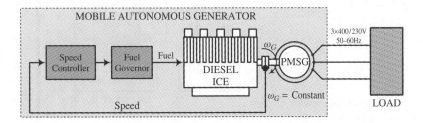

Figure 3.38   A simplified circuit diagram of a diesel-electric mobile gen-set

Fuel consumption of a diesel gen-set depends on several factors, such as the size and operation point of the diesel engine, its load and speed. The typical consumption of a 40 t RTGC is in the range of 15–25 l/h [14].

3.5.3 The Solution

The problem of load power fluctuation, including extreme cases such as negative power when the electric drive brakes, can be solved by the use of energy storage to smooth the load power. Whenever the load is greater than the average, the energy storage is discharged. The energy storage is charged whenever the load is lower than the average including negative power during braking.

Since load fluctuation is quite frequent, with a period in the range of couple of seconds, an ultra-capacitor is the optimal choice. A simplified block diagram of an autonomous power generation system with the UCES is depicted in Figure 3.39. The system is composed of an engine that drives a three-phase PMG. The engine could be a diesel ICE, gas, or steam micro turbine. The generator feeds a three-phase rectifier which provides

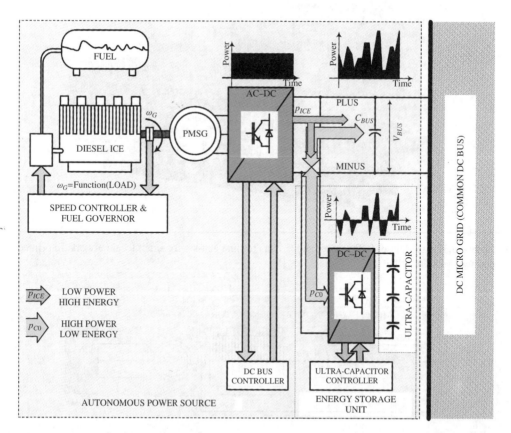

Figure 3.39 Autonomous power generation with a common generator, rectifier and dc bus and distributed inverter-fed motors

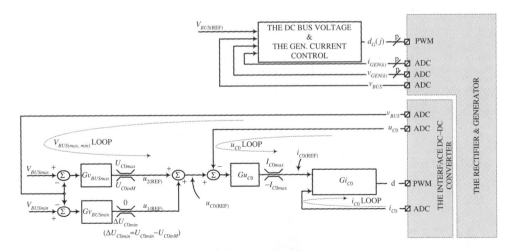

Figure 3.40   One example of the interface dc–dc converter control scheme

a dc bus voltage V_{BUS}. The dc bus is distributed to a decentralized set of three-phase inverters, which drives the ac motor or another type of load. The ultra-capacitor is connected to the dc bus via an interface bi-directional dc–dc converter. The main rectifier and dc–dc converter are controlled in such a way as to provide smooth power of the generator regardless of variation of the dc bus load. The control concept is almost the same as that for variable speed drives and wind turbines. The control objective is to keep the dc bus constant and the generator power smooth without any strong fluctuation. One example of this control is depicted in Figure 3.40.

Simplified waveforms are depicted in Figure 3.41. The ICE power p_{ICE} is smooth, being controlled by the rectifier controller. The ultra-capacitor power is negative whenever the load demand is higher than the ICE power. The ultra-capacitor power p_{C0} is positive whenever the load demand p_{LOAD} is lower than the ICE power p_{ICE}.

As reported in [14], fuel consumption of a 40 t RTGC was reduced from 16 l/h to approximately 8 l/h, which is a 50% reduction in fuel consumption. Regarding hybrid excavator applications, the expected fuel saving varies from 15 to 55%, depending on the concept of the hybrid system [15].

3.6   Energy Transmission and Distribution Applications

3.6.1   STATCOM Applications

Static synchronous compensators (STATCOM) are static power converters used to control reactive power flow in the electric grid (the transmission as well as the distribution network). STATCOM functions as a synchronous voltage source connected to the electric grid via an impedance Z_F. Controlling the voltage source, it is possible to control reactive power flow and therefore control voltage at the PCC. Details are given in Figure 3.42. STATCOM is composed of an input filter impedance Z_F, a three-phase PWM voltage source inverter, and a voltage dc link with voltage V_{BUS}. The dc link is usually a

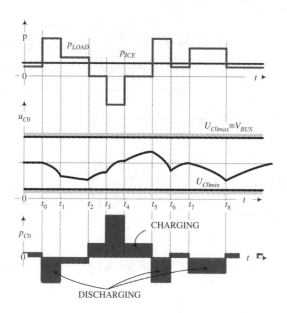

Figure 3.41 Waveforms of an autonomous power generation system equipped with the ultra-capacitor energy storage

bulky electrolytic capacitor C_{BUS}. Using the currently available high power semiconductor switches, such as IGBT or IGCT, the PWM converter can operate at a switching frequency of couple of kilohertz. Thus, high bandwidth can be achieved and reactive power can be controlled in a fraction of the mains period.

3.6.2   Definition of the Problems

We can identify two main issues in the application of traditional STATCOMs. The first is introduced by the load connected to the point of common connection. It could be voltage variation due to fluctuation of the load active power. The second type of issue is caused by faults in the distribution or transmission level of the grid. Since traditional STATCOMs have no capability to provide and absorb active power, only reactive power, the above issues cannot be solved using traditional STATCOMs. A STATCOM with an active power capability is required.

3.6.2.1   Voltage Disturbances and Flickers

Power quality describes the quality of the grid voltage in relation to disturbances such as voltage sags, interruptions, harmonics, and flickers. Voltage flickers are disturbances in lighting systems, which are caused by grid voltage fluctuations [17]. Large nonlinear loads, such as EAFs, welding machines, pumps, and rolling mills are well known as major sources of voltage fluctuations and flickers [19].

EAFs are known as highly nonlinear loads that behave as a variable resistance that models the arc. Figure 3.43 shows detail of the exterior and interior of a large electric arc furnace. Power consumed by the EAF is a constant active and reactive power P_0 and Q,

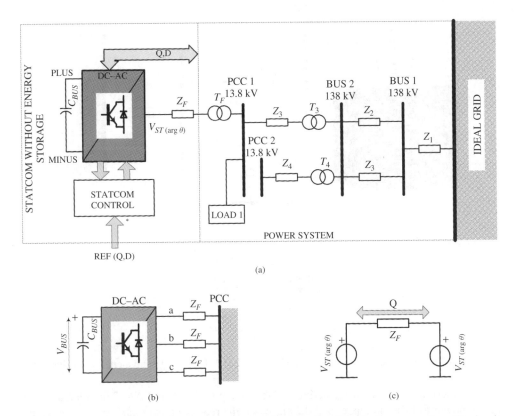

Figure 3.42 (a) STATCOM for reactive power and harmonics distortion compensation, (b) circuit diagram of voltage source STATCOM, and (c) one line equivalent circuit diagram

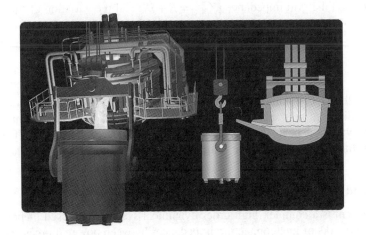

Figure 3.43 Rendering of the exterior and interior of a large electric arc furnace. Copyright © 2011 Uddeholms AB

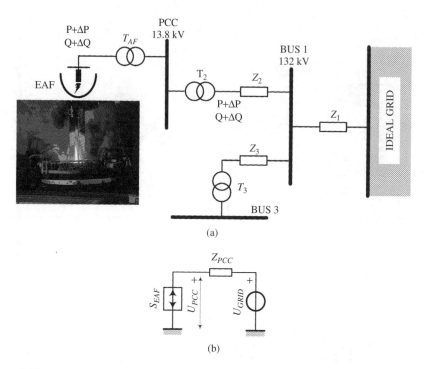

Figure 3.44   (a) An arc furnace connected to 13.8 kV PCC and (b) equivalent circuit diagram

with a superimposed fluctuating active and reactive power ΔP and ΔQ. The fluctuation of active power fails in the range of $\pm 0.2\,pu$, while the reactive power can be up to $\pm -1\,pu$. The frequency is in a broad range, from 1 to 20 Hz [18]. Connecting such a nonlinear load to the grid produces variation of the voltage at the PCC. A simplified one-line circuit diagram and equivalent model are depicted in Figure 3.44. The voltage variation at the PCC depends on the grid impedance ($Z_{PCC} = R + jX$) and the variation of the active (ΔP) and the reactive (ΔQ) power of the arc furnace. Voltage at the PCC is

$$U_{PCC} = \frac{\sqrt{(U_N^2 - (R\Delta P + X\Delta Q))^2 + (X\Delta P - R\Delta Q)^2}}{U_N}, \tag{3.13}$$

where U_N is the PCC nominal voltage.

The normalized current and PCC voltage of a typical EAF are depicted in Figure 3.45. The current variation is in the range of 35–100% of the nominal one. This produces variation of the PCC voltage in the range of 98–102% of the nominal voltage U_N.

3.6.2.2   Grid Stability and Fault Management

A power system is a complex high order dynamical system composed of power sources, loads, and networks of interconnection impedances between power sources and loads. For the sake of simplicity, the power system is usually represented by an ideal grid with zero impedance and constant voltage. A part of the power system of interest is connected to

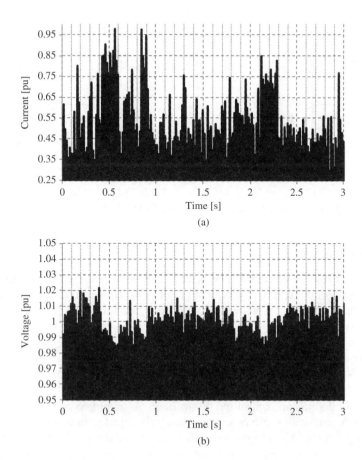

Figure 3.45 (a) Normalized current and (b) the PCC voltage of a typical EAF. The EAF is connected without active/reactive power compensation

the ideal grid. Figure 3.46 shows one example of a single line grid. This simplified circuit diagram can be used to study some particular cases, such as voltage at PCC versus the load variation and fault event.

Faults are unpredictable irregular events that may happen at any time at any point of a power system. Faults can be shunt faults or series faults. Shunt faults are short circuit faults, such as: three-phase short circuits, three-phase to ground short circuits, single-phase to ground short circuits, and two-phase to ground short circuits. Series faults are open circuits between two nodes.

Short circuit faults have two effects on the power system. The first is a large fault current flowing from the generators into the point of the fault. The second effect of the fault is voltage deviation. The voltage drops to an extremely low value during the fault event. The reduction in the voltage RMS value is known as a voltage sag and dip. Voltage sag may cause spurious tripping of variable speed drives and industrial process control systems. As a consequence, it may create power imbalance and system stability issues.

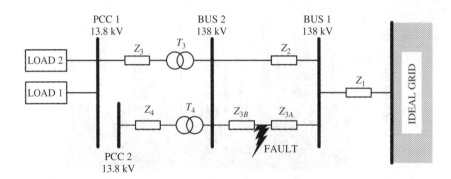

Figure 3.46 A single line circuit diagram of part of a power system of interest

3.6.3 The Solution

As a traditional solution, a STATCOM and similar devices are used to control the grid voltage and support the grid during critical fault events. However, an ordinary STATCOM can only support the system with reactive power. Sometimes, active power support is required as well.

Active power cannot be controlled by a STATCOM due to the lack of a source of active power on the STATCOM dc bus side. To achieve a certain degree of active power control, the STATCOM dc bus voltage must be controlled to be constant by an external power source. Simply speaking, an energy storage device has to be connected to the dc bus, which will define and control the bus voltage. A simplified one-line circuit diagram of a STATCOM with an integrated energy storage is depicted in Figure 3.47. The system consists of a traditional STATCOM and additional energy storage. The STATCOM and the energy storage are connected via an additional interface dc–dc converter. The ordinary

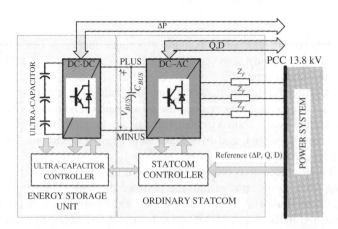

Figure 3.47 A simplified circuit diagram of a STATCOM with integrated energy storage and short-term active power capability

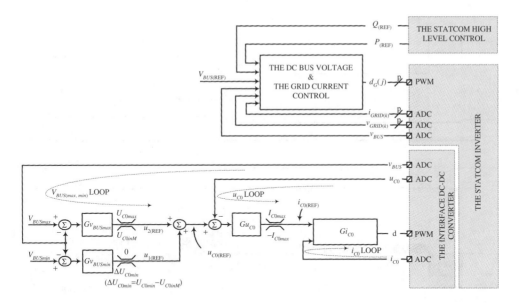

Figure 3.48   One example of the interface dc–dc converter control scheme

STATCOM provides/absorbs reactive energy via the grid connected inverter. The ultra-capacitor provides/absorbs active power via the dc–dc converter and the grid connected inverter.

The grid connected inverter is controlled in such a way as to provide the desired active and/or reactive power. The reference active/reactive power command is generated by the power system upper level control. At same time, the inverter controls the dc bus voltage whenever possible. If the inverter cannot maintain the dc bus voltage at reference value, the interface dc–dc converter takes over control. This happens when active power is absorbed from the grid or provided to the grid. To keep the dc bus voltage constant, the ultra-capacitor is charged or discharged respectively.

One example of an interface converter control scheme is illustrated in Figure 3.48. The control scheme is almost the same as control of the dc–dc converter for controlled electric drive applications, see Section 3.3.4.1. The only difference is that the dc–dc converter control operates in parallel with the control of the grid connected (STATCOM) inverter.

Figure 3.49 shows the power–time profile and the ultra-capacitor voltage when the STATCOM operates in active power mode. At the instant t_0 an event that requires active power support happens. The STATCOM inverter starts supporting the grid with active power. The dc bus capacitor discharges and the voltage starts falling from the steady state set point value. Once the dc bus voltage reaches the minimum reference, the interface dc–dc converter starts supporting the dc bus and discharging the ultra-capacitor. The ultra-capacitor voltage is decreasing toward a minimum U_{C0min}. At the instant t_1, the event is finished and the grid is stabilized. The ultra-capacitor has to be recharged to be ready for the next event. Once the ultra-capacitor is recharged to a set point voltage at the instant t_2, the STATCOM is fully ready for the next discharge event. Before the ultra-capacitor is fully recharged, the STATCOM can absorb active power if required.

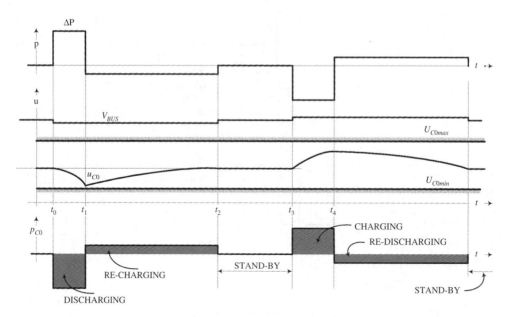

Figure 3.49 Power–time profile and ultra-capacitor voltage and current of a STATCOM with integrated energy storage and short-term active power capability

3.7 Uninterruptible Power Supply (UPS) Applications

3.7.1 UPS System Applications

UPS systems are electric devices whose main role is to provide uninterrupted, reliable, and high-quality power for critical loads. UPS systems protect sensitive loads against power interruptions, as well as over-voltage and under-voltage conditions [26–28]. Applications of UPS systems include medical facilities, data storage and computer systems, emergency equipment, telecommunications, critical industrial processes, and on-line management systems.

The three main families of UPS systems are: static, rotating, and hybrid (static–rotating) UPS. Static UPS systems are composed of static converters and energy storage only. Rotating UPS systems are composed of rotating machines such as ICEs, electric motors, and generators. Hybrid UPS systems are a combination of a static converter with energy storage and a rotating UPS.

3.7.1.1 Static UPS

Static UPS systems are the most common used, especially for short-term applications (up to a couple of tens of minutes) and medium-term applications (up to a couple of hours). Hybrid UPS systems are used in systems that require fast response and long-term autonomy, from a couple of hours up to a couple of days. Pure rotating UPS are not very often used in today's applications.

Static UPS systems can be on-line, off-line, and line interactive.

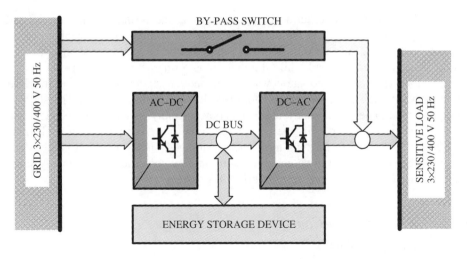

Figure 3.50   Static UPS system

- **On-line UPS system**. The critical load is supplied from the electric grid via an input rectifier and output inverter. The energy storage is fully charged. The transition time from the grid to the energy storage supply is zero. The input rectifier is sized on the full nominal power.
- **Off-line UPS system**. The load is directly supplied from the electric grid via a normally-on switch. The energy storage is fully charged and in stand-by mode. Once the main supply is interrupted, the output inverter starts operating. The load is switched to the inverter output. In contrast to on-line UPS systems, the input rectifier is a charger only and therefore it can be sized on a fraction of the nominal power.
- **Line interactive UPS**. The load and UPS inverter are connected to the grid.

Figure 3.50 shows a simplified one-line circuit diagram of a static UPS system. The UPS is composed of an input rectifier, dc bus, energy storage device, an output inverter, and a bypass switch. The input rectifier and output inverter are voltage sourced PWM converters, mainly of the two-level and three-level topologies.

3.7.1.2   Short-Term versus Long-Term UPS

Depending on the bridging time, we can distinguish three main categories of UPS systems. The first category is short-term UPS systems. The bridging time is very short, up to a couple of tens of seconds. The second category is medium-term UPS systems. The bridging time for this category of UPS systems is from a couple of tens of seconds up to couple of tens of minutes, to a maximum of a couple of hours. The third category is long-term UPS systems. The bridging time for this category of UPS system can be up to a couple of days. For short-term and medium-term UPS systems, electrochemical batteries and flywheels are traditionally used for energy storage. Long-term UPS systems use higher density energy storages, such as ICEs and three-phase generators. Hydrogen fuel cells are also used.

The main issue with long-term energy storages is the response time, the so-called warm-up time, which can be in the range of a couple of seconds up to a couple of tens of seconds. Such a long start-up time is critical and must be bridged by a short-term energy storage that will feed the load while the main long-term energy storage is warming up. Once the long-term energy storage is warmed up and has started to operate at the desired operating point, the short-term energy storage will stop feeding the load. The short-term energy storage can be recharged from the main power supply once this has been recovered. Another drawback of long-term energy storages is response on pulse power. To reduce the effect of pulse power on long-term energy storages, short-term high power density energy storage can be used in combination with the low power long-term energy storage. This category of UPS is a long-term hybrid UPS system.

3.7.2   UPS with Ultra-Capacitor Energy Storage

3.7.2.1   Short-Term UPS

A simplified one-line circuit diagram of a short-term UPS is shown in Figure 3.51. The circuit is similar to the circuit in Figure 3.50. The energy storage device indicated in Figure 3.50 is composed of an ultra-capacitor that is connected to the dc bus via an interface dc–dc power converter. An example of the interface converter control scheme is illustrated in Figure 3.52. The control scheme is almost the same as the control of the dc–dc converter for controlled electric drive applications, see Section 3.3.4.1. The difference is that the dc bus voltage is controlled by just one controller $G_{vBUSmin}$ that

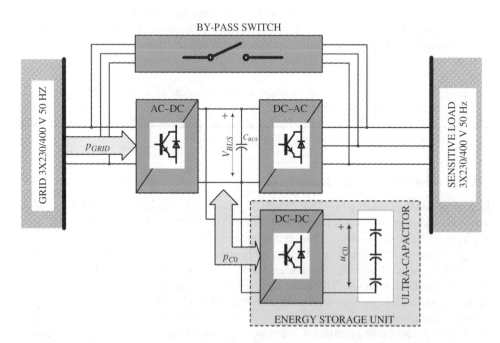

Figure 3.51   Short-term static UPS system with ultra-capacitor-based energy storage

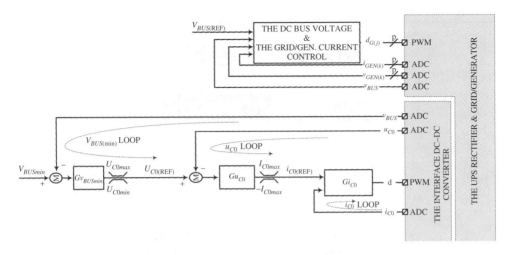

Figure 3.52   One example of the interface dc–dc converter control scheme

regulates the dc bus voltage at the reference V_{BUSmin}. Basically, this is the same case as a controlled electric drive in the ride-through mode. This case is discussed in detail in Section 3.3.4.2.

Figure 3.53 shows power–time profiles and the ultra-capacitor voltage. Let's assume the load is supplied from the grid via the UPS rectifier-inverter. At the instant t_0 the grid is interrupted.

3.7.2.2   Long-Term Hybrid UPS

A circuit diagram of a long-term hybrid UPS is depicted in Figure 3.54. The short-term energy storage is the same as in the short-term UPS. The long-term energy storage can be a hydrogen fuel cell (FC) or ICE and a three-phase generator. The FC is connected to the common dc bus via an interface uni-directional dc–dc converter that is used to control the dc bus voltage regardless of the FC voltage variation. The ICE generator is connected to the common dc bus via an interface uni-directional ac–dc converter. The interface converter controls the dc bus voltage and runs the ICE at an optimal point. Figure 3.55 shows the power–time profiles and the ultra-capacitor voltage of a long-term UPS. Let's assume the load is supplied from the grid via the UPS rectifier-inverter. At the instant t_0 the grid is interrupted and the load is supplied from the ultra-capacitor. At the instant t_1 long-term energy storage (ICE or FC) is started. At the moment t_2, the mains power supply (the grid) is recovered.

3.8   Electric Traction Applications

As we mentioned in Section 3.3, electric traction drives are a special category of controlled electric drives. All traction drives can be split into two main groups:

1. drives for rail vehicles, and
2. drives for road vehicles.

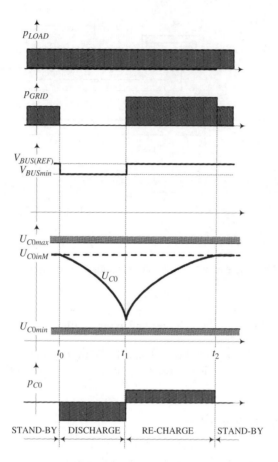

Figure 3.53 Power–time profiles, the ultra-capacitor voltage, and the dc bus voltage of a short-term static UPS system with ultra-capacitor-based energy storage

3.8.1 Rail Vehicles

Rail vehicles are all transportation systems that run on one or two iron rails. Depending on the supply system and the load range, the rail vehicles can be split into three main subgroups: (i) heavy-rail catenary supplied vehicles, (ii) heavy-rail diesel-electric supplied vehicles, and (iii) light-rail rapid transit vehicles.

3.8.1.1 Heavy-Rail Catenary Supplied Vehicles

These are mainly public transportation locomotives, including high speed trains such as the French TGV illustrated in Figure 3.56. A simplified schematic block is shown in Figure 3.57. The traction drive is supplied from an overhead line via a pantograph and the iron rail via the wheels. Overhead supply varies from country to country. Traditionally, in most of Europe, the overhead supply is single phase 25 kV 50 Hz supply, while in Switzerland it is single phase 15 kV $16_{2/3}$ Hz. The on-board equipment is composed of a

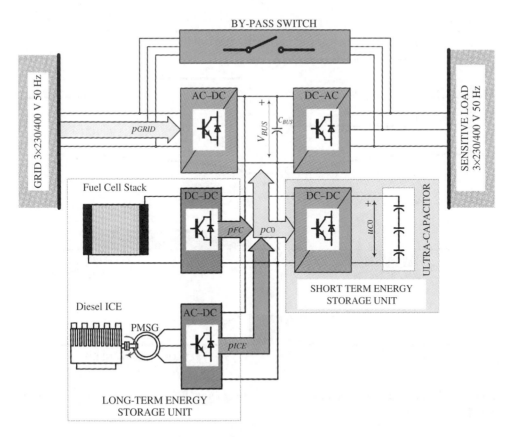

Figure 3.54 Long-term hybrid UPS system with an ultra-capacitor as short-term energy storage

step-down transformer, a rectifier, a dc link (bus capacitor C_{BUS}), and a traction inverter. The inverter drives the parallel connected traction induction motors. A multi-inverter traction system is also possible. The dynamic brake resistor is connected to the dc bus via a brake dc–dc converter. The rectifier, inverter, and dc–dc converter are usually two-level or three-level PWM voltage source converters. Traditionally, GTOs are used as converter switches. Recently, IGCTs and IGBTs have also been used.

3.8.1.2 Heavy-Rail Diesel–Electric Vehicles

Diesel–electric heavy locomotives are traditionally used in North America and in some part of Europe. A GE diesel–electric locomotive is depicted in Figure 3.58.

Figure 3.59 shows a simplified schematic block diagram of a diesel–electric locomotive. The locomotive on-board equipment is composed of an ICE, traditionally a diesel engine or a gas turbine. The ICE drives a three-phase generator that is connected to a three-phase rectifier that provides common dc bus voltage. The rest of the circuit is similar to the traction system of a catenary supplied locomotive.

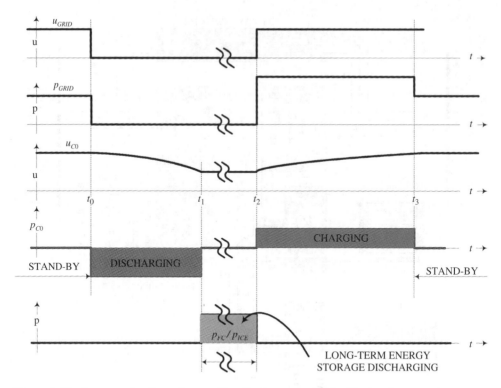

Figure 3.55 Power and voltage time profile of a long term UPS with the ultra-capacitor energy storage

3.8.1.3 Light-Rail Rapid Transit Vehicles

Light-rail traction drives are a commonly adopted solution in urban public transportation. A picture of a light locomotive is given in Figure 3.60. A simplified schematic block diagram is shown in Figure 3.61. The locomotive is supplied by an overhead line via a pantograph and the rail contact via the iron wheels. The overhead line is supplied from a supply substation. The supply voltage varies from country to country. The most common supply is dc voltage, 1.5 and 3 kV.

On-board equipment is similar to the equipment of heavy-traction drives. A common dc bus feeds the traction three-phase inverter that drives the parallel connected three-phase induction motors. An option with a dislocated multi-inverter traction drive is also used.

3.8.2 Road Vehicles

Road vehicles can be split in three main subgroups:

1. public transportation catenary-supplied vehicles,
2. hybrid electric vehicles (HEVs), and
3. pure electric vehicles.

Figure 3.56 Heavy-rail catenary supplied electric locomotive TGV. Image by Smiley.toerist

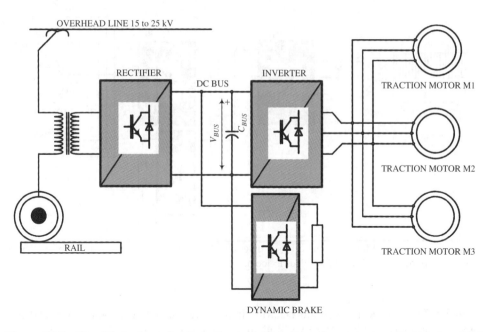

Figure 3.57 Simplified schematic block diagram of a heavy-rail catenary supplied traction drive

3.8.2.1 Public Transportation Catenary-Supplied Vehicles

A trolleybus is an electric bus supplied by overhead wires using spring-loaded trolley poles. The difference between a trolleybus and a light-rail vehicle, such as a tram, is that a tram uses a one pole overhead line and an iron rail as the return path, while a trolleybus must use a two pole overhead line.

Figure 3.58   A Railpower GG20B hybrid gen-set in the Union-Pacific scheme. Copyright © 2007 Bryan Flint

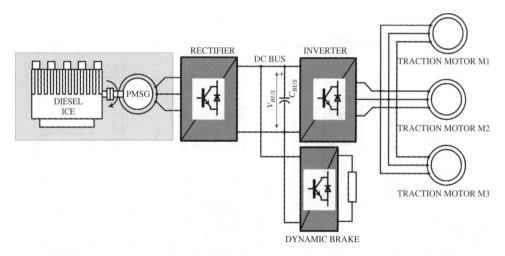

Figure 3.59   A simplified schematic block diagram of a diesel-electric traction drive applied in heavy transportation such as diesel-electric locomotives

Figure 3.63 shows a simplified electric circuit diagram of a trolleybus overhead line and on-board installation. The overhead line voltage is low dc voltage, usually in the range 600–900 V. The voltage is provided by a substation three-phase rectifier connected to the public grid. On-board equipment is similar to the equipment of heavy- and light-rail vehicles. A common dc bus feeds the traction three-phase inverter that drives the parallel connected three-phase induction motors. An option with a dislocated multi-inverter traction drive is also used. A dynamic braking resistor is used as protection in case of the dc bus over-voltage. The dc bus over-voltage may happen when the overhead line cannot accept the vehicle braking power [35].

Figure 3.60 The Hiawatha-Line light-rail in Minneapolis, Minnesota, USA. Copyright © 2006 Sinn

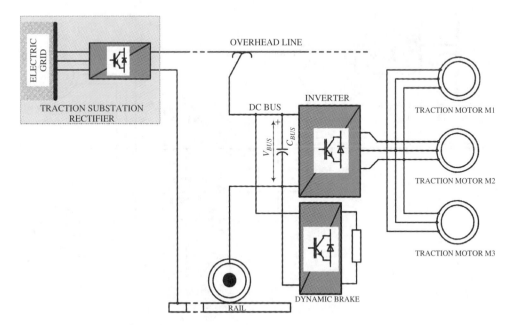

Figure 3.61 A simplified schematic block diagram of dc supplied light-rail traction drive used in tram public transportation

Figure 3.62   Trolleybus Cristalis Line in Lyon, France. Copyright © 2009 Momox de Morteau

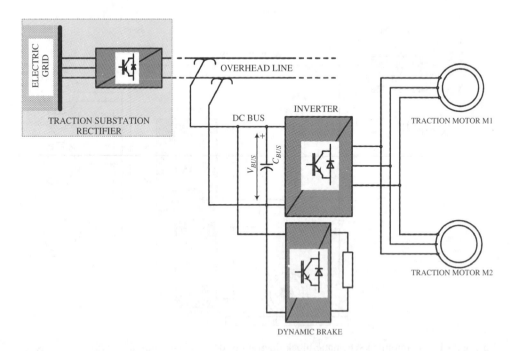

Figure 3.63   Block diagram of a road catenary supplied traction drive applied in public transportation such as a trolleybus

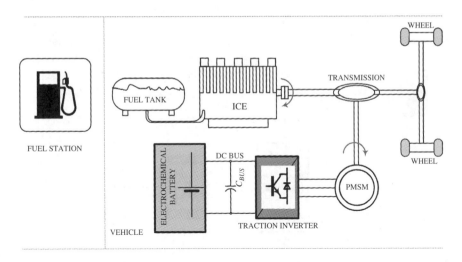

Figure 3.64 Parallel HEV with mechanical coupling

3.8.2.2 Hybrid Electric Vehicles

HEVs are vehicles that combine an ICE and an electric drive. The basic idea behind this concept is to operate the ICE at the maximum efficiency operating point. An HEV basically consists of four main parts: an ICE, an electric traction motor/generator, a traction power converter, and an electric energy storage. Depending on the configuration, HEV can be of series, parallel, and double conversion hybrid configuration. In the series configuration, the ICE and electric drive are attached to a common traction shaft. A simplified circuit diagram is depicted in Figure 3.64.

In parallel configuration, the ICE and the electric drive are connected via a transmission system. A simplified circuit diagram is depicted in Figure 3.65. In the double conversion configuration, the ICE and electric drive are interconnected via an electric link, using the same principle as that used in heavy-rail diesel–electric vehicles (Figure 3.59). A simplified circuit diagram is depicted in Figure 3.66.

3.8.2.3 Electric Vehicles

Pure electric vehicles are vehicles that are driven by an electric drive only. Figure 3.67 shows the Tesla Roadstar Sport 2.5, as an example of a high performance pure electric vehicle. The basic electric drive consists of an energy storage, an electrochemical converter, and an electric traction drive (Figure 3.68). The energy is stored as chemical energy in electrochemical batteries or hydrogen fuel cells. An electrochemical generator is a device that converts stored chemical energy into electrical energy. In the case of electrochemical batteries, the energy storage device and the electrochemical converter are same device: the battery. In contrast to this, in the case of hydrogen fuel cells, the energy is stored as compressed hydrogen in a separate tank. The electrochemical converter is a fuel cell.

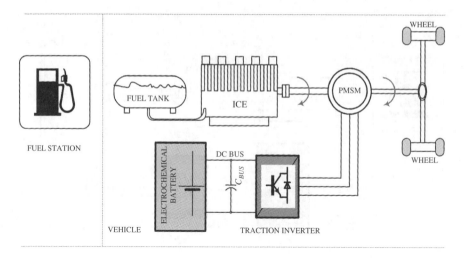

Figure 3.65   Series–parallel HEV with mechanical coupling

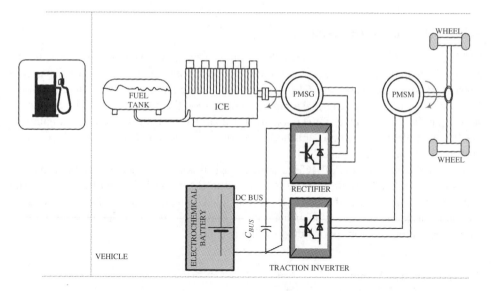

Figure 3.66   Block diagram of a series HEV with electrical coupling

The main difference between the battery and fuel-cell based electric vehicles is the braking process. Electrochemical batteries are bi-directional energy storage. Thus, the vehicle braking energy can be stored in the battery and recovered during the next acceleration cycle. In contrast to this, fuel cells are uni-directional devices. The fuel cell cannot be charged by reversing the terminal current. Therefore, the vehicle braking energy cannot be stored in the fuel cell. The energy is burned on a braking resistor or the vehicle mechanical brakes.

Figure 3.67 Pure electric vehicle, Tesla Roadstar Sport 2.5. Copyright 2010 © Tesla Motors Inc.

3.8.3 A Generalized Traction System

Before we start to discuss the need for energy storage in traction applications, let's briefly present a generalized traction system. Regardless of the category or concept of the traction system, all traction systems have basically the same structure: a generalized traction drive and system. A simplified block circuit diagram of such a system is depicted in Figure 3.69. The system consists of three main elements: (i) the traction motor, (ii) the traction inverter, and (iii) the high energy and high impedance power supply. The power supply impedance is generally nonlinear impedance. It could be internal impedance of an electrochemical battery or hydrogen fuel-cell, impedance of an overhead line, a nonlinear impedance of the substation rectifier or mobile gen-set rectifier, and so on. The traction power is discontinuous, being highly positive and negative, with a very high peak to average power ratio.

3.8.3.1 Identification of the Problem(s)

The role of the traction drive is to move the vehicle from the start-point A to the end-point B, Figure 3.70. Three different phases of the train journey can be identified. The vehicle acceleration phase (position I), the vehicle cruising phase (running at nominal speed) (position II), and the vehicle deceleration phase (position III). The power–time profile of the traction drive is illustrated in Figure 3.70. During the acceleration phase, the vehicle speed increases from zero to a maximum speed. The power quickly increases

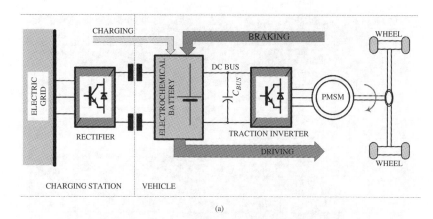

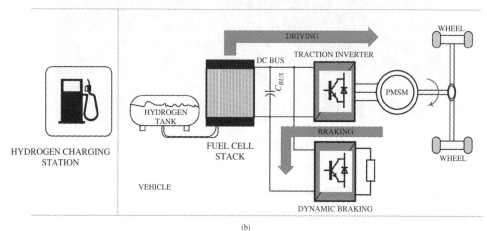

Figure 3.68 An electric vehicle (a) with electrochemical battery storage and (b) with hydrogen fuel cell energy storage

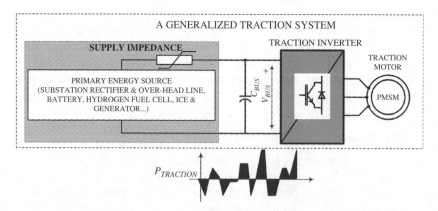

Figure 3.69 A generalized traction converter connected to a high impedance/high energy primary source

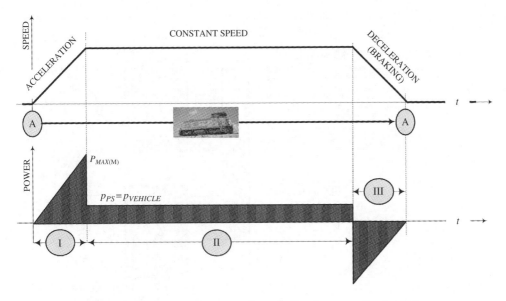

Figure 3.70 Speed and power-time profile of a generalized vehicle

from zero to maximum and stays at maximum during the acceleration phase. During this phase, energy is being stored in the system as the kinetic energy of the moving mass. Once the vehicle has achieved the rated speed, the power drops to the low level. During this phase, the drive power covers the friction losses of the traction system. When the vehicle is approaching its destination, deceleration starts. The vehicle speed decreases from the rated to zero speed. The vehicle kinetic energy is transferred from the vehicle mass to the traction drive system. The drive power is negative. This is the so-called a braking phase of the journey.

The main issue that can be identified from the above case is the variation in the traction power, where the power can be highly positive as well negative. The power fluctuation, especially such a strong variation as is the case with traction, creates some serious issues:

1. supply voltage fluctuation and the system instability [35],
2. additional losses on the main power supply,
3. significant waste of energy due to non-regenerative braking, and
4. heavily oversized and over-rated installations, which lead to the system being over-cost and over-weight.

3.8.3.2 The Solution

A solution to the above described problems is the integration of a short-term energy storage device into the traction drive. As already discussed in previous sections, the short-term energy storage device has to be selected according to the application requirement. The two main parameters are energy storage capability and charging/discharging peak power. Most traction drives require an energy storage device that can handle charging

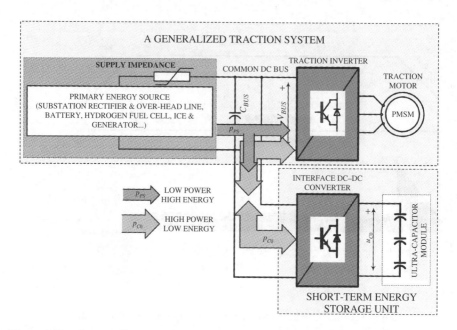

Figure 3.71   A generalized traction system with ultra-capacitor short-term energy storage

and discharging of less than a couple of tens of seconds. Therefore, according to the conclusion of Section 3.2.4, an ultra-capacitor could be the right choice. Figure 3.71 shows a block diagram of a general case traction drive with integrated UCES. The ultra-capacitor is connected to the traction drive dc bus via an interface bi-directional dc–dc converter. More details on interface dc–dc converters are given in Chapter 5.

Figure 3.72 shows an example of a control structure of the interface dc–dc converter. The dc–dc converter controls the ultra-capacitor power via the voltage u_{C0} and current i_{C0} inner control loops. The ultra-capacitor power reference is calculated from the drive

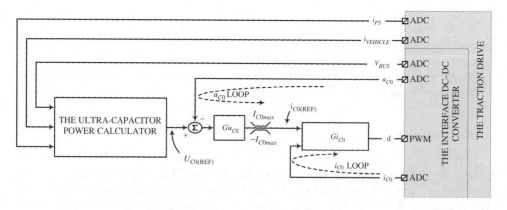

Figure 3.72   An example of a control structure of the interface dc–dc converter

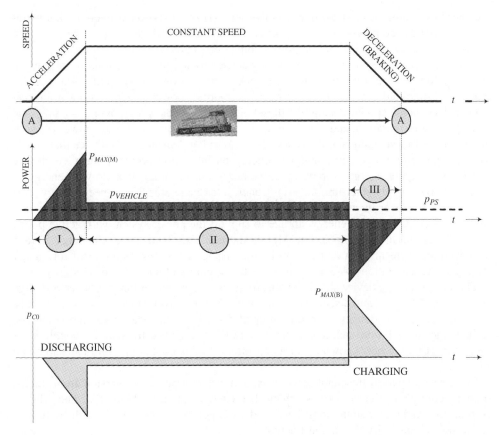

Figure 3.73 Speed and power–time profile of a traction system equipped with ultra-capacitor energy storage

power $p_{VEHICLE}$ and the power of the primary power supply p_{PS}. Different concepts are proposed and analyzed in the literature [36–44].

Figure 3.73 shows the power–time profile of such a traction system. $p_{VEHICLE}$ is the drive power, p_{PS} is the power of the primary power supply, and p_{C0} is the power of the ultra-capacitor. The ultra-capacitor is discharged whenever the drive power is higher than the average power of the primary power source. This mainly happens during drive acceleration. During the drive deceleration, the ultra-capacitor is charged and braking energy is stored in the ultra-capacitor.

3.9 Summary

The fundamentals of power conversion and static power converters are presented in the first part of this chapter.

Static power converters are electric devices whose main role is to modify the presentation of one electric quantity to another, where the quantity is voltage or current. There are four basic families of static power converters: (i) ac–dc power converters, known as rectifiers, (ii) dc–dc power converters, known as choppers, (iii) dc–ac power converters, known as inverters, and (iv) ac–ac power converters, known as cyclo-converters.

Power conversion systems and static power electronics play a significant role in our everyday life. It would be difficult to imagine a power conversion application, such as industrial controlled electric drives, renewable sources, power generation and transmission devices, home appliances, mobile diesel electric gen-sets, earth moving machines and equipment, transportation, and so on, without power electronics, and static power converters. In most of these applications, we are facing growing demands for a device that is able to store and restore a certain amount of energy during short periods. Controlled electric drives may require energy storage to save energy during braking or provide energy in case of power supply interruption. Wind renewable "generators" may need energy storage to smooth power fluctuation caused by wind fluctuation. Power transmission devices, such as STATCOMs, need energy storage to support the power system with active power during faults and unstable operation. Mobile diesel electric gen-sets need energy storage to reduce fuel consumption and CO_2 pollution. There is a strong requirement for energy storage in transportation systems to improve the system's efficiency and reliability.

The energy storage device should be able to quickly store and restore energy at very high power rates. Charging and discharging times are a few seconds up to a few tens of seconds. The charging power density is in the order of 5–10 kW/kg. Today, two energy storage technologies fit well into such requirements: (i) FES and (ii) Electrochemical double-layer capacitors (EDLC), best known as ultra-capacitors. In this book, only ultra-capacitors are addressed.

In the second part of the chapter, we discussed different power conversion applications, such as controlled electric drives, renewable energy sources (Wind, PV, and Marine current, for example), autonomous diesel and NLG gen-sets, STATCOMs with short-term active power capability, UPS, and traction.

All the above mentioned applications face a similar issue; strong variation of the load or input power.

- Controlled electric drive applications:
  - The motor load variation: braking and peak power. The input power should not be affected by the load variation.
  - The input power variation, in the worst case scenario total interruption. The capability of a drive to survive such a case is the ride-through capability. The load must not be affected by the supply interruption.
- Renewable energy sources (Wind, PV, and Marine current):
  - Variation of the input power that should not be transferred to the grid.
- Autonomous diesel and NLG gen-sets:
  - Variation that can be very strong. It should not affect the generator.
- STATCOMs with short-term active power capability:
  - The power system may require short-term active power support.
- UPS:
  - The input power variation, in the worst case scenario total interruption. The load must not be affected by the supply interruption.

- Traction:
  - Variation of the traction motor load: braking and peak power. The primary power source should not be affected by the load variation.

A common solution for all the above listed applications is a short-term energy storage device, in this case an ultra-capacitor with an interface dc–dc converter. Almost the same solution may be applied to all the applications discussed. The control architecture may vary from case to case.

References

1. Erikson, R. and Maksimović, D. (2001) *Fundamentals of Power Electronics*, 2nd edn, Springer, New York.
2. Linder S. (2006) *Power Semiconductors*, Lausanne, EPFL Press (Collection: Electrical Engineering).
3. Wu, B. (2006) *High-Power Converters and AC Drives*, 1st edn, Wiley-IEEE Press, New Jersey.
4. Peng, F.Z. (2002) Z-source inverter. 37th IAS Annual Meeting, October 13–18, 2002, Vol. 2, pp. 775–781.
5. Kolar, J.W. (1993) Dreiphasen-Dreipunkt-Pulsgleichrichter. European Patent Appl.: EP 94 120 245.9-1242, filed Dec. 23, 1993, File No.: A2612/93, entitled Vorrichtung und Verfahren zur Umformung von Drehstrom in Gleichstrom.
6. Grbović, P.J., Delarue, P. and Le Moigne, P. (2011) A novel three-phase diode boost rectifier using hybrid half-DC-BUS-voltage rated boost converter. *IEEE Transactions on Industrial Electronics*, **58**(4), 1316–1329.
7. Marquardt, R. and Lesnicar, A. (2004) *New Concept for High Voltage—Modular Multilevel Converter*, PESC, Aachen.
8. Lorenz, L., G. Deboy, and Zverev I. (2004) Matched pair of CoolMOS transistor with SiC-Schottky Diode—advantages in application, *IEEE Transactions on Industry Applications*, **40**, 5, 1265–1272.
9. Grbović, P.J. (2010) Ultra-capacitor based regenerative energy storage and power factor correction device for controlled electric drives. PhD dissertation. Laboratoire d'Electrotechnique et d'Electronique de Puissance (L2EP) Ecole Doctorale SPI 072, Ecole Centrale de Lille, July 2010, Lille.
10. Grbović, P.J., Delarue, P., Le Moigne, P. and Bartholomeus, P. (2011) The ultra-capacitor based controlled electric drives with braking and ride-through capability: overview and analysis. *IEEE Transactions on Industrial Electronics*, **58**(3), 925–936.
11. Grbović, P.J., Delarue, P. and Le Moigne, P. (2011) Modeling and control of the ultra-capacitor based regenerative controlled electric drive system. *IEEE Transactions on Industrial Electronics*, **58**(8), 3471–3484.
12. Grbović, P.J., Delarue, P. and Le Moigne, P. (2012) The ultra-capacitor based regenerative controlled electric drives with power smoothing capability. *IEEE Transactions on Industrial Electronics*, **59**(12), 4511–4522.
13. Grbović, P.J., Delarue, P. and Le Moigne, P. (2012) A three-terminal ultra-capacitor based energy storage and PFC device for regenerative controlled electric drives. *IEEE Transactions on Industrial Electronics*, **59**(1), 301–316.
14. Kim, S.-M. and Sul, S.-K. (2006) Control of rubber tyred gantry crane with energy storage based on supercapacitor bank. *IEEE Transactions on Power Electronics*, **21**(5), 1420–1427.
15. Kwon, T.-S., Lee, S.-W., Sul, S.-K. *et al.* (2010) Power control algorithm for hybrid excavator with supercapacito. *IEEE Transactions on Industry Applications*, **46**(4), 1447–1455.
16. Lee, J.-H., Lee, S.-H. and Sul, S.-K. (2009) Variable-speed engine generator with supercapacitor: isolated power generation system and fuel efficiency. *IEEE Transactions on Industry Applications*, **45**(6), 2130–2135.
17. Baggini, A. (2008) *Handbook of Power Quality*, 1st edn, John Wiley & Sons, Inc., London.
18. Zhang, Z., Fahmi, N.R., and Norris, W.T. (2001) Flicker analysis and methods for Electric Arc Furnace Flicker (EAF) mitigation (A Survey). Power Tech Proceedings, 2001, Porto, Portugal.
19. Han, C., Huang, A.Q., Bhattacharya, S. *et al.* (2008) Design of an ultra-capacitor energy storage system (UESS) for power quality improvement of electric arc furnaces. *IEEE Industry Application Society Annual Meeting*, 5–9 October, 2008.

20. Ackermann, T. (ed) (2005) *Wind Power in Power Systems*, John Wiley & Sons, Ltd and Royal Institute of Technology, Stockholm.

21. Zhou, Z., Benbouzid, M., Charpentier, J.F. *et al.* (2012) Energy storage technologies for smoothing power fluctuations in marine current turbines. IEEE ISIE 2012, Hangzhou, Chine, pp. 1425–1430.

22. Chen, S.-S., Wang, L., Lee, W.-J. and Chen, Z. (2009) Power flow control and damping enhancement of a large wind farm using a superconducting magnetic energy storage unit. *IET Renewable Power Generation*, **3**(1), 23–38.

23. Cimuca, G., Breban, S., Radulescu, M.M. *et al.* (2010) Design and control strategies of an induction-machine-based flywheel energy storage system associated to a variable-speed wind generator. *IEEE Transactions on Energy Conversion*, **25**(2), 526–534.

24. Rahmann, C., Haubrich, H.-J., Moser, A. *et al.* (2011) Justified fault-ride-through requirements for wind turbines in power systems. *IEEE Transactions on Power Systems*, **26**(3), 1555–1563.

25. Quand, L. and Qiao, W. (2011) Constant power control of DFIG wind turbines with supercapacitorenergy storage. *IEEE Transactions on Industry Applications*, **47**(1), 359–367.

26. Emadi, A., Nasiri, A. and Bekiarov, S. B. (2005) *Uninterruptible Power Supplies and Active Filters*, Boca Raton, CRC Press.

27. Bekiarov, S.B. and Emadi, A. (2002) Uninterruptible power supplies: classification, operation, dynamics and control. *Proceedings of IEEE APEC*, **1**, 597–604.

28. Tao, H., Duarte, J.L. and Hendrix, M.A.M. (2008) Line-interactive UPS using a fuel cells the primary source. *IEEE Transactions on Industrial Electronics*, **55**(8), 3012–3021.

29. Rufer, A. (2003) Power-electronic interface for a supercapacitor-based energy-storage substation in DC-transportation networks. Proceedings of EPE Conference, Toulouse.

30. Grbović, P.J., Delarue, P. and Le Moigne, P. (2012) Interface converters for ultra-capacitor applications in power conversion systems. EPE PEMC (ECCE Europe), IEEE Energy Conversion Congress and Exposition, Novi Sad, Serbia, September 2–6, 2012.

31. Beck, H.-P. and Hesse, R., (2007) Virtual synchronous machine. Proceedings of the 9th International Conference EPQU, pp. 1–6.

32. Torres, M. and Lopes, L.A.C. (2009) Virtual synchronous generator control in autonomous wind-diesel power systems. IEEE Electrical Power and Energy Conference.

33. Van, T.V., K. Visscher, J. Diaz, et al. (2010) Virtual synchronous generator: an element of future grids. Innovative Smart Grid Technologies Conference Europe (ISGT Europe), 2010 IEEE PES.

34. Zhong, Q.-C. and Weiss, G. (2011) Synchronverters: inverters that mimic synchronous generators. *IEEE Transactions on Industrial Electronics*, **58**(4), 1259–1267.

35. Rufer, A., Hotellier, D. and Barrade, P. (2004) A supercapacitor-based energy storage substation for voltage compensation in weak transportation networks. *IEEE Transactions on Power Delivery*, **19**(2), 629–636.

36. Camara, M.B., Gualous, H., Gustin, G. *et al.* (2010) DC-DC converter design for supercapacitor and battery management in hybrid vehicle applications-polynomial control strategy. *IEEE Transactions on Industrial Electronics*, **57**(2), 587–597.

37. Allegre, A.L., Bouscayrol, A., Delarue, P. *et al.* (2010) Energy storage system with supercapacitor for an innovative subway. *IEEE Transactions on Industrial Electronics*, **57**(12), 4001–4012.

38. Jih-Sheng, L. and Nelson, D.J. (2007) Energy management power converters in hybrid electric and fuel cell vehicle. *Proceedings of IEEE*, **95**(4), 766–777.

39. Lhomme, W., Delarue, P., Barrade, et al. (2005) Design and control of a supercapacitor storage system for traction applications. Proceedings of IAS 2005, pp. 2013–2020.

40. Gao, L., Dougal, R.A. and Liu, S. (2005) Power enhancement of an actively controlled battery/ultracapacitor hybrid. *IEEE Transactions on Power Electronics*, **20**(1), 236–243.

41. Ortúzar, M.O., Moreno, J. and Dixon, J. (2007) Ultracapacitor-based auxiliary energy system for an electric vehicle: implementation and evaluation. *IEEE Transactions on Industrial Electronics*, **54**(4), 2147–2156.

42. Jahns, T.M. and Blasko, V. (2001) Recent advances in power electronics technology for industrial and traction machine drives, Invited paper. *IEEE Proceedings*, **89**(6), 963–975.

43. Emadi, L., Lee, Y.J. and Rajashekare, K. (2008) Power electronics and motor drives in electric, hybrid electric, and plug-in hybrid electric vehicle. *IEEE Transactions on Industrial Electronics*, **55**(6), 2237–2245.

44. Solero, L., Lidozzi, A. and Pomilio J. A., (2005) Design of multiple-input power converter for hybrid vehicle, *IEEE Transactions on Power Electronics*, **20**(5), 1007–1016.

4

Ultra-Capacitor Module Selection and Design

4.1 Introduction

Before we start discussing design aspects of ultra-capacitor cells and modules, let's consider a general case power conversion system with integrated ultra-capacitor energy. A one-line circuit diagram is shown in Figure 4.1. The system consists of four main subsystems numbered 1 to 4: (1) the main power converter that feeds the load, (2) the ultra-capacitor energy storage, (3) the interface bi-directional dc–dc converter that interconnects the ultra-capacitor and the main dc bus, and (4) the input rectifier.

The main power converter feeds the load. The topology of the main converter and load depends on the application. It could be a three-phase inverter in applications such as variable speed drives (VSDs), traction drives, power quality equipment, and so on. In addition, the converter could be a dc–dc converter in applications such as automotive high voltage and low voltage systems. As this part of the conversion system is outside the scope of this book, we will not discuss design details further.

Design aspects of the interface dc–dc converter will be addressed in Chapter 5 of this book. In this chapter, we will address and discuss design aspects of the ultra-capacitor as the energy storage device. The ultra-capacitor design and selection criterion are discussed in the first part of the chapter. In the second, we address the efficiency of the ultra-capacitor and the entire power conversion system. Efficiency as a function of the ultra-capacitor size and cost is developed. Some aspects of the ultra-capacitor module, such as thermal design, parallel–series connection, and voltage balancing issues are addressed in detail.

4.1.1 The Analysis and Design Objectives

The ultra-capacitor design objective(s) could be summarized as follows: design and select an ultra-capacitor module according to the application requirement (Figure 4.2 shows an

Ultra-Capacitors in Power Conversion Systems: Applications, Analysis and Design from Theory to Practice,
First Edition. Petar J. Grbović.
© 2014 John Wiley & Sons, Ltd. Published 2014 by John Wiley & Sons, Ltd.

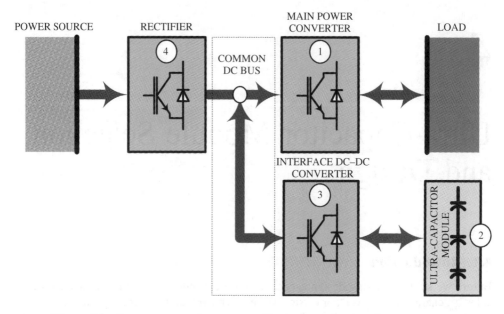

Figure 4.1   A power conversion system with ultra-capacitor based energy storage

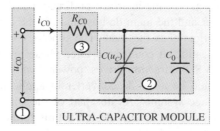

Figure 4.2   The ultra-capacitor model with main parameters highlighted

ultra-capacitor module with the highlighted main parameters to be analyzed and selected):

1. the module voltage rating,
2. the module capacitance, and
3. the module internal (parasitic) resistance that defines the conversion losses, thermal design, and efficiency.

4.1.2   Main Design Steps

The design of an ultra-capacitor for a power conversion application can be split into four fundamental steps.

- The first design step is selection of the ultra-capacitor voltage. Voltage rating depends on the application's requirement and the topology of the interface converter.

- The second step is selection of the module capacitance. The module capacitance is selected according to the application requirements, such as energy storage capability and conversion efficiency.
- The third step is the calculation of conversion losses and thermal stress of the ultra-capacitor module.
- The final step is the design of the string voltage balancing circuit.

4.1.3 The Ultra-Capacitor Model

Most of the ultra-capacitor models presented in the literature consider a nonlinear (voltage dependent) transmission line or a finite ladder RC network. For simplicity of analysis, the transmission line effect is neglected, and a first-order nonlinear model depicted in Figure 4.3 is used. The equivalent series resistance (ESR) R_{C0} is assumed as the frequency-independent resistance. The ultra-capacitor total capacitance is the voltage-controlled capacitance defined as

$$C_{C0}(u_C) = C_0 + k_C \cdot u_C, \tag{4.1}$$

where

C_0 is the initial capacitance that represents the electrostatic capacitance of the capacitor and

k_C is a coefficient that represents effects of the diffused layer of the ultra-capacitor.

Another form of the ultra-capacitor $C(u_C)$ characteristic can be derived from Equation 4.1 using the nominal capacitance C_N and the initial capacitance C_0. The nominal capacitance is the capacitance at the nominal voltage U_{CON}. The coefficient k_C can be defined as

$$k_C = \frac{C_N}{U_{CON}} \left(1 - \frac{C_0}{C_N} \right) = \frac{C_N}{U_{CON}} (1 - k_0), \tag{4.2}$$

where

C_N is the nominal capacitance at the nominal voltage U_{CON} and
k_0 is the normalized initial capacitance.

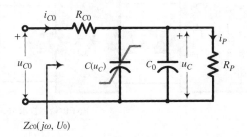

Figure 4.3 First order RC model of the ultra-capacitor

Substituting Equation 4.2 into Equation 4.1 yields

$$C_{C0}(u_C) = \left(\frac{C_0}{C_N}\right) C_N + \frac{C_N}{U_{C0N}}(1 - k_0) \cdot u_C = C_N \left(k_0 + \frac{(1 - k_0)}{U_{C0N}} \cdot u_C \right). \tag{4.3}$$

The capacitance definition (Equation 4.3) is more convenient from a practical perspective. Ultra-capacitor manufacturers give nominal capacitance and nominal voltage as parameters. The normalized initial capacitance is more or less constant for a family of ultra-capacitor cells. Typical values are $k_0 = 0.7$–0.8.

4.2 The Module Voltage Rating and Voltage Level Selection

The analysis and design objective is to select the ultra-capacitor module with proper voltage rating. The ultra-capacitor module is connected to the power conversion system via an interface dc–dc converter, as illustrated in Figure 4.4. The system dc bus voltage and the interface converter input voltage is V_{BUS}. The ultra-capacitor voltage and the interface converter output voltage is u_{C0}.

When we say the voltage rating of a device, the design usually has a straightforward meaning: select a device with the rated voltage. In the case of the ultra-capacitor, the situation is, however, a little bit different. Let's refer to Figure 4.5 that illustrates the definition of the different voltage levels of an ultra-capacitor energy storage device. Identify the following four voltage levels:

1. U_{C0max}, maximum operating voltage,
2. U_{C0min}, minimum operating voltage,
3. U_{C0inM}, intermediate operating voltage, and
4. U_{C0N}, the ultra-capacitor rated voltage.

Each of the listed voltage levels will be defined and the selection procedure described in the following sections.

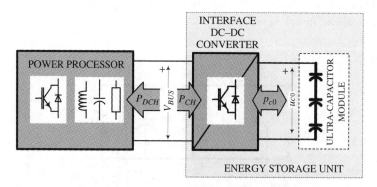

Figure 4.4 The ultra-capacitor connected to "the power processor" via an interface dc–dc converter

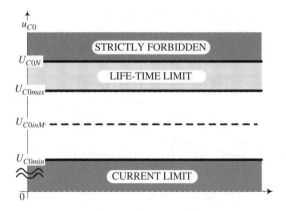

Figure 4.5   The ultra-capacitor voltage level definition

4.2.1   Relation between the Inner and Terminal Voltages

From the ultra-capacitor model, Figure 4.3, we can distinguish two different voltages; the inner voltage u_C and the terminal voltage u_{C0}. These voltages may significantly differ, depending on the current and the internal resistance. The internal resistance voltage drop is

$$\Delta u_{C0} = i_{C0} R_{C0}. \tag{4.4}$$

The ultra-capacitance ESR R_{C0} is a function of the capacitance $f(C_N)$. The function will be discussed in Section 4.3.2. For the moment, assume that the ESR is

$$R_{C0} \cong \left(0.68 \cdot 10^{-3} + \frac{1}{C_N} \right). \tag{4.5}$$

The current i_{C0} can be defined from the carbon loading factor k_{CL} as

$$i_{C0} = C_N k_{CL}. \tag{4.6}$$

The carbon loading factor determines the ultra-capacitor current density per farad of capacitance. Typical values are in the range 50–150 mA/F. For more details please see Section 2.4.4.

Substituting Equations 4.5 and 4.6 into Equation 4.4 yields the normalized voltage drop

$$\frac{\Delta u_{C0}}{U_{CON}} = \frac{(C_N 0.68 \cdot 10^{-3} + 1)k_{CL}}{U_{CON}} \tag{4.7}$$

Figure 4.6 shows the ultra-capacitor normalized voltage drop versus the capacitance and the carbon loading factor k_{CL}. Please note that the voltage drop increases with the capacitance and the loading factor. The voltage drop could be as high as 25% if the ultra-capacitor is loaded with 150 mA/F. Such a high voltage drop is significant and we have to differentiate the internal voltage u_C from the terminal voltage u_{C0}.

As discussed in Section 2.5, ultra-capacitor analysis is carried out with the internal voltage u_C as the variable. A logical question that arises is: what voltage shall we take into consideration when designing the ultra-capacitor module?

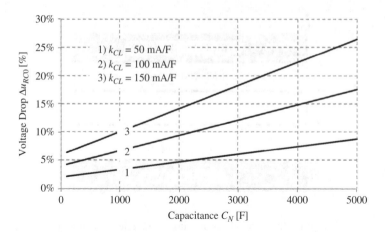

Figure 4.6 The ultra-capacitor relative voltage drop versus the capacitance and the carbon loading factor

Let the ultra-capacitor be charged to an initial voltage. As the current is zero (steady state), the internal and terminal voltages are the same. Hence, we can take the internal and terminal voltages as the same. Now, let's charge the ultra-capacitor to the maximum operating voltage. Once the terminal voltage reaches the maximum operating voltage, the charging is stopped. The internal voltage is lower than the terminal voltage. The difference depends on the charging current. The energy stored is not determined by the maximum operating voltage, but the internal one. The difference can be significant. To resolve this issue, we have to determine the internal and terminal maximum operating voltage U_{Cmax} and U_{C0max} respectively. The relation between these two is

$$U_{C0\,max} = U_{C\,max} + i_{C0} R_{C0}.$$ (4.8)

The same applies to the minimum operating voltage U_{C0min} and U_{Cmin}. We have to use the internal one (U_{Cmin}) that is linked to the terminal voltage by the equation

$$U_{C0\,min} = U_{C\,max} - i_{C0} R_{C0}.$$ (4.9)

4.2.2 Maximum Operating Voltage

The ultra-capacitor module's maximum operating voltage is the maximum voltage that the ultra-capacitor will be exposed to during charging. This limit is not strict and it can be exceeded. If so, there is no immediate destructive mechanism, but excursion above the limit will strongly influence the ultra-capacitor's life span.

The maximum operating voltage depends on the interface dc–dc converter topology. In this analysis, we have assumed that the dc–dc converter is a non-isolated direct dc–dc converter with a voltage gain not greater than unity ($u_{C0}/v_{BUS} \leq 1$). Therefore, the ultra-capacitor voltage cannot be greater than the dc bus voltage. As the ultra-capacitor is fully charged when the dc bus voltage is at maximum, the ultra-capacitor maximum operating voltage U_{C0max} is

$$U_{C0\max} \leq V_{BUS\max},\tag{4.10}$$

where V_{BUSmax} is the maximum operating dc bus voltage.

The internal maximum operating voltage U_{Cmax} can take any value in the range

$$U_{C\max} = [U_{C0\max}, U_{C0\max} - i_{C0}R_{C0}],\tag{4.11}$$

depending on the charging current i_{C0}.

The maximum operating dc bus voltage depends on the application voltage rating. Table 4.1 shows a summary of some possible applications, voltage rating and related maximum dc bus voltage. This table does not include MV traction applications, such as 1.5–3 kV trams and 15–25 kV heavy duty locomotives.

4.2.3 Minimum Operating Voltage

The ultra-capacitor minimum operating voltage is determined by two criteria. The first is the current capability I_{C0max} of the interface dc–dc converter and the ultra-capacitor itself.

$$U_{C0\min} \geq \frac{P_{C0}}{I_{C0\max}},\tag{4.12}$$

where P_{C0} is the conversion power.

The internal minimum operating voltage U_{Cmin} can take any value in the range

$$U_{C\min} = [U_{C0\min}, U_{C0\min} + i_{C0\max}R_{C0}].\tag{4.13}$$

The second criterion that determines the minimum operating voltage U_{C0min} is the so-called "constant power stability limit." This was intensively discussed in Chapter 3. The minimum internal operating voltage U_{Cmin} determined by the stability criterion is

$$U_{C\min} \geq \sqrt{P_{C0}4R_{C0}}\tag{4.14}$$

where R_{C0} is the ultra-capacitor's internal resistance.

The minimum terminal operating voltage U_{C0min} is determined from the worst case scenario of the above criteria (Equations 4.12 and 4.14),

$$U_{C0\min} = \max\left[\left(\sqrt{P_{C0}4R_{C0}} - i_{C0}R_{C0}\right), \frac{P_{C0}}{I_{C0\max}}\right].\tag{4.15}$$

Table 4.1 Power conversion systems' operating voltage and related maximum dc bus voltage

System	Low Voltage DC	Single Phase 110 V	Single/Three Phase 230 V	Three Phase 400 V	Three Phase 690 V
V_{BUSmax} (V)	16 to 56	150 to 450	350 to 450	700 to 900	1000 to 1200
Application area	Telecom, UPS, automotive	Domestic	Domestic, industry in Japan	Domestic and industry	Industry

Very often, the minimum voltage is limited to 40–50% of the rated voltage ($U_{C0min} = 0.4/0.5 \ U_{C0max}$).

4.2.4 The Ultra-Capacitor Intermediate Voltage

4.2.4.1 Definition of Intermediate Voltage

The ultra-capacitor intermediate voltage U_{C0inM} is defined as the long-term average voltage of the ultra-capacitor. The energy storage system is controlled in such a way as to have the tendency to keep the ultra-capacitor voltage as close as possible to the intermediate voltage. The intermediate terminal voltage U_{C0inM} and internal voltage U_{CinM} can be assumed to be the same.

To explain the importance of the intermediate voltage U_{C0inM} let's analyze two examples. The first is short-term UPS (uninterruptible power supply), while the second is an industrial controlled electric drive.

Figure 4.7 shows the ultra-capacitor voltage when the ultra-capacitor is used in UPS application. The ultra-capacitor is charged to the intermediate voltage U_{C0inM} and UPS is in stand-by mode. At the instant t_0 the mains supply is interrupted and the ultra-capacitor is discharged. The voltage decreases toward the minimum operating voltage U_{C0min}. At the instant t_1 the mains supply has recovered and the ultra-capacitor is re-charged. The voltage is increasing toward the intermediate voltage U_{C0inM}. At the instant t_2 the voltage reaches U_{C0inM} and re-charging is finished.

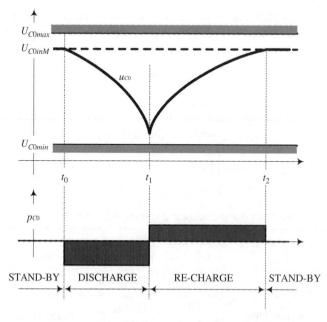

Figure 4.7 UPS application. Discharging and recharging of the ultra-capacitor in the case of power interruption

From this simple example it is obvious how to define the intermediate voltage U_{C0inM}. The intermediate voltage U_{C0inM} should be as high as possible in order to maximize the utilization of the ultra-capacitor. The top limit for the intermediate voltage U_{C0inM} is the maximum operating voltage. However, how we should select the intermediate voltage U_{C0inM} if the load profile is different?

A typical example of such an application is a controlled electric drive. As we have already discussed in Chapter 3, the ultra-capacitor can be used as short-term energy storage in controlled electric drive applications. In such applications, the ultra-capacitor may have two roles. The first is to store and recover the drive braking energy, while the second is to provide energy in the case of power interruption. Figure 4.8 shows the ultra-capacitor voltage and current for such an example.

Figure 4.8a illustrates an example when the ultra-capacitor is being charged while the drive operates in braking mode. The ultra-capacitor voltage is regulated on the intermediate value U_{C0inM}. At the instance t_0, the drive starts a braking operation and the ultra-capacitor is charged. The voltage u_{C0} is rising toward the maximum U_{C0max}. At the instance t_1, the braking has finished. The ultra-capacitor remains in stand-by mode. The voltage is constant. As soon as the drive operates in motoring mode, at the instance t_2, the ultra-capacitor is discharged toward the intermediate value U_{C0inM}. Once the voltage reaches the intermediate value U_{C0inM}, at the instance t_3, discharge is stopped and the ultra-capacitor remains in stand-by mode, awaiting the next braking sequence.

Figure 4.8b illustrates an example where the ultra-capacitor is being discharged while the drive operates in ride-through mode. The ultra-capacitor voltage is regulated on the intermediate value U_{C0inM}. At the instance t_0, the drive lost the mains power supply. The ultra-capacitor is discharged in order to supply the drive load. The voltage u_{C0} is decreasing toward the minimum U_{C0min}. At the instance t_1, the main power supply is recovered and takes over the drive load. At the same time the interface dc–dc converter starts recharging the ultra-capacitor toward the intermediate value U_{C0inM}. Once the voltage reaches the intermediate value U_{C0inM}, at the instance t_2, recharging is stopped and the ultra-capacitor remains in stand-by mode, awaiting the next braking or ride-through sequence.

From this example, it is not so obvious what the value of the intermediate voltage is. How does it depend on the braking and ride-through energy? How does it depend on the ultra-capacitor minimum and maximum voltage? Responses to these questions follow in the next section.

4.2.4.2 Determination of the Intermediate Voltage

Let the ultra-capacitor defined by Equation 4.1 be charged from the intermediate voltage U_{CinM} to maximum voltage U_{Cmax}. The energy stored in the ultra-capacitor is

$$E_{CH} = C_N \left[\frac{k_0}{2} \left(U_{C\,max}^2 - U_{CinM}^2 \right) + \frac{2}{3} \frac{(1 - k_0)}{U_{C0N}} \left(U_{C\,max}^3 - U_{CinM}^3 \right) \right]. \tag{4.16}$$

The energy realized from the ultra-capacitor when the ultra-capacitor is discharged from the intermediate voltage U_{CinM} to the minimum voltage U_{Cmin} is

$$E_{DCH} = C_N \left[\frac{k_0}{2} \left(U_{CinM}^2 - U_{C\,min}^2 \right) + \frac{2}{3} \frac{(1 - k_0)}{U_{C0N}} \left(U_{CinM}^3 - U_{C\,min}^3 \right) \right]. \tag{4.17}$$

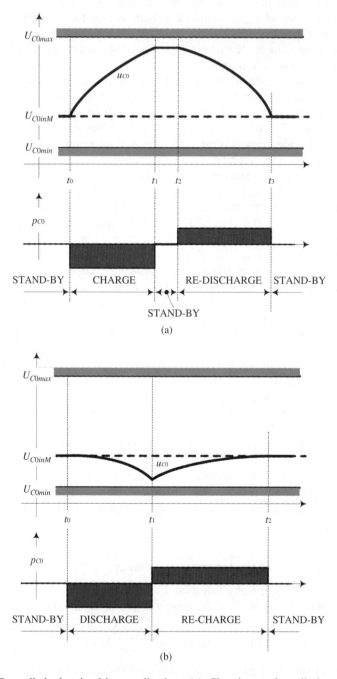

Figure 4.8 Controlled electric drive application. (a) Charging and re-discharging the ultra-capacitor when the drive is braking. (b) Discharging and recharging of the ultra-capacitor in the case of ride-through mode

From Equations 4.16 and 4.17 we can write the following third-order equation

$$\frac{2}{3}\frac{(1-k_0)}{U_{C0N}}(E_{CH}+E_{DCH})U_{inM}^3 + \frac{k_0}{2}(E_{CH}+E_{DCH})U_{CinM}^2 - A = 0, \qquad (4.18)$$

where A is a constant that depends on the capacitor parameters (U_{C0N} and k_0) and minimum/maximum operating voltage (U_{Cmax} and U_{Cmin}).

$$A = \frac{k_0}{2}(E_{DCH}U_{C\,max}^2 + E_{CH}U_{C\,min}^2) + \frac{2}{3}\frac{(1-k_0)}{U_{C0N}}(E_{DCH}U_{C\,max}^3 + E_{CH}U_{C\,min}^3). \qquad (4.19)$$

Theoretically, the intermediate voltage U_{CinM} can be computed as a function of the maximum and minimum voltages (U_{Cmax} and U_{Cmin}) and the capacitor coefficient k_0 from Equation 4.18. However, since Equation 4.18 is a cubical equation, it would be difficult to find a closed form analytical solution of U_{CinM}. It could be done numerically using the Matlab symbolic toolbox or Excel solver tool. However, it would not be practical in most cases and therefore a simplification of Equation 4.18 is necessary.

Let's assume that the ultra-capacitor is a linear capacitor with $k_C \cong 0$ and therefore $k_0 \cong 1$. Now we can easily compute the intermediate voltage U_{C0inM} from Equation 4.18

$$U_{C0inM} \cong \sqrt{\frac{E_{DCH}U_{C\,max}^2 + E_{CH}U_{C\,min}^2}{E_{DCH}+E_{CH}}}. \qquad (4.20)$$

The charge energy E_{CH} is energy stored in the ultra-capacitor when the ultra-capacitor is charged from the intermediate voltage U_{CinM} to the maximum voltage U_{Cmax}.

$$E_{CH} = \int_0^{T_{CH}} P_{CH}(t)dt - \int_0^{T_{CH}} R_{C0}(t)i_{C0(CH)}^2 dt, \qquad (4.21)$$

where

$P_{CH}(t)$ is the system charging power at the input terminals of the interface converter (Figure 4.4) and

T_{CH} is charging time.

The discharge energy E_{DCH} is the energy realized from the ultra-capacitor when the ultra-capacitor is discharged from the intermediate voltage U_{CinM} to the minimum voltage U_{Cmin}.

$$E_{DCH} = \int_0^{T_{DCH}} P_{DCH}(t)dt + \int_0^{T_{CH}} R_{C0}(t)i_{C0(DCH)}^2 dt, \qquad (4.22)$$

where

T_{DCH} is the discharge time and

$P_{DCH}(t)$ is the system conversion power at the interface converter input terminals, Figure 4.4.

However, this is not the end of the story. Please note from Equations 4.21 and 4.22 that the charge and discharge energies are functions of the ultra-capacitor charge/discharge current $i_{C0(CH)}$ and $i_{C0(DCH)}$. Additionally, the charge/discharge current is a function of the ultra-capacitor initial voltage, in this case the intermediate voltage U_{Cmin}. For more details see Section 2.5.3 and Equations 2.66, 2.68 and 2.79. Therefore, the charge/discharge energy is also a function of the intermediate voltage. This comes back to the fact that the simplified Equation 4.20 is actually much more complicated that it looks. To overcome this issue, we can compute the intermediate voltage in two iterations.

The first iteration

In the first iteration we will assume that the ultra-capacitor is an ideal one without resistance ($R_{C0} = 0$). Under such an assumption, we can compute the first iteration charge/discharge energy

$$E_{CH(1)} = \int_0^{T_{CH}} P_{CH}(t)dt, \quad E_{DCH(1)} = \int_0^{T_{DCH}} P_{DCH}(t)dt. \tag{4.23}$$

Substituting Equation 4.23 into Equation 4.20 yields the first iteration intermediate voltage

$$U_{C0inM(1)} \cong \sqrt{\frac{E_{DCH(1)}U_{C\max}^2 + E_{CH(1)}U_{C\min}^2}{E_{DCH(1)} + E_{CH(1)}}}. \tag{4.24}$$

The second iteration

In the second iteration we will use the first iteration intermediate voltage (Equation 4.24) to compute the charge/discharge current from Equations 2.68 and 2.79. Then, compute the charge and discharge energy from Equations 4.21 and 4.22 and substitute the results into Equation 4.20 to obtain the second iteration intermediate voltage. The second iteration intermediate voltage is accurate enough and can be used as the final one.

Figure 4.9 illustrates the ultra-capacitor intermediate voltage U_{CinM} versus the charge energy E_{CH} and minimum discharge voltage U_{Cmin}. The ultra-capacitor voltages are normalized on the rated voltage U_{C0max}, while the charge energy is normalized on the discharge energy E_{DCH}. For example, if the discharge energy is 25% of the charge energy ($E_{DCH}/E_{CH} = 0.25$ on the x axis) and the minimum discharge voltage is 50%, the intermediate voltage is approximately 63%.

4.2.5 The Ultra-Capacitor Rated Voltage

The ultra-capacitor rated voltage U_{CON} determines the ultra-capacitor life span. The expected life span T_{exp} of the ultra-capacitor is

$$T_{exp}(u_{C0}, \theta) = k_1 \exp\left(\frac{u_{C0}}{U_{CON}}k_2 + \theta k_3\right), \tag{4.25}$$

where u_{C0} is the ultra-capacitor operating voltage, θ is the ultra-capacitor operating temperature, and k_1, k_2, and k_3 are coefficients. These coefficients are usually given in the ultra-capacitor data sheet or provided by the supplier on demand.

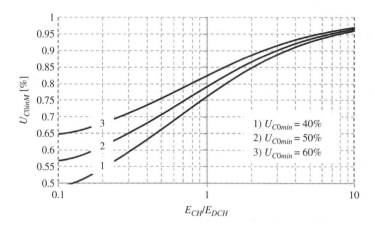

Figure 4.9 The ultra-capacitor intermediate voltage u_{C0inM} versus relative charge energy (ratio of the discharge energy E_{DCH} to the charge energy E_{CH}) and relative minimum ultra-capacitor voltage (determined as the minimum to maximum ultra-capacitor voltage ratio)

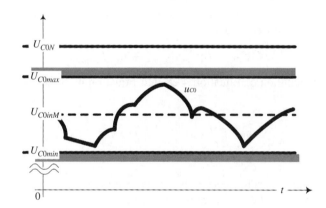

Figure 4.10 The ultra-capacitor voltage profile over a long time

However, the ultra-capacitor operating voltage is not constant over time. It rather varies between minimum and maximum, depending on the load profile. An example of an ultra-capacitor voltage profile is indicated in Figure 4.10. Thus, the expected life span equation (Equation 4.25) cannot be used directly. Average expected life span has to be computed as

$$T_{AV}(U_{CN}, \theta) = \psi(u_{C0}(t), \theta(t)). \tag{4.26}$$

From the previous equation, one can compute the rated voltage as a function of the given average expected life span and temperature

$$U_{C0N} = f(T_{AV}, \theta(t)). \tag{4.27}$$

4.2.6 Exercises

Exercise 4.1

A power conversion system with the following parameters is given: constant conversion power $P_{C0} = 22$ kW, maximum dc bus voltage $V_{BUSmax} = 750$ V and ultra-capacitor resistance $R_{C0} = 2\,\Omega$.

1. Calculate the maximum and minimum operating voltage of the ultra-capacitor.
2. Calculate the required current capability and power rating of the interface dc–dc converter that interconnects the ultra-capacitor and the power conversion system.

Solution

1. Since the ultra-capacitor is connected to the dc bus via a non-isolated dc–dc converter, the maximum operating voltage is determined by the dc bus voltage (Equation 4.10), $U_{C0max} = V_{BUSmax} = \mathbf{750\,V}$.

   The minimum operating voltage is determined by the system stability (Equation 4.14). For a given resistance $R_{C0} = 2\,\Omega$ and discharge power $P_{C0} = 22$ kW we have the minimum internal voltage $U_{Cmin} = \mathbf{419.5\,V}$.

   The terminal voltage is given by Equation 4.15. The ultra-capacitor current can be computed from the power balance

$$P_{C0} = U_{C\,min} I_{C0\,max} - R_{C0} I_{C0\,max}^2. \tag{4.28}$$

Substituting the solution of Equation 4.28 into Equation 4.15 yields the terminal minimum voltage

$$U_{C0\,min} = U_{C\,min} - R_{C0} I_{C0\,max} = \frac{U_{C\,min}}{2}.$$

$$U_{C0\,min} = 209.75\,\text{V}. \tag{4.29}$$

2. The maximum ultra-capacitor current is achieved at the minimum voltage U_{Cmin}. From the power balance equation (Equation 4.28) we have

$$I_{C0\,max} = \frac{U_{C\,min}}{2R_{C0}}. \tag{4.30}$$

The current capability of the interface dc–dc converter is therefore $I_{C0max} = \mathbf{104.88\,A}$. The converter power rating is $P_{DC} = \mathbf{78.66\,kW}$.

   Please note that the dc–dc converter power rating is much higher than the conversion power ($P_{DC} = 3.5\,P_{C0}$). This is an example of one of the major disadvantages of the ultra-capacitors as energy storage devices. The lower the minimum operating voltage the higher the power rating of the dc–dc converter is.

Exercise 4.2

A UPS system with the following parameters is given: output power $P_{C0} = 22$ kW, the UPS inverter efficiency $\eta_{INV} = 98\%$, maximum dc bus voltage $V_{BUSmax} = 750$ V, minimum ultra-capacitor voltage $U_{C0min} = 400$ V, and maximum discharge (bridging) time $T_{DCH} = 10$ seconds. The ultra-capacitor is connected to the UPS dc bus via a non-isolated dc–dc converter.

1. Calculate the ultra-capacitor intermediate voltage U_{C0inM}.
2. Calculate the ultra-capacitor minimum internal and terminal voltage that guarantee stability of the system. Assume that the capacitor resistance is $R_{C0} = 1\,\Omega$.

Solution

1. The intermediate voltage is given by Equation 4.20. Since there is no need to charge the ultra-capacitor above the intermediate voltage, the intermediate voltage is equal to the maximum voltage. As in the previous case, the ultra-capacitor is connected to the dc bus via a non-isolated dc–dc converter. Hence, the maximum operating voltage and the intermediate voltage is $U_{C0max} = V_{BUSmax} = 750\,V$ and $U_{CinM} = U_{C0max} = 750\,V$.
2. The minimum operating voltage is determined by the system's stability (Equation 4.14). For a given resistance $R_{C0} = 1\,\Omega$ and discharge power $P_{C0} = 22\,kW$ we have the minimum internal voltage

$$U_{Cmin} = 296.6\,V.$$

The ultra-capacitor current and terminal voltage are computed from Equations 4.29 and 4.30 $U_{C0min} = 148.3\,V$ and $I_{C0max} = 148.3\,A$.

Exercise 4.3

A variable speed drive (VSD) system with the following parameters is given: output power $P_0 = 15\,kW$, inverter efficiency $\eta_{INV} = 98\%$, maximum dc bus voltage $V_{BUSmax} = 750\,V$, and minimum ultra-capacitor voltage $U_{C0min} = 400\,V$. The VSD requires a ride-through capability of 2 seconds and braking of 10 seconds both at the full load. The ultra-capacitor is connected to the VSD dc bus via a non-isolated dc–dc converter.

1. Calculate the ultra-capacitor intermediate voltage U_{C0inM}.
2. Calculate the ultra-capacitor minimum internal voltage that guarantees stability of the system. Assume that the capacitor resistance is $R_{C0} = 1\,\Omega$.
3. Let's assume that the ultra-capacitor nominal voltage is $U_{CON} = 800\,V$ and the same voltage rating is selected for the ultra-capacitor as the previous example. Which ultra-capacitor will have the longer life span? The one used in the UPS system or the VSD system?

Solution

1. The intermediate voltage is computed from Equation 4.20. The discharge and charge energies are approximated by

$$E_{DCH} \cong \frac{1}{\eta_{INV}} \int_0^{T_{DCH}} P_{DCH}(t)dt = \frac{P_0 T_{DCH}}{\eta_{INV}},$$

$$E_{CH} \cong \eta_{INV} \int_0^{T_{CH}} P_{CH}(t)dt = \eta_{INV} P_0 T_{CH}. \tag{4.31}$$

Substituting Equation 4.31 and the minimum/maximum operating voltage into Equation 4.20 yields the intermediate voltage $U_{C0inM} = \mathbf{461.9\,V}$.

2. The minimum operating voltage is determined by the system stability criterion (Equation 4.14). For a given resistance $R_{C0} = 1\,\Omega$ and discharge power $P_{C0} = 15.3\,kW$ the minimum internal voltage is $U_{Cmin} = \mathbf{247.4\,V}$.

3. As explained in Section 4.2.5, Equations 4.25 and 4.27, the ultra-capacitor's nominal and long-term average voltages have a strong influence on the ultra-capacitor life span.

   In the UPS application considered in the example above, the intermediate voltage is the same as the maximum operating voltage. Hence, the intermediate to nominal voltage is close to unity $U_{C0inM}/U_{CON} = \mathbf{0.9375}$.

   In contrast to this, the intermediate voltage of the VSD system is low and therefore the intermediate to nominal voltage is low $U_{C0inM}/U_{CON} = \mathbf{0.577}$. Therefore, we can conclude that the life span of the ultra-capacitor used in the VSD system will be longer than that of the UPS one.

4.3   The Capacitance Determination

The capacitance of the ultra-capacitor can be determined based on two criteria: (i) the energy storage/recovery capability and (ii) the conversion efficiency. In the following section we will explore these two criteria in more detail.

4.3.1   Energy Storage/Recovery Capability

Let the ultra-capacitor defined by Equation 4.1 be charged/discharged between the minimum and maximum voltage. The energy is

$$\Delta E_C = C_N \left[\frac{k_0}{2} \left(U_{C\,max}^2 - U_{C\,min}^2 \right) + \frac{2}{3} \frac{(1 - k_0)}{U_{CON}} (U_{C\,max}^3 - U_{C\,min}^3) \right], \tag{4.32}$$

where U_{Cmax} is the ultra-capacitor internal maximum voltage and U_{Cmin} is the ultra-capacitor minimum voltage.

The total energy variation when the ultra-capacitor is charged from the minimum to the maximum voltage is the sum of the discharge energy E_{DCH} and charge energy E_{CH},

$$\Delta E_C = E_{CH} + E_{DCH}. \tag{4.33}$$

The charge and discharge energies are defined in Section 4.2.4.2, Equations 4.21 and 4.22. The capacitance C_N can be computed from Equation 4.32 as

$$C_N = \frac{E_{CH} + E_{DCH}}{\left[\frac{k_0}{2} \left(U_{C\,max}^2 - U_{C\,min}^2 \right) + \frac{2}{3} \frac{(1 - k_0)}{U_{CON}} (U_{C\,max}^3 - U_{C\,min}^3) \right]}. \tag{4.34}$$

4.3.2   Conversion Efficiency

4.3.2.1   A General Case Analysis

As discussed in Chapter 2, ultra-capacitor losses depend on a few parameters: the series resistance, capacitance, and the ultra-capacitor initial voltage. Is it possible to select the ultra-capacitor for the losses given as a design parameter? A short analysis follows.

Let the ultra-capacitor be charged from U_{C0inM} toward U_{C0max} with an arbitrary power or current profile. The ultra-capacitor losses are

$$P_{C0(\varsigma)}(t) = R_{C0}(t)i_{C0}^2(t), \tag{4.35}$$

where

$i_{C0}(t)$ is the ultra-capacitor current profile and
$R_{C0}(t)$ is the ultra-capacitor ESR that is generally time-dependent resistance.

For the sake of simplicity, let's assume that the current profile is such that the frequency spectrum falls outside of the ultra-capacitor critical frequency range. The critical frequency range has been discussed in Section 2.3.1.1. Under such an assumption, we can write that the ESR is a constant resistance R_{C0}.

The energy dissipated on the ESR R_{C0} during charging and discharging is

$$E_{\varsigma(CH)} = R_{C0} \int_0^{T_{CH}} i_{C0(CH)}^2(t)dt = R_{C0} \int_0^{T_{CH}} F_1^2(R_{C0}, C_N, K_0, P_{C0(CH)}, t)dt, \tag{4.36}$$

and

$$E_{\varsigma(DCH)} = R_{C0} \int_0^{T_{DCH}} i_{C0(DCH)}^2(t)dt = R_{C0} \int_0^{T_{DCH}} F_2^2(R_{C0}, C_N, K_0, P_{C0(DCH)}, t)dt, \tag{4.37}$$

where the ultra-capacitor charging and discharging current are functions F_1 and F_2 of the ultra-capacitor parameters and conversion power. The function strongly depends on the conversion power profile and the conversion method. At the end of this section we will give an example of the constant current conversion mode.

The ultra-capacitor resistance depends on the capacitance. The greater the capacitance the smaller the resistance

$$\frac{\partial R_{C0}}{\partial C_N} = \frac{\partial [R_{C0}(C_N)]}{\partial C_N} < 0. \tag{4.38}$$

The internal resistance R_{C0} is a strongly nonlinear function of the ultra-capacitor rated capacitance C_N, that can be approximated by a function

$$R_{C0} = \left(k_{RC(0)} + \frac{k_{RC(1)}}{C_N} + \frac{k_{RC(2)}}{C_N^2} \right), \tag{4.39}$$

where the coefficients $k_{RC(0)}$, $k_{RC(1)}$, and $k_{RC(2)}$ depend on the ultra-capacitor cells type.

Figure 4.11a illustrates the ultra-capacitor series resistance R_{C0} versus the capacitance C_N. The module rated voltage is $U_{CON} = 800$ V. The module is composed of 286 series connected ultra-capacitor cells LSUC 2.8 V [1]. From the example in Figure 4.11a we can identify the coefficients of Equation 4.39 and define the resistance as

$$R_{C0} \cong \left(-0.046 + \frac{1.42}{C_N} - \frac{0.17}{C_N^2} \right). \tag{4.40}$$

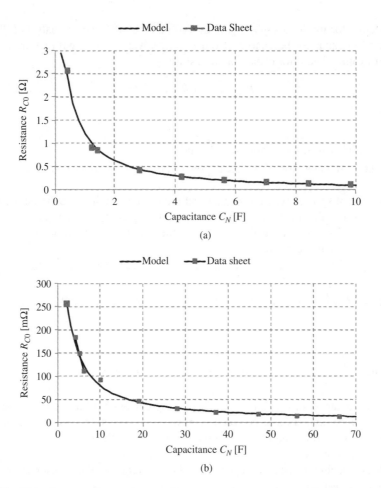

Figure 4.11 The ultra-capacitor resistance R_{C0} versus the capacitance C_0. (a) A small module composed of 286 LSUC 2.8 V cells. (b) A large module composed of 320 K2 Maxwell Technology cells. The module rated voltage is $U_{CON} = 800$ V

Figure 4.11b illustrates another example. The ultra-capacitor series resistance R_{C0} versus the capacitance C_N is shown. The module rated voltage is $U_{CON} = 800$ V. The module is composed of 320 series connected ultra-capacitor cells K2 2.5 V [2]. From the example in Figure 4.11b we can identify the coefficients of Equation 4.39 and define the resistance as

$$R_{C0} \cong \left(1.37 + \frac{856}{C_N} - \frac{688}{C_N^2} \right). \tag{4.41}$$

Substituting Equation 4.39 into Equation 4.36 yields charging and discharging energy losses

$$E_{S(CH)} \cong \left(k_{RC(0)} + \frac{k_{RC(1)}}{C_N} + \frac{k_{RC(2)}}{C_N^2} \right) \left[\int_0^{T_{CH}} F_1^2 \left(R_{C0}, C_N, K_0, P_{C0}, t \right) dt \right], \tag{4.42}$$

and

$$E_{\varsigma(DCH)} \cong \left(k_{RC(0)} + \frac{k_{RC(1)}}{C_N} + \frac{k_{RC(2)}}{C_N^2}\right)\left[\int_0^{T_{DCH}} F_2^2\left(R_{C0}, C_N, K_0, P_{C0(DCH)}, t\right)dt\right].$$

(4.43)

According to the discussion conducted in Section 2.4.2, the charge/discharge and round trip efficiency is

$$\eta_{RTP} = \frac{\Delta E_{C(DCH)}}{\Delta E_{C(CH)}}\left[\frac{\Delta E_{C(DCH)} - E_{\varsigma(DCH)}}{\Delta E_{C(CH)} + E_{\varsigma(CH)}}\right]$$

$$= \left[\frac{\Delta E_{C(CH)} - \left(k_{RC(0)} + \frac{k_{RC(1)}}{C_N} + \frac{k_{RC(2)}}{C_N^2}\right)\left[\int_0^{T_{DCH}} F_2^2\left(R_{C0}, C_N, K_0, P_{C0(DCH)}, t\right)dt\right]}{\Delta E_{C(CH)} + \left(k_{RC(0)} + \frac{k_{RC(1)}}{C_N} + \frac{k_{RC(2)}}{C_N^2}\right)\left[\int_0^{T_{CH}} F_1^2\left(R_{C0}, C_N, K_0, P_{C0(CH)}, t\right)dt\right]}\right],$$

(4.44)

where the energy losses are given by Equations 4.42 and 4.43. If stand-by losses can be neglected, the charging and discharging energy variations are the same, defined as

$$\Delta E_{C(CH)} = \Delta E_{C(DCH)} = C_N\left[\frac{k_0}{2}\left(U_{C\,\max}^2 - U_{C\,\min}^2\right) + \frac{2\left(1 - k_0\right)}{3}\frac{U_{C\,\max}^3 - U_{C\,\min}^3}{U_{CON}}\right].$$

(4.45)

It is obvious that the efficiency (Equation 4.44) depends on the capacitance C_N. It could be proven that the first derivate of the efficiency over the capacitance is strictly positive,

$$\frac{\partial \eta_{RTP}}{\partial C_N} > 0,$$

(4.46)

the greater the capacitance C_N the higher the efficiency η_{RTP}.

From the efficiency equation (Equations 4.44 and 4.45) and the capacitor resistance characteristic (Equation 4.39) we can compute the capacitance and internal resistance for given efficiency η_{RTP} as the design parameter,

$$C_N = C_N(\eta_{RTP}), \quad R_{C0(\eta_{RTP})} = R_{C0}(\eta_{RTP})$$

(4.47)

4.3.2.2 Constant Current Charging/Discharging Case

To illustrate the above conducted analysis, let's take the example of a constant current conversion system.

Discharging

The ultra-capacitor is fully charged to the maximum internal voltage U_{Cmax}. At some instant, the ultra-capacitor is discharged from U_{Cmax} toward the minimum voltage U_{Cmin} with a constant current $I_{C0(DCH)}$. According to the analysis conducted in Section 2.5.2.1,

Equation 3.21, the discharge efficiency is

$$\eta_{DCH} = 1 - I_{0(DCH)} \left(k_{RC(0)} + \frac{k_{RC(1)}}{C_N} + \frac{k_{RC(2)}}{C_N^2} \right)$$

$$\times 6 \frac{\left(\frac{1-k_0}{U_{CON}} \left(U_{C\,max}^2 - U_{C\,min}^2 \right) + k_0 (U_{C\,max} - U_{C\,min}) \right)}{4 \frac{1-k_0}{U_{CON}} (U_{C\,max}^3 - U_{C\,min}^3) + 3 k_0 (U_{C\,max}^2 - U_{C\,min}^2)} \qquad (4.48)$$

where U_{CON} is the ultra-capacitor rated voltage. The ultra-capacitor resistance R_{C0} is approximated by the function (Equation 4.39).

As discussed in Section 2.5.2.1, two different cases can be distinguished: the load profile is defined and constant; and the maximum/minimum voltages are defined and constant.

If the load profile is defined and constant, the ultra-capacitor minimum voltage is a function of the rated capacitance C_N. The minimum voltage is defined from Equation 2.52 as

$$U_{C\,min} = \sqrt{\left(\frac{U_{CON}}{2} \frac{k_0}{(1-k_0)} \right)^2 + \frac{U_{CON}}{(1-k_0)} \left(U_{C\,max} k_0 + \frac{(1-k_0)}{U_{CON}} U_{C\,max}^2 - \frac{I_{0(DCH)} T_{DCH}}{C_N} \right)}$$
$$- \frac{U_{CON}}{2} \frac{k_0}{(1-k_0)}. \qquad (4.49)$$

Substituting Equation 4.49 into Equation 4.48 yields the discharge efficiency as a function on the load profile only. From that function it would be possible to find the capacitance C_N for the efficiency as the parameter. However, the equation is difficult to solve analytically. At the end of this section we will give a couple of examples and graphics of efficiency versus capacitance.

The ultra-capacitor minimum voltage can be defined and is constant, while the load profile is not constant. This case is not often seen in real applications and therefore we will not discuss it in more detail.

To illustrate the above conducted analysis, an example is given. The ultra-capacitor is discharged from maximum to minimum voltage with a constant current. The maximum voltage is $U_{Cmax} = V_{BUSmax} = 750\,V$. The discharge current is $I_{C0} = 10\text{--}50\,A$, while the discharge time is $T_{DCH} = 10$ seconds.

Figure 4.12 shows the calculated minimum voltage versus the discharge current and the rated capacitance C_N of the ultra-capacitor. The efficiency versus the capacitance and discharge current was also computed. The results are shown in Figure 4.13.

As predicted by Equation 4.46, the discharge efficiency strongly depends on the capacitance at the charge profile given as a parameter. The bigger the capacitor the higher efficiency is. This efficiency is, however, not free. The higher the efficiency required the bigger and more expensive the capacitor is. The conversion efficiency versus the

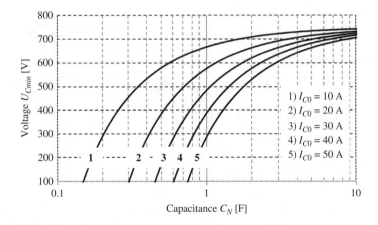

Figure 4.12 The ultra-capacitor minimum voltage versus rated capacitance C_N. The ultra-capacitor module voltage $U_{Cmax} = 750$ V, discharge time $T_{DCH} = 10$ seconds, the discharge current $I_{C0} = 10$ A to $I_{C0} = 50$ A

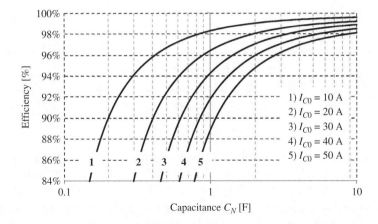

Figure 4.13 The charge efficiency versus the ultra-capacitor rated capacitance C_N. The ultra-capacitor module rated voltage $U_{C0max} = 750$ V, discharge time $T_{DCH} = 10$ seconds, the discharge current $I_{C0} = 10$ A to $I_{C0} = 50$ A

ultra-capacitor module cost is computed using the prediction of 2 US\$/kJ for the ultra-capacitor module. The graph is depicted in Figure 4.14.

Charging

The ultra-capacitor initial voltage is the minimum internal voltage U_{Cmin}. At some instant, the ultra-capacitor is charged from U_{Cmin} toward the maximum voltage U_{Cmax} with a constant current $I_{C0(CH)}$. The charge efficiency is computed from Equations 2.58 and 4.39 as

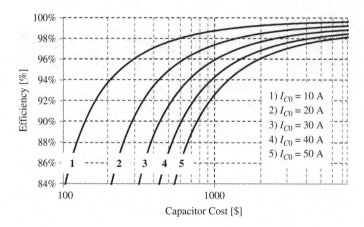

Figure 4.14 The charge efficiency versus the ultra-capacitor cost. The ultra-capacitor module rated voltage $U_{COmax} = 750\,\text{V}$, discharge time $T_{DCH} = 10\,\text{seconds}$, the discharge current $I_{C0} = 10\,\text{A}$ through $I_{C0} = 50\,\text{A}$

$$\eta_{CH} = \frac{1}{1 + \left(R_0 + \dfrac{k_{RC(1)}}{C_N} + \dfrac{k_{RC(2)}}{C_N^2}\right) I_{0(CH)} 6 \dfrac{\left(\dfrac{1-k_0}{U_{CON}}\left(U_{C\,max}^2 - U_{C\,min}^2\right) + k_0(U_{C\,max} - U_{C\,min})\right)}{4\dfrac{1-k_0}{U_{CON}}\left(U_{C\,max}^3 - U_{C\,min}^3\right) + 3k_0(U_{C\,max}^2 - U_{C\,min}^2)}}.$$

$$(4.50)$$

The same discussion from the previous section about the load profile and minimum/maximum voltage applies to charge efficiency. If the load profile is defined, the minimum and/or maximum voltages are variables. Let's assume that the maximum voltage is constant, while the minimum voltage is a function of the charge profile $(I_{0(CH)}, T_{CH})$ and capacitance. The minimum voltage is computed from Section 2.5.2.2 and Equation 2.61 as

$$U_{C\,min} = \sqrt{\left(\frac{U_{CON}}{2}\frac{k_0}{(1-k_0)}\right)^2 + \frac{U_{CON}}{(1-k_0)}\left(\begin{array}{c}\dfrac{(U_{C0\,max} - R_{C0}I_{0(CH)})\,k_0}{(1-k_0)} \\ + \dfrac{(1-k_0)}{U_{CON}}(U_{C0\,max} - R_{C0}I_{0(CH)})^2 \\ - \dfrac{I_{0(DCH)}T_{DCH}}{C_N}\end{array}\right)}$$

$$-\frac{U_{CON}}{2}\frac{k_0}{(1-k_0)}.$$

$$(4.51)$$

The second case is when the minimum and maximum voltages are constant while the charge profile $(I_{0(CH)}, T_{CH})$ is variable. This case is also uncommon and is not discussed further here.

Round Trip Efficiency

The conversion round trip efficiency can be computed as

$$\eta_{RTP} = \eta_{CH}\eta_{DCH},\tag{4.52}$$

where the charge and discharge efficiencies are given by Equations 4.48 and 4.50. The variables can be the load/charge profile or minimum/maximum voltages.

4.3.3 End-of-Life Effect on the Capacitance Selection

The ultra-capacitor End-of-Life (EOL) is characterized by a reduction in the capacitance and an increase of the capacitor resistance and leakage current. According to [2], the EOL capacitance is 80% and the resistance and leakage current are 200% of the initial (nominal) value. The initial values will be termed Start of Life (SOL) values in further discussion.

The ultra-capacitor parameter deviation must be considered at the beginning of the design. Otherwise, the performance of the conversion system cannot be ensured. The capacitance computed from the energy storage capability (Equation 4.34), has to be increased for a factor that takes into account the EOL capacitance reduction. Hence, the designed (SOL) capacitance is

$$C_N = \frac{1}{0.8}\frac{E_{CH} + E_{DCH}}{\left[\frac{k_0}{2}\left(U_{C\,\text{max}}^2 - U_{C\,\text{min}}^2\right) + \frac{2}{3}\frac{(1-k_0)}{U_{CON}}\left(U_{C\,\text{max}}^3 - U_{C\,\text{min}}^3\right)\right]}.\tag{4.53}$$

The EOL increase in the capacitor resistance has to be adequately taken into account in the design. Let the ultra-capacitor SOL and EOL resistance be $R_{C0(SOL)}$ and $R_{C0(EOL)}$. From the EOL definition it follows that

$$R_{C0(EOL)} = 2R_{C0(SOL)}.\tag{4.54}$$

Substituting the EOL condition, Equitation 4.54 into the ESR-to-capacitance function, Equation 4.39 yields

$$C_N = -\frac{k_{RC(1)}}{2k_{RC(0)} - R_{C0(\eta_{RTP})}} \pm \sqrt{k_{RC(1)}^2 - 2k_{RC(2)}(2k_{RC(0)} - R_{C0(\eta_{RTP})})},\tag{4.55}$$

where $R_{C0}(\eta_{RTP})$ is the module resistance that gives the required efficiency at the EOL,

$$R_{C0(\eta_{RTP})} = \left(k_{RC(0)} + \frac{k_{RC(1)}}{C_N(\eta_{RTP})} + \frac{k_{RC(2)}}{C_N^2(\eta_{RTP})}\right)\tag{4.56}$$

the parameters $k_{RC(0)}$, $k_{RC(1)}$, and $k_{RC(2)}$ are determined from the ultra-capacitor $R_{C0}(C_N)$ characteristic (Equation 4.39). The capacitance $C_N(\eta_{RTP})$ is the capacitance determined from the efficiency characteristic (Equation 4.44).

Finally, the ultra-capacitor rated capacitance is

$$C_N = \max \times \left\{ \frac{1}{0.8} \frac{E_{CH} + E_{DCH}}{\left[\frac{k_0}{2} \left(U_{C\,max}^2 - U_{C\,min}^2 \right) + \frac{2}{3} \frac{(1-k_0)}{U_{CON}} \left(U_{C\,max}^3 - U_{C\,min}^3 \right) \right]} , \underbrace{C_N = C_N(\eta_{RTP})}_{Equation\ 4.47} \right\}.$$

(4.57)

If the capacitance is selected according to Equation 4.57, it can be guaranteed that the conversion system performances at the EOL of the ultra-capacitor are as required.

4.3.4 Exercises

Exercise 4.4

A UPS system with the parameters from Exercise 4.2 is given. The ultra-capacitor nominal voltage is $U_{CON} = 800$ V. Assume the ultra-capacitor is a nonlinear capacitor with the initial capacitance coefficient $k_0 = 0.8$.

1. Calculate the nominal capacitance C_N of the ultra-capacitor energy storage. Take into account the EOL effect on the capacitance selection considering that the ultra-capacitor resistance is approximated by Equation 4.40.
2. Estimate the discharge efficiency.

Solution

1. The nominal capacitance is computed from Equation 4.34,

$$C_N = \frac{E_{DCH}}{\left[\frac{k_0}{2} \left(U_{C\,max}^2 - U_{C\,min}^2 \right) + \frac{2}{3} \frac{(1-k_0)}{U_{CON}} \left(U_{C\,max}^3 - U_{C\,min}^3 \right) \right]}$$

(4.58)

where the discharge energy is given by Equations 4.21 and 4.22.

However, to calculate the discharge energy, we need first to calculate the conversion losses. The losses can be computed according to analysis conducted in Section 2.5.3.1 and Equation 2.71. The conversion losses are a function of the ultra-capacitor resistance and capacitance. If we include this in the capacitance equation, we will have an equation that cannot be solved analytically. To avoid this issue, we can calculate the capacitance in several iterations.

The first iteration:
Let's calculate the discharge energy neglecting the conversion losses (an ideal capacitor is assumed). From $P_0 = 22$ kW, $\eta_{INV} = 98\%$, $T_{DCH} = 10$ seconds we have $E_{DCH} = 224.5$ kJ. From Equation 4.58 and the discharge energy E_{DCH} we compute the capacitance $C_N = 1.02$ F. Now we have a rough idea of the ultra-capacitor size, and we can go to the second iteration.

The second iteration:
For the capacitance computed, we computed the resistance from Equation 4.40 $R_{C0} = 1.16\,\Omega$. From Equation 2.72 and the capacitance and resistance computed above, we compute the discharge energy losses $E_{\varsigma(DCH)} = 8.83\,\text{kJ}$ and the discharge energy $E_{DCH} = 233.4\,\text{kJ}$.

Now having the discharge energy and the energy losses, we compute a new value of the capacitance $C_N = 1.06\,\text{F}$. Please notice the difference (error) between the first and second iteration capacitance is approximately 0.04 F, which is 4%.

To take into account the EOL effect on the capacitance selection, the initial (SOL) capacitance should be higher than the computed one. As the EOL capacitance is 80% of the initial one, the ultra-capacitor should be selected with the initial capacitance $C_N = 1.325\,\text{F}$.

2. The discharge efficiency is computed from Equation 2.77 as

$$\eta_{DCH} = 1 - \frac{E_{\varsigma(DCH)}}{\Delta E_C}. \tag{4.59}$$

For the selected ultra-capacitor $C_N = 1.325\,\text{F}$ we compute the initial and the EOL resistance $R_{C0} = 0.934\,\Omega$ and $R_{C0(EOL)} = 1.87\,\Omega$. From the above parameters we can compute the discharge efficiency at the SOL and EOL $\eta_{DCH} = 97.5\%$ and $\eta_{DCH(EOL)} = 95\%$.

4.4 Ultra-Capacitor Module Design

In Section 2.8 we addressed some aspects of high power ultra-capacitor modules. Here, in this section, we will briefly address some aspects of the module design.

4.4.1 Series/Parallel Connection

The ultra-capacitor module's nominal parameters, nominal voltage U_N, and capacitance C_N were selected in Sections 4.2.5 and 4.3.3. Now we have to design the module according to the selected parameters. The design steps can be listed as:

1. an elementary ultra-capacitor cell rated voltage $U_{N(cell)}$,
2. number of series connected cells N,
3. the capacitance of a cell $C_{N(cell)}$,
4. the number of parallel connected cells M.

4.4.1.1 Number of Series and Parallel Cells

From a list of available cells we select the cell nominal voltage $U_{N(cell)}$. Usually, it is 2.5–2.8 V [1, 2]. The number of the series connected cells is

$$N = floor\left(\frac{U_{C0N}}{U_{C0N(cell)}}\right). \tag{4.60}$$

The capacitance of an individual cell or a bank of parallel connected cells is

$$C_{N1} = C_N N. \tag{4.61}$$

If necessary, M cells will be connected in parallel. This is usually the case if capacitance (Equation 4.61) is from the list of available ultra-capacitor cells. It also may happen that an optimal solution is to use M paralleled small cells instead of a large one. The number of paralleled cells is

$$M = floor\left(\frac{C_N}{C_{N(cell)}}\right)N, \qquad (4.62)$$

where $C_{N(cell)}$ is the capacitance of the selected ultra-capacitor cell [1, 2].

4.4.1.2 The Cells' Inter-Connection Arrangement

In the previous section, we defined the number of series connected (N) and parallel connected (M) ultra-capacitor cells. Is there a unique way to connect the cells into a module or are there more combinations? If there is more than one way to connect the cells into the module, which one is optimal?

For given number of series and parallel connected cells, there are two extreme ways to connect the cells into a module. The first one is the series–parallel connection (Figure 4.15a). The second is the parallel–series connection (Figure 4.15b).

In the first case, N cells are series connected into a submodule. Then M submodules are parallel connected. The submodules are parallel connected only into two points, plus and minus, of the entire module. There is no internal parallel connection. In the second case, M cells are parallel connected into a submodule. Then, N submodules are series connected.

Is there some significant difference between these two connections? These two connections are electrically identical, N series and M parallel connected cells. However, there are two differences that may become important in some cases.

Averaging of the Capacitance Deviation
Let's consider the second case; M cells parallel connected into a submodule. If the capacitance of each individual cell is C_j, the capacitance of the submodule k and average cell

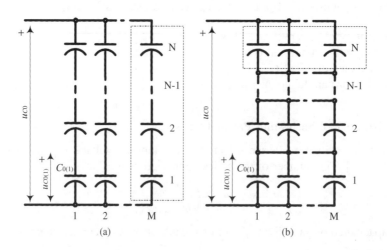

Figure 4.15   (a) Series–parallel connection of cells and (b) parallel–series connection

capacitance are

$$C_k = \sum_{j=1}^{m} C_j, \quad C_{cell(AV)} = \frac{1}{M} \sum_{j=1}^{m} C_j.$$ (4.63)

The capacitance deviation between each individual cell can be $\pm 20\%$ of the rated capacitance. However, if the number of parallel connected cells into one submodule is high, the capacitance deviation between N submodules will be lower than the deviation between the individual cells. Hence, N series connected submodules will have better capacitance distribution and therefore better voltage sharing between the submodules. More detail about voltage sharing is given in Section 4.4.3.

It was shown in [3] that voltage variation in the first case (series–parallel) is much higher than in the second case (parallel–series). An obvious conclusion from the above analysis and results presented in [3] is that a preferred combination is M-parallel N-series connected ultra-capacitor cells. But, this is not the end of the story.

Fault Tolerance

What will happen in the case of failure of one cell? How does the connection scheme affect the module fault tolerance if it affects it at all? Yes, the connection may affect the fault tolerance significantly. The ultra-capacitor cell may fail in short circuit or open circuit faults. Due to the technology of the ultra-capacitor, the open circuit fault is rarely seen in the field. Hence, we will not consider it further in this discussion.

Let's assume that the Nth cell of the Mth column is short circuited, as shown in Figure 4.16a. In the first case (series–parallel connection) the cell that is short circuited is not connected in parallel to any other cell. This failure is reflected as the loss of only one cell in the module. The voltage on the remaining cells in the Mth column is

$$u_{C01} = \cdots = u_{C0(N-1)} = \frac{u_{C0}}{N-1},$$ (4.64)

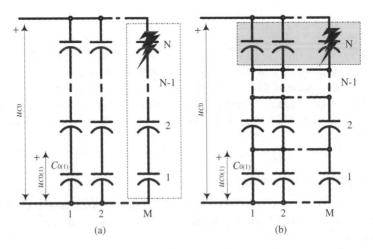

(a)   (b)

Figure 4.16 Short circuit fault of one cell of the module. (a) Series–parallel connection and (b) parallel–series connection

where

u_{C0} is voltage of the entire module and
u_{C01} are the remaining cells in the Mth column.

The voltage rise normalized on the cell nominal voltage is

$$\Delta u = \frac{\Delta u_{C01}}{\frac{u_{C0}}{N}} = \frac{1}{N-1}, \tag{4.65}$$

where $\frac{u_{C0}}{N}$ is the cell nominal voltage.

Assuming that the number of series connected cells N is high and selected with some margin, the voltage rise (Equation 4.65) is not significant. The voltage on the remaining cells is within specification.

The module energy capability is reduced, where the energy reduction is

$$\Delta e = \frac{1}{NM}. \tag{4.66}$$

Since the number of cells (NM) is large, the factor (Equation 4.66) is very small. Let's take an example. An ultra-capacitor module is composed of $N = 50$ series connected cells and $M = 20$ parallel connected cells. The voltage rise is $\Delta u = 2\%$, while the energy reduction factor is $\Delta e = 0.1\%$.

Let's now consider the second case (parallel–series connection). The Nth cell of the Mth column is short circuited, as shown in Figure 4.16b. The cell that is short circuited is connected in parallel with $M - 1$ cells, M parallel connected in total. Thus, the failure is reflected as a loss of one submodule of the total of N submodules. The voltage on the remaining cells in the module is the same as in the first case (Equations 4.64 and 4.65). There is no difference. However, there is a significant difference in module energy capability before and after the fault. The energy reduction is

$$\Delta e = \frac{1}{N}. \tag{4.67}$$

Let's take the same example as in the first case. The voltage rise is the same, $\Delta u = 2\%$, while the energy reduction factor is much higher than in the first case, $\Delta e = 2\%$.

From the previous short analysis we have seen that the first case (series–parallel connection) is a better option if the module fault tolerance is a design objective. In contrast to this, the second case (parallel–series connection) is a better solution if module voltage balancing is the design objective. In the end, the connection selection depends on the design objective.

4.4.2   Current Stress and Losses

Conversion losses were intensively discussed in Section 2.6. Here, in this section, we will summarize the results from Section 2.6 and give new results that take into account the EOL resistance.

The ultra-capacitor is not an ideal device (having an internal resistance R_{C0}). Hence, whenever the ultra-capacitor is charged and discharged, a fraction of the stored/restored energy is realized on the internal resistance R_{C0}. As addressed in Section 2.3, the internal resistance is the frequency dependent resistance.

Instantaneous power dissipated on the resistance R_{C0} is generally given by

$$p(t) = \frac{1}{4\pi^2} \int_{-\infty}^{+\infty} I(j\omega)e^{j\omega t}d\omega \cdot \int_{-\infty}^{+\infty} R_{C0}(j\omega)I(j\omega)e^{j\omega t}d\omega, \qquad (4.68)$$

where R_{C0} is the frequency dependent resistance.

Let $R_{C0(cell)}$ be the ESR of the selected ultra-capacitor cell. The total ESR of the ultra-capacitor module consisting of N series connected and M parallel connected cells is

$$R_{C0}(j\omega) = \sum_{k=1}^{N} \frac{1}{\displaystyle\sum_{p=1}^{M} \frac{1}{R_{C0(cell)(kp)}(j\omega)}} \cong R_{C0(cell)}(j\omega)\frac{N}{M}, \qquad (4.69)$$

where $R_{C0(cell)(kp)}$ is the resistance of the cell in the pth row and k^{th} column of the module.

Usually the cells are of same type and size, and therefore all resistance of cells is the same $R_{C0(cell)}$.

The relation between the cell and module current is

$$i_{C0(cell)}(t) = \frac{i_{C0}(t)}{M}. \qquad (4.70)$$

The module resistance given by Equation 4.69 is the initial (SOL) resistance. According to the standard and the EOL definition, the EOL resistance is 200% of the initial one. Such a significant increase in the resistance has to be considered at the beginning of the design in order to guarantee the performance of the conversion system at the EOL of the ultra-capacitor. Hence, the conversion losses have to be computed with the EOL resistance,

$$R_{C0}(j\omega) = 2R_{C0(cell)}(j\omega)\frac{N}{M}. \qquad (4.71)$$

4.4.2.1   Periodic Current at Moderate Frequency

If the ultra-capacitor current is a periodic function with the frequency falling within the range of critical frequencies of the ultra-capacitor, the losses at the EOL of the ultra-capacitor cell are computed as

$$P_{C0(\varsigma)(cell)} \cong 2\left[\frac{1}{2}\sum_{k=0}^{+\infty} R_{C0(cell)}(k\omega_0) \cdot \frac{I_{0(k)}^2}{M^2}\right],$$

$$P_{C0(\varsigma)(MOD)} \cong 2\frac{N}{M}\left[\frac{1}{2}\sum_{k=0}^{+\infty} R_{C0(cell)}(k\omega_0) \cdot I_{0(k)}^2\right]. \qquad (4.72)$$

$R_{C0(cell)}(k\omega_0)$ is the frequency dependent initial (SOL) internal resistance of the ultra-capacitor cell and $I_{0(k)}$ is magnitude of the kth harmonics of the ultra-capacitor module current.

4.4.2.2   Periodic Current at High Frequency

If the frequency of the ultra-capacitor current is above the critical frequency, the conversion losses at the EOL of the ultra-capacitor are

$$P_{C0(\varsigma)(cell)} = 2R_{C0(cell)(HF)}\frac{I_{0(RMS)}^2}{M^2},$$

$$P_{C0(\varsigma)(MOD)} = 2\frac{N}{M}R_{C0(cell)(HF)}I_{0(RMS)}^2 \qquad (4.73)$$

where $I_{0(RMS)}$ is the ultra-capacitor module RMS current and $R_{C0(cell)(HF)}$ is the high frequency resistance of the ultra-capacitor.

4.4.2.3   Periodic Current at Very Low Frequency or Long Single Pulse Current

Let the ultra-capacitor current be a periodic function at a very low frequency or a long single pulse. If the frequency falls within the range of the ultra-capacitor thermal frequency, the losses have to be computed as instantaneous losses

$$P_{C0(\varsigma)(cell)}(t) = 2R_{C0(cell)(LF)}\frac{i_{C0}^2(t)}{M^2},$$

$$P_{C0(\varsigma)(MOD)}(t) = 2\frac{N}{M}R_{C0(cell)(LF)}i_{C0}^2(t), \qquad (4.74)$$

where i_{C0} is the ultra-capacitor module's current profile and $R_{C0(cell)(LF)}$ is low frequency resistance of the ultra-capacitor.

4.4.2.4   A Short Single Pulse Current

Let the ultra-capacitor current be a single pulse with a pulse width that falls within the range of the ultra-capacitor at time constant. In such a case, the instantaneous losses are computed from the general form equation and the EOL resistance,

$$p_{C0\varsigma(cell)}(t) = 2\frac{1}{M^2}\frac{1}{4\pi^2}\int_{-\infty}^{+\infty}I(j\omega)e^{j\omega t}d\omega \cdot \int_{-\infty}^{+\infty}R_{C0(cell)}(j\omega)I(j\omega)e^{j\omega t}d\omega,$$

$$p_{C0\varsigma(MOD)}(t) = 2\frac{N}{M}\frac{1}{4\pi^2}\int_{-\infty}^{+\infty}I(j\omega)e^{j\omega t}d\omega \cdot \int_{-\infty}^{+\infty}R_{C0(cell)}(j\omega)I(j\omega)e^{j\omega t}d\omega, \qquad (4.75)$$

where $I(j\omega)$ is the current spectrum.

4.4.3   String Voltage Balancing

4.4.3.1   Definition of the Problem

The ultra-capacitor is not a perfect device. Some parameters of the ultra-capacitor may vary significantly from cell to cell. Variation and dispersion of parameters such as capacitance, ESR, and leakage current may cause dispersion of the voltage distribution among series connected cells.

Dispersion of the Cells' Capacitance
The ultra-capacitor cell capacitance may vary $\pm 20\%$ of the nominal value. The capacitance dispersion creates dynamic variation in the ultra-capacitor voltage when the ultra-capacitor is charged and discharged. Voltages of individual cells are defined as

$$\frac{1}{C_{0(1)}} \int i_{C0}dt \neq \frac{1}{C_{0(2)}} \int i_{C0}dt \dots \neq \frac{1}{C_{0(N)}} \int i_{C0}dt, \tag{4.76}$$

where $C_{0(1)}$ to $C_{0(N)}$ is the capacitance of the cell 1 to N respectively.

Dispersion of the Cells' ESR
Dispersion of the cells' ESR also creates dynamic variation in the voltage when the ultra-capacitor is charged and discharged. Voltages of individual cells are defined as

$$R_{C0(1)}i_{C0} \neq R_{C0(2)}i_{C0} \dots \neq R_{C0(N)}i_{C0}, \tag{4.77}$$

where $R_{C0(1)}$ to $R_{C0(N)}$ is the ESR of the cell 1 to N respectively.

Dispersion of the Cells' Leakage Current
The leakage current of individual cells may significantly vary from cell to cell. This variation creates long-term (static) dispersion of the voltage distribution. Voltages of individual cells are defined as

$$\frac{1}{C_{0(1)}} \int i_{\varsigma(1)}dt \neq \frac{1}{C_{0(2)}} \int i_{\varsigma(2)}dt \dots \neq \frac{1}{C_{0(N)}} \int i_{\varsigma(N)}dt, \tag{4.78}$$

where $i_{\varsigma 1}$ to $i_{\varsigma N}$ is the leakage of the cell 1 to N respectively

4.4.3.2　Balancing Circuits

We can distinguish four balancing methods and circuits; passive resistive balancing, passive clamp, switched resistor balancing, and switched mode balancing circuit [4, 5].

Passive Resistive Balancing Circuit
The passive resistive balancing circuit is the simplest balancing circuit (Figure 4.17a). The same circuit is broadly used to balance voltage of series connected electrolytic capacitors [6]. The balancing circuit consists of resistors R_1 to R_N, each connected in parallel with a corresponding ultra-capacitor cell C_1 to C_N. The resistance is selected according to the ultra-capacitor cell leakage current.

Let the ultra-capacitor cell leakage current be $I_{C0\varsigma}$ and the variation of the nominal leakage current be ΔI_ς. To keep the voltage dispersion within a limit, the balancing resistor current must be a multiple of the worst case scenario leakage current dispersion. The resistor current is

$$I_{R\varsigma} \geq 10\Delta I_{C0\varsigma}M, \tag{4.79}$$

where Δ is the relative leakage current, usually in the range $\pm 20\%$, $I_{C0\varsigma}$ is the cell nominal leakage current, and M is the number of parallel connected cells.

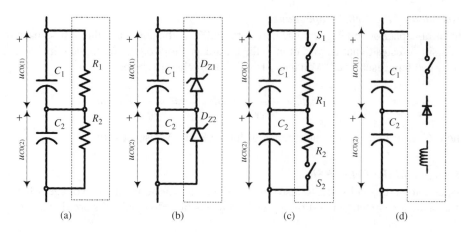

Figure 4.17 (a) Passive resistive balancing circuit, (b) passive clamping circuit, (c) switched resistor balancing circuit, and (d) switched mode loss-free balancing circuit

The resistance of the balancing resistors is computed from

$$R \leq \frac{U_N}{10 \Delta I_{C0_S} M}.$$ (4.80)

Stand-by losses of the balancing resistor and entire balancing circuit are

$$P_R \geq U_N 10 \Delta I_{C0_S} M$$

$$P_{R(MOD)} \geq U_N 10 \Delta I_{C0_S} N M$$ (4.81)

where N is the number of series connected cells.

The passive resistive balancing circuit is a robust and simple solution. However, because of some disadvantages, such as additional stand-by losses and no dynamic balancing capability, this solution is not used very often in industrial applications.

Passive Voltage Clamping Circuit

The second option is a passive voltage clamping circuit. The circuit consists of low voltage zener diodes D_{Z1} to D_{ZN} connected in parallel with the series connected ultra-capacitor cells C_1 to C_N (Figure 4.17b). The cell voltage is limited to the clamping diode breakdown voltage. Thus, there are no stand-by losses if the cell voltage is below the limit. The main disadvantage of this kind of balancing circuit is the lack of dynamic voltage balancing capability. In fact, it would be possible to achieve dynamic limiting of the voltage, but it would require high power clamping diodes.

Switched Resistor Balancing Circuit

The third solution is a switched resistor balancing circuit. The circuit consists of resistors R_1 to R_N and switches S_1 to S_N. The resistors are connected in parallel with the capacitor cells via the switches, as illustrated in Figure 4.17c. The cells' voltages are measured and actively controlled by acting on the switches. If the cell voltage has a tendency to increase above the upper reference, the switch is turned on and the resistor is connected in parallel with the capacitor. The capacitor is discharged. Once the voltage is reduced to the lower reference the switch is turned off. The voltage is actively controlled with dynamic capability. Efficiency is the main drawback of the circuit.

Let the capacitors C_1 and C_2 be defined as

$$C_1 = C_N(1 + \Delta_1), \quad C_2 = C_N(1 + \Delta_2), \tag{4.82}$$

where C_N is the nominal capacitance and Δ_1 and Δ_2 are the normalized deviations of the capacitance of capacitors C_1 and C_2 respectively.

Let the ultra-capacitor module be charged with current I_{C0}. If the capacitor C_1 is bigger than C_2, $\Delta_1 > \Delta_2$, the resistor R_2 has to be connected in parallel with the C_2 capacitor. The maximum resistance of the resistor R_2 that will guarantee the balanced voltages is

$$R_2 \leq \frac{U_N}{I_{C0}} \frac{1 + \Delta_1}{\Delta_1 - \Delta_2}. \tag{4.83}$$

The worst case scenario is $\Delta_1 = \Delta_{max}$ and $\Delta_2 = -\Delta_{max}$, where maximum capacitance deviation is $\Delta_{max} = 0.2$ (pu) (20%). Substituting the worst case condition into Equation 4.83 yields

$$R_2 \leq \frac{U_N}{I_{C0}} 3. \tag{4.84}$$

The same applies if the capacitor C_2 is bigger than C_1 ($\Delta_2 > \Delta_1$). The resistor R_1 is the same as the resistor R_2,

$$R_1 = R_2 \leq \frac{U_N}{I_{C0}} 3. \tag{4.85}$$

Switch Mode Balancing Circuit

In order to provide very efficient and dynamic balancing of the cells' voltages, a switching mode loss-free balancing circuit is developed (Figure 4.17d). The voltage is actively controlled with good balancing capability, high dynamic, and high efficiency. High complexity is a drawback.

4.4.3.3 Switching Mode Balancing Circuit

Decentralized Half-Bridge Balancing Circuit

An example of a three-cell decentralized half-bridge balancing circuit is shown in Figure 4.18a. The circuit consists of two pairs of switches S_{1A}/S_{2B} and S_{2A}/S_{3B}, and two inductors L_{12} and L_{23}. The switches and inductors are connected as indicated in Figure 4.18a. Each pair of switches are driven by complementary driving signals with a duty cycle of approximately 50%. Since the duty cycle is 50%, the voltages of the two capacitors are equal

$$u_{C01} = u_{C02} \text{ and } u_{C02} = u_{C03} \Rightarrow u_{C01} = u_{C02} = u_{C03}. \tag{4.86}$$

The inductor current depends on the ultra-capacitor charge/discharge current, the capacitance dispersion (Equation 4.82), and the number of series connected cells. Let the module

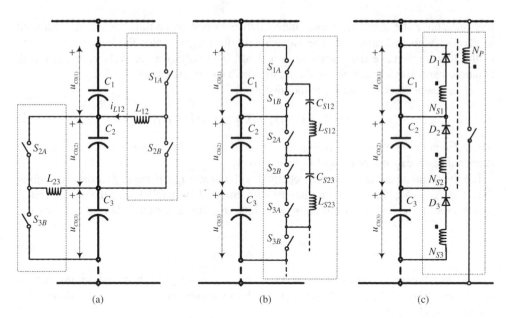

(a)   (b)   (c)

Figure 4.18   (a) Decentralized half-bridge balancing circuit, (b) decentralized resonant half-bridge balancing circuit, and (c) centralized multi-output fly-back balancing circuit

be composed of N series connected capacitors $C_{(1)}$ to $C_{(N)}$, all of them having different capacitance. The capacitances are defined by Equation 4.82. The balancing circuit can be described by a system of linear equations in matrix form

$$
\begin{bmatrix}
-1 & 1 & . & 0 & . & 0 \\
0 & -1 & 1 & 0 & \cdots & 0 \\
 & & & & \vdots & \\
 & & & -1 & 1 & 0 \\
0 & 0 & \cdots & 0 & -1 & 1
\end{bmatrix}
\begin{bmatrix}
I_{C0}\Delta_1 \\
I_{C0}\Delta_2 \\
\vdots \\
I_{C0}\Delta_j \\
\vdots \\
I_{C0}\Delta_N
\end{bmatrix}
=
\begin{bmatrix}
-1 & \frac{1}{2} & 0 & \cdots & 0 \\
\frac{1}{2} & -1 & \frac{1}{2} & 0 & \vdots \\
0 & \frac{1}{2} & -1 & \frac{1}{2} & 0 \\
\vdots & \vdots & & & \vdots \\
0 & \cdots & 0 & \frac{1}{2} & -1
\end{bmatrix}
\begin{bmatrix}
i_{12} \\
\vdots \\
i_{(j)(j+1)} \\
\vdots \\
i_{(N-1)(N)}
\end{bmatrix}, \quad (4.87)
$$

where

$\Delta_{(j)}$ is the capacitance deviation of the jth capacitor in the stack and
$i_{(j)(j+1)}$ is current of the inductor $L_{(j)(j+1)}$.

It can be proven that the worst case scenario inductor current is

$$
I_L = 2\Delta_{max} I_{C0(max)} \leq 0.4 I_{C0(max)}, \quad (4.88)
$$

where Δ_{max} is the maximum deviation of the capacitance.

Ultra-capacitor manufacturers guarantee a maximum capacitance dispersion of $\Delta_{max} \leq 20\%$.

As the balancing half-bridge circuit operates at around 50% of the duty cycle, the current ripple is close to the maximum. The inductance L is

$$L \geq \frac{U_{N(cell)}}{4 f_{SW} \Delta i_L}, \tag{4.89}$$

where

$U_{N(cell)}$ is the cell nominal voltage,
f_{SW} is the switching frequency, and
Δi_L is the inductor current ripple.

The current ripple is usually 30% of the inductor current. Hence, the inductance is

$$L \geq \frac{U_{N(cell)}}{4 f_{SW} 0.12 I_{C0(max)}}. \tag{4.90}$$

The switching frequency can be moderately high to very high since very low voltage MOSFETs (metal oxide semiconductor field effect transistors) are used as the switches. However, very high switching frequency is followed by significant stand-by losses, which is not desirable in a design where the ultra-capacitor is used to store energy over a long time.

Let the number of series connected ultra-capacitors be N. The number of switches and inductors are $2N - 2$ and $N - 1$ respectively. The switches are low voltage MOSFETs. The voltage rating of the MOSFETs is

$$U_M = 2u_{C0} + \Delta u_M \tag{4.91}$$

where

Δu_M is the commutation over-voltage,

$$\Delta u_M = L_\xi \frac{di_{SW}}{dt} = L_\xi \frac{0.8 I_L}{t_F}. \tag{4.92}$$

L_ξ is the commutation loop inductance,
t_F is the switch fall time,
and I_L is the inductor current.

Since the cell voltage u_{C0} is very low ($\cong 2.5$ V), the over-voltage Δu_M can be dominant and it will determine the MOSFET voltage rating. Two MOSFETs and the corresponding gate driver can be integrated into a switch-cell.

Let the current of the inductor $L_{(j)(j+1)}$ be defined by a linear function of the capacitor's deviation $\Delta_{(1)} \ldots \Delta_{(j)} \ldots \Delta_{(N)}$ and the ultra-capacitor charge/discharge current I_{C0},

$$i_{(j)(j+1)} = I_{C0} \sum_{k=1}^{N} k_k \Delta_{(k)} \tag{4.93}$$

where the coefficients k_k are obtained as a solution of matrix equation (Equation 4.87). Conduction losses of jth half-bridge switches are

$$P_{SW(j)A} = P_{SW(j+1)B} = R_{DS} \frac{I_{C0}^2}{2} \left(\sum_{k=1}^{N} k_{jk} \Delta_{(k)} \right)^2 \le R_{DS} \frac{(0.4 I_{C0(max)})^2}{2}, \tag{4.94}$$

where R_{DS} is the switch on-state resistance.

Total conduction losses of the entire balancing circuit are

$$P_{TOTAL} = 2 R_{DS} \frac{I_{C0(max)}^2}{2} \sum_{j=1}^{N-1} \left(\sum_{k=1}^{N} k_{jk} \Delta_{(k)} \right)^2. \tag{4.95}$$

Total conduction losses cannot be computed as a simple sum of the worst case scenario losses (Equation 4.94). The reason is non-uniform distribution of the capacitance deviation and therefore balancing current and losses. It is possible to find the worst case scenario of Equation 4.95, but it is outside the scope of this book.

$$P_{SW(j)A} = P_{SW(j+1)B} = f_{SW} E_{SW} \frac{2 U_{C0(cell)}}{U_{REF} I_{REF}} \underbrace{I_{C0} \sum_{k=1}^{N} k_{jk} \Delta_{(k)}}_{The\_inductor\_current}$$

$$\le f_{SW} E_{SW} \frac{2 U_{C0(cell)}}{U_{REF} I_{REF}} 0.4 I_{C0(max)}, \tag{4.96}$$

where

f_{SW} is the switching frequency,

E_{SW} is the total switching energy at rated voltage U_{REF} and current I_{REF} [7].

The MOSFETs switching losses can be neglected if fast switching is preserved and a reasonable "low" switching frequency is used. However, due to the issue of commutation over-voltage, the switching speed has to be reduced to keep the over-voltage reasonably low. More details are given in Exercise 4.6.

Decentralized Series Resonant Half-Bridge Balancing Circuit

Figure 4.18b shows a circuit diagram of a decentralized series resonant half-bridge balancing circuit. An example of three cells is given. Let the number of series connected cells be N. The balancing circuit consists of N pairs of switches, and $N-1$ resonant circuits. Each pair of switches is connected in parallel with the corresponding capacitor cell. The switches are denoted as $S_{1A}/S_{1B} \ldots S_{NA}/S_{NB}$. The resonant circuits are connected between the output terminals. Each resonant circuit consists of series connected resonant inductor Ls and capacitor Cs. Each pair of switches is driven with complementary signals at a duty cycle of approximately 50%. The switching frequency has to be slightly below the resonant frequency in order to ensure zero current switching.

A series resonant converter that operates in discontinuous conduction mode (DCM), type 1, is used to balance the voltage of the two series connected dc bus capacitors

[8]. Because the resonant converter operating in DCM, type 1, has the input to output voltage ratio of one, the voltage of the top and bottom capacitor must be equal. This is an inherent feature of such kinds of resonant converters. It is not possible to have a difference between the input and output voltage. If for some reason the voltages are unbalanced, the resonant circuit current will increase in a way to eliminate the voltage imbalance. Thus, the voltages will be automatically balanced and feedback control is not required.

Let's analyze capacitors C_1 and C_2. If the cell voltage u_{C01} is regarded as the input voltage, the output voltage is u_{C02}. The same applies for the capacitors C_2 and C_3. The voltage u_{C02} is regarded as the input voltage and u_{C03} is the output voltage. Since the input to output voltage ratio must be one, it is obvious that all the voltages must be balanced,

$$u_{C01} = u_{C02} \quad \text{and} \quad u_{C02} = u_{C03} \Rightarrow u_{C01} = u_{C02} = u_{C03} = \ldots = u_{C0}. \tag{4.97}$$

The switches are low voltage MOSFETs. Voltage rating of the MOSFETs is

$$U_M = u_{C0} + \Delta u_M \tag{4.98}$$

where Δu_M is the commutation over-voltage.

Unlike the half-bridge balancing circuit, the resonant circuit is characterized by zero current switching. Hence, the commutation over-voltage is very low, in same range as the cell voltage u_{C0}. Two MOSFETs and the corresponding gate driver can be integrated into a switch-cell.

In fact, the balancing circuit of Figure 4.18b is a switched capacitor converter with resonant commutation. The mean of balancing is the switched capacitor C_S. The capacitor is switched from one cell to another and due to the charge transfer the voltage on the capacitors C_1 and C_2 is balanced. The role of the resonant inductor L_S is to minimize the conduction losses and ensure the zero current switching condition. To ensure zero current switching, the inductance and capacitance must satisfy the condition

$$L_S C_S \leq \left(\frac{T_S}{2\pi}\right)^2; \tag{4.99}$$

where T_s is the switching period.

The capacitor and inductor can be selected for minimum size, cost, and loss of the LC balancing circuit. The total volume of the capacitor and inductor can be expressed in a general form

$$W = F_1(C_S) + F_2\left(\left(\frac{T_S}{2\pi}\right)^2 \frac{1}{C_S}\right), \tag{4.100}$$

where functions $F1$ and $F2$ depend on the capacitor and the inductor technology. From the minimum volume condition

$$\frac{\partial W}{\partial C_S} = \frac{\partial F_1(C_S)}{\partial C_S} + \frac{\partial F_2\left(\left(\frac{T_S}{2\pi}\right)^2 \frac{1}{C_S}\right)}{\partial C_S} = 0, \tag{4.101}$$

we can determine the capacitance C_S and inductance L_S. The capacitor C_S is a low voltage ceramic chip capacitor while inductor L_S is a small air-core inductor. If the switching frequency is sufficiently high, the stray inductance of the switches and capacitor can be used as the resonant inductor.

Centralized Multi-Output Fly-Back Balancing Circuit

A centralized multi-level fly-back circuit (Figure 4.18c), can also be used as a loss-free balancing circuit. The circuit is basically am N-output fly-back converter. The fly-back transformer primary winding is connected to the module full voltage u_{C0}. Each output is connected to the corresponding ultra-capacitor cell. If the secondary windings have an equal number of turns, the cell voltages u_{C01} to u_{C0N} must be equal regardless of the capacitor charge/discharge current and the deviation of the capacitor's parameters.

Although the circuit is simple and does not require closed loop control, it is not practical if a high number of capacitors are series connected. The transformer becomes complicated and connection between the transformer and the cells impractical.

4.4.4 Exercises

Exercise 4.5

A high power UPS system is equipped with an ultra-capacitor module as short-term energy storage. The UPS parameters are: dc bus voltage $V_{BUS} = 700$ V, nominal power $P_N = 400$ kW, the inverter efficiency $\eta_{INV} = 98\%$, and discharge time $T_{DCH} = 20$ seconds.

1. Select the ultra-capacitor module maximum operating voltage, minimum operating voltage, and intermediate operating voltage if the module is connected to the UPS dc bus via a bi-directional dc–dc converter with a maximum current capability of $I_{C0max} = 900$ A.
2. Design the ultra-capacitor module according to the above given parameters. The design should include the capacitance C_N, selection of the cells, number of series N and parallel M connected cells, and the module equivalent resistance R_{C0}. Assume the coefficient $k_0 = 0.8$.

Solution

1. Maximum operating voltage of the ultra-capacitor module is the UPS dc bus voltage $U_{C0max} = V_{BUS} = 700$ V.

   The minimum operating voltage is defined by the current capability of the dc–dc converter or the system stability (Equation 4.15). Since we do not know the module resistance, we assume the minimum voltage is determined by the current capability only. From Equation 4.12 we compute the minimum voltage $U_{C0min} = 444.5$ V.

   The intermediate voltage can be directly determined. Because the UPS does not need to store energy from the load, but only provide energy to the load, it is obvious the intermediate voltage has to be as high as possible. In this case $U_{C0inM} = U_{C0max} = 700$ V.
2. The nominal capacitance C_N is computed from

$$C_N = \frac{E_{DCH}}{\left[\frac{k_0}{2} \left(U_{C\,max}^2 - U_{C\,min}^2 \right) + \frac{2}{3} \frac{(1-k_0)}{U_{CON}} \left(U_{C\,max}^3 - U_{C\,min}^3 \right) \right]} \qquad (4.102)$$

where the discharge energy is given by Equations 4.21 and 4.22.

   However, to calculate the discharge energy, we need first to calculate the conversion losses. The losses can be computed according to the analysis conducted in Section 2.5.3.1 and Equation 2.72. The conversion losses are a function of the ultra-capacitor resistance and capacitance. If we include this in the capacitance equation, we will have an equation that cannot be solved analytically. Another issue is that we need the internal minimum voltage, which depends on the ultra-capacitor resistance (Equation 4.13). To avoid this issue, we can calculate the capacitance in two iterations.

The first iteration:

Let's calculate the discharge energy neglecting the ultra-capacitor resistance. Hence, there are no discharge losses and the internal and terminal minimum voltages are the same. From Equation 4.31 and the UPS data we have $E_{DCH} = 8.163\,\text{MJ}$. Substituting the discharge energy E_{DCH} and the module voltages into Equation 4.102 yields the capacitance $C_N = 51.326\,\text{F}$. Now we have a rough idea of the ultra-capacitor size, we can go on to the second iteration.

The second iteration:

For the capacitance computed above, we can preliminarily estimate the resistance from the capacitance-to-resistance characteristics (Equation 4.41), $R_{C0} = 18\,\text{m}\Omega$. From Equation 4.13 we compute the minimum internal voltage $U_{C0min} = 460.7\,\text{V}$. From the Equation 2.72 and the capacitance, resistance, and minimum internal voltage computed above, we compute the discharge energy losses $E_{\zeta(DCH)} = 85.77\,\text{kJ}$ and the discharge energy $E_{DCH} = 8.25\,\text{MJ}$.

Now having the discharge energy and the energy losses, we compute a new value of the capacitance $C_N = 51.80\,\text{F}$. Please note the difference (error) between the first and second iteration capacitance is approximately 0.44 F, which is 0.85% of the rated capacitance.

To take into account the EOL effect on the capacitance selection, the initial (SOL) capacitance should be higher than the computed one. As the EOL capacitance is 80% of the initial one, the ultra-capacitor should be selected with the initial capacitance $C_N = 64.75\,\text{F}$.

Number of series and parallel connected cells

Before we define the number of series connected ultra-capacitor cells, we have to select the cells' family. For such large ultra-capacitor module we will select large cells type K2 from Maxwell Technologies or LSUC 2.8 from LS Mtron Ultra-capacitors. The cell nominal voltage is $U_{CON(cell)} = 2.5\,\text{V}$ and $U_{CON(cell)} = 2.8\,\text{V}$ respectively.

The number of series connected cells is

$$N = floor\left(\frac{U_{CON}}{U_{CON(cell)}}\right). \qquad (4.103)$$

Substituting the above data into Equation 4.103 yields $N = 320$ for K2 cells and $N = 286$ for LSUC 2.8.

The capacitance of an individual cell or a bank of parallel connected cells is

$$C_{N(cell)} = C_N N. \qquad (4.104)$$

From the module capacitance C_N and the number of series connected cells we compute $C_{NI} = 20720\,\text{F}$ for Maxwell K2 cells and $C_{NI} = 18518.5\,\text{F}$ for LS Mtron LSUC 2.8 cells. Such a large ultra-capacitor cell does not exist. Hence, M cells have to be connected in parallel. The number of paralleled cells is

$$M = floor\left(\frac{C_N}{C_{N(cell)}}\right) N. \qquad (4.105)$$

Table 4.2 Summary of the ultra-capacitor cell selection

Ultra-capacitor cell	N	M	$Q = NM$	R_{C0} (mΩ)
K2 2000 F/2.5 V	320	11	3520	9.54
K2 3000 F/2.5 V	320	7	2240	12.5
LSUC 2.8 1700 F/2.8 V	286	11	3146	13
LSUC 2.8 3000 F/2.8 V	286	7	2002	11

Preliminarily, we can select 2000 and 3000 F cells of the K2 and LSUC 2.8 series. Substituting the data given into Equation 4.105 yields the results summarized in Table 4.2.

Exercise 4.6

The ultra-capacitor for the above exercise is given. The leakage current of the selected cells is $I_{0\zeta} = 5$ mA.

1. Design a passive resistive balancing circuit.
2. Design an active balancing circuit based on decentralized half-bridge balancing topology. The switching frequency is $f_{SW} = 25$ kHz. The selected switches have switching energy $E_{SW} = 100\,\mu$J at referent conditions $U_{REF} = 25$ V, $I_{REF} = 100$ A, and $t_F = 20$ ns.

Solution(s)

1. Since $M = 7$ cells are parallel connected, the total leakage current per sub-module set is $I_{0\zeta} = 35$ mA. The balancing resistance computed from Equation 4.80 is $R = 40\,\Omega$. In total, $N = 286$ resistors. Power dissipation per resistor and entire module computed from Equation 4.81 is $P_R = 196$ mW and $P_{R(module)} = 56$ W.
2. The first step in the design of half-bridge balancing circuit is the calculation of the inductor current I_L (Equation 4.88). The maximum ultra-capacitor current is achieved at minimum voltage and rated power,

$$I_{C0\,max} = \frac{P_N}{\eta_{INV}\, U_{C0\,min}}. \tag{4.106}$$

Substituting the data given into Equations 4.106 and 4.88 yields the ultra-capacitor maximum current and the balancing inductor current $I_{C0max} = 918$ A and $I_L = 367$ A. The inductance of the balancing inductor is computed from Equation 4.90 as $L = 0.227$ uH.

Let's now select the balancing switches. The maximum current rating is determined by the inductor current previously calculated. The switch blocking voltage is defined in Equations 4.91 and 4.92. Let's assume the commutation inductance of $L_\zeta = 10$ nH and the current fall time of $t_F = 20$ ns. Substituting these numbers and the inductor current into Equations 4.91 and 4.92 yields the switch maximum voltage of $U_M = 151.8$ V! This is an extremely high voltage rating compared to the steady state blocking voltage of only 5 V. According to these results, it seems we need to select a 200 V/400 A rated MOSFET. Such a MOSFET is very big and expensive. If take into account the number of series connected ultra-capacitor cells $N = 286$, it follows that we need $N_{SW} = 2N - 2 = 286 = 570$ of these MOSFETS! Does this make sense? Do we have another solution? The only way to reduce the switch rating is to reduce the voltage rating. Since the voltage rating is determined mainly by the switching over-voltage, we have to reduce this. The over-voltage is determined by two factors: the commutation inductance and the current fall time. The commutation inductance can be reduced, but not significantly. Hence, the only way to reduce the over-voltage is to increase the current fall time.

Let the objective be to use a 25 V rated MOSFET. The over-voltage is therefore

$$\Delta u_M = \frac{U_M}{k_M} - 2u_{C0}. \tag{4.107}$$

Where k_M is the design margin, usually 1.25. From Equations 4.92 and 4.107 we can define the current fall time

$$t_F = L_\xi \frac{0.8 I_L}{\frac{U_M}{k_M} - 2u_{C0}}. \tag{4.108}$$

Substituting the data given into Equation 4.108 yields $t_F > \mathbf{195.7\,ns}$.

We can select MOSFET BSB013NE2LXI from Infineon technologies. The MOSFET rating is 25 V/163 A. Since the balancing inductor current is $I_L = 367\,A$, three of these MOSFETs have to be parallel connected.

The MOSFET conduction losses are

$$P_{\xi(1)} = \left(\frac{I_L}{2M_{SW}}\right)^2 R_{DS} = \left(\frac{I_L}{6}\right)^2 R_{DS}, \tag{4.109}$$

where

M_{SW} is the number of paralleled MOSFETs and
R_{DS} is the MOSFET on-state resistance.

Substituting the data given into Equation 4.109 yields conduction losses of $P_{\xi(1)} = \mathbf{4.86\,W}$ per single MOSFET and $P_\xi = \mathbf{29.16\,W}$ per half-bridge. Switching losses are computed from Equation 4.96. The switching energy is given at $t_F = 20\,ns$. As we have approximately 10 times the fall time, the switching energy will be also 10 times the referent one, $E_{SW} = \mathbf{1\,mJ}$ **(25 V, 100 A, 200 ns)**. Substituting this into Equation 4.96 yields switching losses of $P_{SW(1)} = \mathbf{6.17\,W}$ per switch and $P_{SW} = \mathbf{36.7\,W}$ per half-bridge!

From this extreme example, it is obvious that the balancing of a string of series connected ultra-capacitor cells is not a trivial issue. Design of an appropriate balancing circuit is a very important part of the module design and it must not be considered a trivial task.

4.5 The Module's Thermal Management

Although the ultra-capacitor is characterized by relatively low internal resistance and therefore low losses, the losses cannot be neglected in the analysis and design. The losses cause a temperature rise in the ultra-capacitor cells. As addressed in Section 4.2.5, elevated temperature has a strong negative impact on the ultra-capacitor's life span. Each $10°$ of temperature increase will reduce the life span by a factor of 2. The temperature inside the ultra-capacitor core must be limited to the maximum temperature defined by the electrolyte boiling point. Hence, the ultra-capacitor will be destroyed if the core temperature reaches $\theta_{max} = 81\,°C$. Therefore, it is obvious that the ultra-capacitor's thermal analysis, cooling system design, and management are very important steps in the module design process.

Once the ultra-capacitor losses have been determined for the current profile given in Equations 4.72–4.75, the thermal model of the ultra-capacitor module has to be determined. The cooling system has to be designed from the thermal model and the loss profile. The design objective is to keep the ultra-capacitor hot-spot temperature below the limit, where the limit is specified by the application requirements, mainly life expectancy.

4.5.1 The Model's Definition

The cell temperature is determined by the cell's losses and the cell-to-ambient thermal impedance. A simplified equivalent model is depicted in Figure 4.19a. If the cell is a standalone cell (not surrounded by other cells), the model, Figure 4.19a, is correct. However, if several cells are placed in a module and enclosed in the module housing, the model of Figure 4.19a is not valid any more. In this case, the cell ambient temperature is no longer constant and independent. The temperature around the cell strongly depends on the cell losses and thermal impedance from the cell ambient to the module ambient. Moreover, there is thermal coupling between the cell and the rest of the system. As a consequence of thermal coupling, the cell ambient temperature depends on the losses of the rest of the cells in the module. This effect is taken into account when considering the cell ambient temperature as a state variable, as illustrated in Figure 4.19b.

Let's assume that a module is made of $Q = NM$ cells enclosed in the module housing. The core temperature of the nth cell is

$$\theta_{(cell)(n)} = Z_{(cell)(n)} P_{C0\varsigma(cell)(n)} + \theta_{AMB(cell)(n)}, \qquad (4.110)$$

where $\theta_{(cell)(n)}$ is the ambient temperature of the nth cell, which is a function of the losses of all the cells in the module and cross-coupling thermal impedances between the cells.

$$\theta_{AMB(cell)(n)} = \theta(P_1 \ldots P_n \ldots P_Q, Z), \qquad (4.111)$$

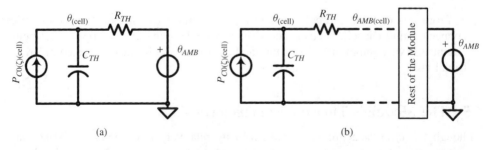

(a)   (b)

Figure 4.19   The ultra-capacitor cell thermal model. (a) A stand alone cell. (b) A cell inside the module

The entire module can be described by a matrix equation

$$
\begin{bmatrix}
\theta_{(cell)(1)} \\
\cdot \\
\theta_{(cell)(n)} \\
\cdot \\
\theta_{(cell)(Q)}
\end{bmatrix}
=
\begin{bmatrix}
Z_{11} & \cdot & Z_{1n} & \cdot & Z_{1Q} \\
\cdot & \cdot & \cdot & \cdot & \cdot \\
Z_{n1} & \cdot & Z_{nn} & \cdot & Z_{nQ} \\
\cdot & \cdot & \cdot & \cdot & \cdot \\
Z_{Q1} & \cdot & Z_{Qn} & \cdot & Z_{QQ}
\end{bmatrix}
\begin{bmatrix}
P_{C0\varsigma(cell)(1)} \\
\cdot \\
P_{C0\varsigma(cell)(n)} \\
\cdot \\
P_{C0\varsigma(cell)(Q)}
\end{bmatrix}
+
\begin{bmatrix}
1 \\
\cdot \\
1 \\
\cdot \\
1
\end{bmatrix}
\theta_{AMB},
\qquad (4.112)
$$

where thermal impedances Z_{nn} are self (driving point) impedances, while Z_{nQ} are mutual impedances. The cells losses $P_{(cell)(n)}$ are given by Equations 4.72–4.74. The model (4.112) is schematically illustrated in Figure 4.20a.

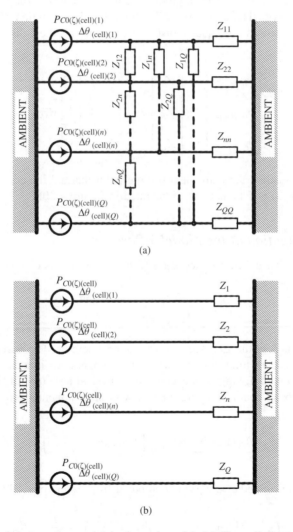

(a)

(b)

Figure 4.20 The ultra-capacitor module's thermal model. (a) The cells losses are different and (b) the cells losses are same

Usually, the cell's resistances are equal and therefore the losses of all the cells are also the same. In some extreme cases, the resistances and therefore losses can vary from cell to cell. An extreme case is low ambient temperatures. As discussed in Section 2.3.1.4, Equation 2.19, the ultra-capacitor resistance strongly increases at extremely low temperatures. At a minimum temperature of $-40\,°C$, the resistance can be as high as 170% of the resistance at room temperature. If the ambient temperature is very low, the temperature of the outer cells of the module will be significantly lower than that of the inner cells. As a consequence, the resistance and losses of the outer cells will be higher than those of the inner cells. If this is the case the model (4.59) is not correct, strictly mathematically speaking. The correct model is a nonlinear model having temperature dependent losses instead of having constant losses as assumed in Equation 4.59. Analysis of such a nonlinear system of very high order, sometimes 300th order (300 cells), is very difficult in closed form. Therefore, we will assume equal distribution of the losses among all the cells. If the losses are equally distributed among all the cells, the model (4.112) can be reduced to

$$
\begin{bmatrix} \theta_{(cell)(1)} \\ \cdot \\ \theta_{(cell)(n)} \\ \cdot \\ \theta_{(cell)(Q)} \end{bmatrix} = \begin{bmatrix} Z_1 \\ \cdot \\ Z_n \\ \cdot \\ Z_Q \end{bmatrix} P_{C0\varsigma(cell)} + \begin{bmatrix} 1 \\ \cdot \\ 1 \\ \cdot \\ 1 \end{bmatrix} \theta_{AMB},
\tag{4.113}
$$

where Z_n is the so-called the equivalent driving point thermal impedance of the nth cell. The model (4.113) is schematically illustrated in Figure 4.20b.

4.5.2   Determination of the Model's Parameters

The driving point thermal impedance $Z_n(s)$ (Equation 4.113) is defined as

$$
Z_n(s) = \frac{\Delta\theta_n(s)}{P_{C0(cell)}(s)} = s\frac{L(\Delta\theta_n(t))}{P_{C0(cell)}},
\tag{4.114}
$$

where the operator L is the Laplace transformation operator. The variable $\Delta\theta_n(t)$ is the temperature response in time domain, where the temperature is measured on the ultra-capacitor cell terminal. Taking into account the physical structure of the ultra-capacitor module, this response could be approximated with an exponential series having a finite number of terms.

$$
\Delta\theta_n(t) \cong \sum_{k=1}^{R} \Delta\Theta_{n(k)}\left(1 - \exp\left(-\frac{t}{\tau_{n(k)}}\right)\right).
\tag{4.115}
$$

The driving point impedance $Z_n(s)$ is computed from Equations 4.114 and 4.115 as

$$
Z_n(s) \cong sL\left(\sum_{k=1}^{R} \frac{\Delta\Theta_{n(k)}}{P_{C0(cell)}}\left(1 - \exp\left(-\frac{t}{\tau_{n(k)}}\right)\right)\right) = \sum_{k=1}^{R} \frac{R_{n(k)}}{1 + s\tau_{n(k)}}.
\tag{4.116}
$$

Time domain impedance, so-called transient impedance, is defined as the time response of the impedance on a unity step excitation.

$$Z_n(t) \cong \sum_{k=1}^{R} R_{n(k)} \left(1 - \exp \left(-\frac{t}{\tau_{n(k)}} \right) \right) = \sum_{k=1}^{R} \frac{\Delta \Theta_{n(k)}}{P_{C0(cell)}} \left(1 - \exp \left(-\frac{t}{\tau_{n(k)}} \right) \right). \quad (4.117)$$

The thermal impedances can be approximated by low order, usually first or second order, transfer functions. In the case of a first-order model, the transfer functions are the first-order low pass filters,

$$Z_n(s) = \frac{R_n}{1 + s\tau_n}, \quad (4.118)$$

where R_n and τ_n are the thermal resistance and time constant of the nth cell. If the model is of second order, the transfer functions are second-order low pass filters with real and different roots.

$$Z_n(s) = \frac{R_{n(1)}}{1 + s\tau_{n(1)}} + \frac{R_{n(2)}}{1 + s\tau_{n(2)}}, \quad (4.119)$$

where $\tau_{n(1)}$ and $\tau_{n(2)}$ are the thermal time constants of the nth cell.

Now we have defined the thermal model of an ultra-capacitor module. Is this sufficient to design the module cooling system? No, it is not sufficient at all. What we are still missing are the parameters of the model, more precisely thermal resistances $(R_{n(1)}, R_{n(2)})$ and time constants $(\tau_{n(1)}, \tau_{n(2)})$ of the thermal impedance Z_n (Equations 4.118 and 4.119). In the following section we will address this issue in more detail.

4.5.3 The Model's Parameters—Experimental Identification

There are three methods to identify the parameters of the thermal model of a thermal system: (i) experimental identification, (ii) 3-D finite element (FEM) simulation, and (iii) analytical solution. In the following sections we will discuss these three methods in more detail.

In this book we will address the experimental method only. As an example, let's refer to Figure 4.21 that shows the structure of a large ultra-capacitor module 96 V/135 F, 622 kJ. The module consists of 40 series connected large ultra-capacitor cells, each cell 2.5 V/5400 F. An objective of the identification process is to find the parameters (Equation 4.116) of the module thermal model (4.113) for different cooling options. The cooling options are natural convection and forced air-cooling.

Experimental identification of the thermal model parameters consists of the following four steps.

- **Step A** The first step is to design a mock-up prototype of the module with the selected ultra-capacitor cells (Figure 4.21). The module is equipped with a set of thermal probes. The probes are attached to the cell's case and terminal. The probes could also be placed in the air between the cells. The inlet and outlet temperatures are measured too. For the sake of clarity, only a few probes are depicted in Figure 4.21.

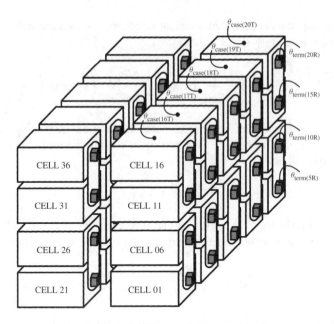

Figure 4.21   A large ultra-capacitor module for heavy duty applications: 96 V/135 F, 622 kJ

- **Step B** The second step is the loading of the ultra-capacitor module. As the ultra-capacitor losses have to be constant, charge/discharge with a constant current is used. The charge/discharge current has to be set at a value that gives losses in the same range as the losses expected in the application (computed in the previous chapter). More details of test set-up are given in Section 4.6.
- **Step C** The third step is the measurement of temperature of selected points inside the ultra-capacitor module. The temperature is measured with an appropriate data logger. The sampling time should be in the range 1–10 seconds. The measured temperatures are recorded as a data file that can be imported in the Matlab or Excel tool for further analysis.
- **Step D** The final step in the experimental identification of the thermal model is analysis of the experimental data. The objective is to calculate the model parameters ($R_{n(1)}$, $R_{n(2)}$ and $\tau_{n(1)}, \tau_{n(2)}$ of Z_n) based on the temperature response $\Delta\theta_n(t)$ of each cell of the module.

The model parameters can be easily computed using least-square method tools, such as Matlab Curve Fitting Toolbox or Excel Solver. The input in the algorithm is a time series $\Delta\theta_n(t)$ while the outputs are parameters of the model of defined order (first or second).

4.5.4   The Cooling System Design

Once the thermal model of the ultra-capacitor module has been determined and parameters identified, the cooling system has to be designed. In this section we will briefly discuss the design objective(s) and design methodology.

4.5.4.1   Design Objective(s)

Let the following parameters be defined and specified as the design inputs:

1. **The ultra-capacitor load profile.** The load is usually in the form of power or current time profile, $P_{C0}(t)$ or $I_{C0}(t)$. If the power is given as an input, the current is computed from Equations 2.66 and 2.67, Section 2.5.3.
2. **Maximum temperature of an ultra-capacitor cell.** The absolute maximum cell temperature, the point of destruction, is $\theta_{max} = 81\,^{\circ}\mathrm{C}$. To keep the life expectancy as required by the application, the cell maximum temperature is usually limited to $\theta_{max} = 60\text{--}65\,^{\circ}\mathrm{C}$. If a long life span is expected, the maximum temperature is even lower, being limited to $\theta_{max} = 50\text{--}60\,^{\circ}\mathrm{C}$. As the ultra-capacitor module consists of $Q = NM$ cells, the temperature distribution within the module is not uniform. The temperature of the hottest ultra-capacitor cell is the module's worst case scenario hot-spot temperature. This temperature must not exceed the limit θ_{max}.
3. **Maximum ambient temperature.** Maximum ambient temperature strongly depends on the area of application. Usually, the design criterion is $\theta_{AMB} = 40\,^{\circ}\mathrm{C}$.

Now we can formalize the design objective as: "Design the module cooling system in such a way as to keep the module hot-spot temperature below the limit θ_{max}, assuming the maximum ambient temperature $\theta_{AMB} = 40\,^{\circ}\mathrm{C}$ or as specified and the load profile given. The deigned cooling must provide the required performance over the entire life span including the end of life period."

4.5.4.2   Design Methodology

Let the nth cell be the hot-spot cell having the worst case scenario temperature. The hot-spot temperature is maximum at an instance $t = T_P$, where T_P depends on the load profile. The hot-spot maximum temperature can be computed in a general form

$$\Theta_{n(\max)} = L^{-1}\left[P_{C0(cell)}(s)\sum_{k=1}^{R}\frac{R_{n(k)}}{1+s\tau_{n(k)}}\right]\Bigg|_{t=T_P} + \Theta_{AMB}, \qquad (4.120)$$

where the thermal impedance $Z_n(s)$ is defined by Equation 4.116 or approximated by Equations 4.118 and 4.119. $P_{C0(cell)}(s)$ is the cell losses in the Laplace domain and L^{-1} is the inverse Laplace transformation. The losses are computed at a given current profile and the EOL ultra-capacitor internal resistance (Equation 4.74).

From Equation 4.120 we can find the losses to thermal impedance ratio

$$\underbrace{L^{-1}\left[P_{C0(cell)}(s)\sum_{k=1}^{R}\frac{R_{n(k)}}{1+s\tau_{n(k)}}\right]\Bigg|_{t=T_P}}_{\Delta\Theta_{n(\max)}} \leq \Theta_{n(\max)} - \Theta_{AMB}. \qquad (4.121)$$

Equation 4.121 is a very general equation. Thermal impedance that satisfies the condition 4.121 strongly depends on the load profile and the instance T_P. For better understanding of the design methodology and Equation 4.121, we should look at a few different examples.

Low to High Frequency Periodic Load Profile

Let the ultra-capacitor load profile be such that the losses of a cell are a periodic function

$$P_{C0(cell)}(t) = P_{C0(cell)(AV)} + \sum_{i=1}^{\infty} P_{C0(cell)(i)} \sin(i\omega_0 t + \psi_{(i)}), \quad (4.122)$$

where the fundamental repetition frequency ω_0 is much higher than the module's thermal model dominant frequency.

$$\omega_0 >> \frac{1}{\tau_n}, \quad (4.123)$$

where τ_n is the driving point thermal time constant of the nth cell.

The average losses are

$$P_{C0(cell)(AV)} = \frac{1}{T_0} \int_0^{T_0} P_{C0(cell)}(t)dt. \quad (4.124)$$

The temperature of the nth cell can be computed as a steady state temperature

$$\Theta_{n(max)} = P_{C0(cell)(AV)} \sum_{k=1}^{R} R_{n(k)} + \Theta_{AMB} \cong P_{C0(cell)(AV)}(R_{n(1)} + R_{n(2)}) + \Theta_{AMB}, \quad (4.125)$$

that is determined by the average losses and the cell thermal resistance.

The thermal resistances are computed from Equations 4.121 and 4.125 as

$$(R_{n(1)} + R_{n(2)}) \leq \frac{\Theta_{n(max)} - \Theta_{AMB}}{P_{C0(cell)(AV)}}. \quad (4.126)$$

Single Pulse of Moderate Duration

Let the ultra-capacitor load profile be such that the losses can be represented by a pulse of moderate duration T_P, where T_P is the same order of magnitude as the thermal time constant.

General case losses are defined as

$$P_{C0(cell)}(t) = \begin{cases} P_{C0(cell)}(t) & 0 \leq t \leq T_P \\ 0 & T_P < t < \infty \end{cases}. \quad (4.127)$$

In most real applications, the losses are a segment of a periodic function (Equation 4.122),

$$P_{C0(cell)}(t) = \begin{cases} P_{C0(cell)(AV)} + \sum_{i=1}^{\infty} P_{C0(cell)(i)} \sin(i\omega_0 t + \psi_{(i)}) & 0 \leq t \leq T_P \\ 0 & T_P < t < \infty \end{cases}. \quad (4.128)$$

If the condition 4.123 is valid, Equation 4.128 can be approximated by

$$P_{C0(cell)}(t) = \begin{cases} P_{C0(cell)(AV)} & 0 \leq t \leq T_P \\ 0 & T_P < t < \infty \end{cases}, \qquad (4.129)$$

where the average losses are given by Equation 4.124.

Substituting Equation 4.129 into Equation 4.120 yields the hot-spot maximum temperature

$$\theta_{n(max)} = P_{C0(cell)(AV)} \sum_{k=1}^{R} R_{n(k)} \left(1 - \exp\left(-\frac{T_P}{\tau_{n(k)}}\right)\right) + \theta_{AMB} \qquad (4.130)$$

$$\cong P_{C0(cell)(AV)} \left[R_{n(1)} \left(1 - \exp\left(-\frac{T_P}{\tau_{n(1)}}\right)\right) + R_{n(2)} \left(1 - \exp\left(-\frac{T_P}{\tau_{n(2)}}\right)\right) \right] + \theta_{AMB}.$$

The thermal impedance is computed from Equation 4.130 as

$$\left[R_{n(1)} \left(1 - \exp\left(-\frac{T_P}{\tau_{n(1)}}\right)\right) + R_{n(2)} \left(1 - \exp\left(-\frac{T_P}{\tau_{n(2)}}\right)\right) \right] = \frac{\theta_{n(max)} - \theta_{AMB}}{P_{C0(cell)(AV)}}. \qquad (4.131)$$

4.5.5 Exercises

Exercise 4.7

An ultra-capacitor module is made of 40 series connected large ultra-capacitor cells with the following parameters: $C_N = 3000\,\text{F}$, $U_{CON} = 2.5\,\text{V}$, $R_{C0} = 0.3\,\text{m}\Omega$. The module is tested under constant current charge/discharge conditions. The temperature of the inner cells is measured. The ambient temperature is a constant $\theta_{AMB} = 30\,°\text{C}$. The charge/discharge current is $I_{C0} = 100\,\text{A}$.

1. Identify the cell thermal impedance if the hottest cell temperature is a time series given in Table 4.3. Assume the first order RC thermal model.

Table 4.3 The ultra-capacitor cell's temperature measured by an experiment

t (s)	θ (°C)	t (s)	θ (°C)	t (s)	θ (°C)	t (s)	θ (°C)
0	30.00	480	43.00	960	44.22	1440	44.62
60	34.20	540	43.29	1020	44.29	1500	44.65
120	37.08	600	43.51	1080	44.35	1560	44.68
180	39.07	660	43.68	1140	44.41	1620	44.70
240	40.45	720	43.83	1200	44.46	1680	44.73
300	41.42	780	43.95	1260	44.50	1740	44.75
360	42.12	840	44.05	1320	44.54	1800	44.77
420	42.62	900	44.14	1380	44.58		

2. Calculate the hot-spot maximum temperature at the EOL of the ultra-capacitor module. The load profile is given by Equations 4.122–4.124 and the charge/discharge current $I_{C0} = 100$ A.
3. Calculate the hot-spot maximum temperature at the EOL of the ultra-capacitor module if the load profile is given by Equations 4.128 and 4.129. The pulse width is $T_p = 30$ seconds.

   Calculate the maximum charge/discharge current if the load profile is given by Equations 4.128 and 4.129. The pulse width is $T_p = 30$ seconds. The maximum hot-spot temperature is limited to $\theta_{n(max)} = 60\,^{\circ}\text{C}$.

Solution

1. The first step of the analysis and the model parameter identification is to compute the cell driving point transient thermal impedance given by Equation 4.117. However, first we need to compute the cell losses. The cell losses are

$$P_{C0(\varsigma)(cell)}(t) = R_{C0(cell)(LF)} I_{C0}^2. \tag{4.132}$$

Substituting the exercise date into Equation 4.132 yields $P_{C0(\varsigma)(cell)} = 3\,\text{W}$.

   The driving point thermal impedance can be approximated by a low pass filter of first or second order (Equations 4.118 and 4.119). In this exercise we will use the first-order approximation only.

   The next step is to identify the thermal resistance and time constant of the cell driving point thermal impedance. For this we will use Excel Solver to find the model parameters. The identification process steps are described below.

   (a) Insert the experimental results into the Excel worksheet. The first column is discrete time kT_S, while the second is temperature measured $\theta_{(cell)}(kT_S)$ (Figure 4.22a).
   (b) In the third column, calculate the driving point transient impedance using Equation 4.117 (Figure 4.22a).
   (c) Define parameters R_n and τ_n. Set some initial vales, for example, $R_n = 6$ and $\tau_n = 100$ (Figure 4.22a).
   (d) To declare parameters in Excel, open Formulas, Create from selection. Now, R_n and τ_n are declared as parameters (Figure 4.22b).
   (e) In the fourth column, calculate the transient impedance using the parameters R_n and τ_n. For the moment, the computed value of the transient impedance is very different from the one measured because our first guess is far from the optimal one.

$$\widehat{Z}_n(kT_S) \cong R_n \left(1 - \exp\left(-\frac{kT_S}{\tau_n}\right)\right). \tag{4.133}$$

   (f) In the fifth column calculate the impedance error

$$Error = (Z_n(kT_S) - \widehat{Z}_n(kT_S))^2. \tag{4.134}$$

   (g) Calculate sum of the errors

$$\sum (Z_n(kT_S) - \widehat{Z}_n(kT_S))^2. \tag{4.135}$$

RC Model		Time	Temperature Measured	Measured Impedance	Identified Impedance	(Error)^2	Σ((error)^2)
R_n	6	0	30.00	0.00	0.000	0.000	58.875
τ_n	100	60	34.20	1.40	2.707	1.706	
		120	37.08	2.36	4.193	3.358	
		180	39.07	3.02	5.008	3.945	
		240	40.45	3.48	5.456	3.893	
		300	41.42	3.81	5.701	3.587	
		360	42.12	4.04	5.836	3.229	
		420	42.62	4.21	5.910	2.897	
		480	43.00	4.33	5.951	2.616	
		540	43.29	4.43	5.973	2.386	
		600	43.51	4.50	5.985	2.199	
		660	43.68	4.56	5.992	2.046	
		720	43.83	4.61	5.996	1.921	
		780	43.95	4.65	5.998	1.816	
		840	44.05	4.68	5.999	1.728	
		900	44.14	4.71	5.999	1.653	
		960	44.22	4.74	6.000	1.588	

(a)

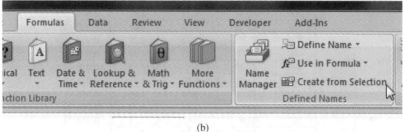

(b)

(c)

Figure 4.22 Parameter identification using the Excel Solver tool. (a) The results before running the Solver, (b) declaration of the coefficients, (c) Solver window, (d) identification is finished, and (e) the results after running the Solver

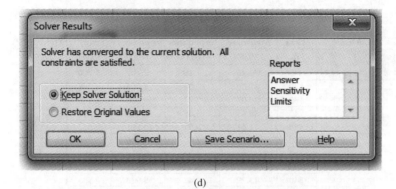

(d)

RC Model		Time	Temperature Measured	Measured Impedance	Identified Impedance	(Error)^2	Σ((error)^2)
R_n	4.81681	0	30.00	0.00	0.00	0.000	0.165
τ_n	191.68	60	34.20	1.40	1.29	0.011	
		120	37.08	2.36	2.24	0.014	
		180	39.07	3.02	2.93	0.008	
		240	40.45	3.48	3.44	0.002	
		300	41.42	3.81	3.81	0.000	
		360	42.12	4.04	4.08	0.002	
		420	42.62	4.21	4.28	0.005	
		480	43.00	4.33	4.42	0.008	
		540	43.29	4.43	4.53	0.010	
		600	43.51	4.50	4.61	0.011	
		660	43.68	4.56	4.66	0.010	
		720	43.83	4.61	4.70	0.009	
		780	43.95	4.65	4.73	0.007	
		840	44.05	4.68	4.76	0.005	
		900	44.14	4.71	4.77	0.004	
		960	44.22	4.74	4.78	0.002	

(e)

Figure 4.22   (*continued*)

(h) Open Date, and Solver on right side.
   i. In the window "Set target cell" put the cell with the sum of errors (Equation 4.135).
   ii. In the field "Equal to" check the button "Minimum."
   iii. In the field "By changing cells" insert cells with the parameters R_n and τ_n.
   iv. Press the button "Solve" in the top right corner.
   v. The Solver has found the value of the parameters R_n and τ_n that gives minimum error.
   vi. When the next window pops up, check the button "Keep solver solution."
(i) The model parameters are identified as $R_n = 4.81\,\text{K/W}$ and $\tau_n = 191.68\,\text{seconds}$.
   Figure 4.22 shows some detail of the Excel worksheet and the Solver window. Figure 4.23 shows the measured and estimated thermal impedance.

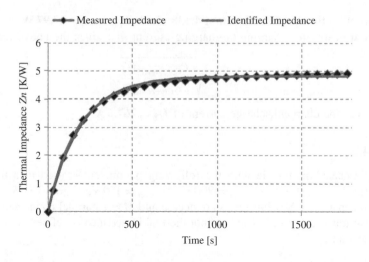

Figure 4.23 The ultra-capacitor hot-spot cell thermal transient impedance

2. The ultra-capacitor losses at the EOL are computed from Equation 4.74, where we have assumed that the ultra-capacitor resistance is doubled at EOL. From the exercise date and Equation 4.74 we have $P_{C0(\varsigma)(cell)} = \mathbf{6\,W}$.

   The hot-spot cell temperature at the EOL, assuming constant and very long term load, is (Figure 4.23)

$$\Theta_{n(\max)} \cong P_{C0(\varsigma)(cell)} R_n + \Theta_{AMB}. \tag{4.136}$$

   Substituting the thermal resistance, losses, and ambient temperature into Equation 4.136 yields $\theta_{n(max)} = \mathbf{58.86\,^\circ C}$. Please note that the hot-spot temperature at the EOL is significantly higher than that at the beginning of exploitation (SOL).

3. If the load profile is a pulse of duration T_P, where T_P falls in the same range as the thermal time constant, the cell temperature has to be computed from Equation 4.130 using the first term only.

$$\theta_{n(\max)} \cong P_{C0(cell)(AV)} R_n \left(1 - \exp\left(-\frac{T_P}{\tau_n} \right) \right) + \theta_{AMB}. \tag{4.137}$$

   Substituting the thermal resistance and time constant, the pulse width T_P, the EOL losses, and the ambient temperature into Equation 4.137 yields $\theta_{n(max)} = \mathbf{34.62\,^\circ C}$. Please note that the hot-spot temperature in the case of pulse load is very low. The temperature rise is only $\mathbf{4.62\,^\circ C}$ even at the EOL of the ultra-capacitor.

4. From Equation 4.137 we can compute the ultra-capacitor cell losses that give a maximum temperature of $\theta_{n(max)} = \mathbf{60\,^\circ C}$ at the end of a pulse of $T_P = 30$ seconds. The losses are

$$P_{C0(cell)(AV)} = \frac{\theta_{n(\max)} - \theta_{AMB}}{R_n \left(1 - \exp\left(-\frac{T_P}{\tau_n} \right) \right)}. \tag{4.138}$$

Substituting the data into Equation 4.138 yields $P_{C0(\zeta)(cell)} = \mathbf{42.97\,W}$.

The ultra-capacitor charging/discharging current that gives the above losses at the EOL is

$$I_{C0} = \sqrt{\frac{P_{C0(\varsigma)(cell)}(t)}{2R_{C0(cell)(LF)}}}, \tag{4.139}$$

which gives the charge/discharge current of $I_{C0} = \mathbf{267.6\,A}$.

Exercise 4.8

A large ultra-capacitor module with the following parameters is used in a high power UPS system: $U_{CON} = 800\,V$, $U_{C0max} = 700\,V$, $U_{C0min} = 350\,V$, $C_N = 40\,F$, $R_{C0} = 38\,m\Omega$. The module consists of $N = 290$ series connected and $M = 7$ parallel connected cells. The module is naturally air-cooled. The hot-spot thermal impedance is defined as $R_n = 6\,K/W$, $\tau_n = 1800$ seconds.

1. Compute the hot-spot temperature at the EOL of the ultra-capacitor if the load profile is a single pulse determined by $P_{C0} = 400\,kW$ and $T_P = 20$ seconds. Assume that the ultra-capacitor is discharged with rated power and then recharged with 10% of the rated power. The ambient temperature is $\theta_{AMB} = 40\,°C$.
2. Compute the hot-spot temperature if the ultra-capacitor module is recharged with nominal power.

Solution

1. The temperature of the nth cell of the module is governed by the equation

$$\theta_{n(max)} = L^{-1}\left[P_{C0(cell)}(s)\sum_{k=1}^{R}\frac{R_{n(k)}}{1 + s\tau_{n(k)}}\right] + \theta_{AMB}. \tag{4.140}$$

As the cell thermal impedance is a first-order impedance defined by R_n and τ_n, the cell temperature can be described by the following first order differential non-homogeneous equation.

$$\frac{d\theta}{dt} + \frac{\theta}{\tau_n} = \frac{R_n}{\tau_n}P_{C0(cell)}(t) + \frac{\theta_{AMB}}{\tau_n}. \tag{4.141}$$

"Exact" calculation

A general solution of Equation 4.141 can be written in the form

$$\theta(t) = \underbrace{\left(\frac{R_n}{\tau_n}\int_0^t \exp\left(\frac{x}{\tau_n}\right)P_{C0(cell)}(x)dx\right)\exp\left(-\frac{t}{\tau_n}\right)}$$

$$+ (\theta_0 - \theta_{AMB})\exp\left(-\frac{t}{\tau_n}\right) + \theta_{AMB}$$

$$= \frac{R_n}{\tau_n}J(t)\exp\left(-\frac{t}{\tau_n}\right) + (\theta_0 - \theta_{AMB})\exp\left(-\frac{t}{\tau_n}\right) + \theta_{AMB}, \tag{4.142}$$

where

θ_0 is the cell initial temperature and

$P_{CO(cell)}(x)$ is the cell losses where x is the time variable (in fact the time), and

$J(t)$ is an integral to be solved.

The cell time dependent cell losses are computed from Equations 4.74 and 2.67 as

$$P_{CO(\varsigma)(cell)}(t) = 2R_{CO(cell)} i_{CO(cell)}^2(t) = \frac{1}{NM} 2R_{CO} \frac{P_{CO(DCH)}^2 C_N}{C_N U_{CO\,max}^2 - 2P_{CO(DCH)} t}. \quad (4.143)$$

Please note that the resistance is doubled in order to take into account the EOL of the ultra-capacitor.

To find the solution of Equation 4.142 we have to solve the following integral

$$J(t) = \int_0^t \exp\left(\frac{x}{\tau_n}\right) P_{CO(cell)}(x) dx$$

$$= \frac{1}{NM} 2R_{CO} P_{CO(DCH)}^2 C_N \int_0^t \frac{\exp\left(\frac{x}{\tau_n}\right)}{C_N U_{CO\,max}^2 - 2P_{CO(DCH)} x} dx. \quad (4.144)$$

The integral $J(t)$ (Equation 4.144) can be rewritten in the following form

$$J(t) = -\frac{1}{NM} R_{CO} P_{CO(DCH)} C_N \exp\left(\frac{C_N U_{CO\,max}^2}{2P_{CO(DCH)} \tau_n}\right) \int_{x_0}^{x_1(t)} \frac{\exp(x)}{x} dx, \quad (4.145)$$

where the integral limits are

$$x_0 = -\frac{C_N U_{CO\,max}^2}{2P_{CO(DCH)} \tau_n}, \quad x_1(t) = \frac{2P_{CO(DCH)} t - C_N U_{CO\,max}^2}{2P_{CO(DCH)} \tau_n}. \quad (4.146)$$

The remaining issue is to solve the integral

$$J_1(t) = \int_{x_0}^{x_1(t)} \frac{\exp(x)}{x} dx. \quad (4.147)$$

Unfortunately, the integral (Equation 4.147) cannot be solved in closed form. Expanding Equation 4.147 in a Taylor series yields

$$J_1(t) = \int_{x_0}^{x_1(t)} \sum_{k=0}^{\infty} \frac{x^{k-1}}{k!} dx = \ln\left(\frac{x_1(t)}{x_0}\right) + \left(\sum_{k=1}^{\infty} \frac{x^k}{kk!}\right)\Bigg|_{x_0}^{x_1(t)}. \quad (4.148)$$

Substituting Equation 4.146 into Equations 4.148 and 4.145 yields

$$J(t) = \frac{1}{NM} R_{CO} P_{CO(DCH)} C_N \exp\left(\frac{C_N U_{CO\,max}^2}{2P_{CO(DCH)} \tau_n}\right) \left[\ln\left(\frac{x_0}{x_1(t)}\right) - \left(\sum_{k=1}^{\infty} \frac{x^k}{kk!}\right)\Bigg|_{x_0}^{x_1(t)}\right].$$

$$(4.149)$$

Substituting the integral $J(t)$ (Equation 4.149) and the condition that initial temperature equals the ambient temperature ($\theta_0 = \theta_{AMB}$) into Equation 4.142 yields the cell temperature at the end of the load pulse T_P

$$\theta_{n(\max)} = \theta(t = T_P) = \frac{R_n}{\tau_n} J(t = T_P) \exp\left(-\frac{T_P}{\tau_n}\right) + \theta_{AMB}$$

$$= \frac{R_n}{\tau_n} \left\{ \frac{1}{NM} R_{C0} P_{C0(DCH)} C_N \exp\left(\frac{C_N U_{C0\,\max}^2}{2 P_{C0(DCH)} \tau_n}\right) \left[\ln\left(\frac{x_0}{x_1(t)}\right) - \left(\sum_{k=1}^{\infty} \frac{x^k}{kk!}\right) \Big|_{x_0}^{x_1(t)} \right] \right\}$$

$$\exp\left(-\frac{T_P}{\tau_n}\right) + \theta_{AMB}. \tag{4.150}$$

Taking only the first element of the series expansion ($k = 1$) (Equation 4.148) and substituting the data given into Equation 4.150 yields the cell maximum temperature $\theta_{n(max)} = \mathbf{40.747\,°C}$.

An approximation and simplified calculation

Since the module time constant τ_n is almost two orders of magnitude higher than the load pulse TP, we can assume the thermal process is purely adiabatic. That means that the heat generated in the cell is stored in the thermal capacity and not transferred to the ambient. A differential equation describing such an adiabatic system is derived from Equation 4.141 as

$$\lim_{R_n \to \infty} \left[\frac{d\theta}{dt} + \frac{\theta}{\tau_n} = \frac{R_n}{\tau_n} P_{C0(cell)}(t) + \frac{\theta_{AMB}}{\tau_n} \right] = \underbrace{\left[\frac{d\theta}{dt} = \frac{1}{C_{n(TH)}} P_{C0(cell)}(t) \right]}_{ADIABATIC\_SYSTEM}, \tag{4.151}$$

where $C_{n(TH)}$ is the cell thermal time capacitance

$$C_{n(TH)} = \frac{\tau_n}{R_n}. \tag{4.152}$$

From the data given we have $C_{n(TH)} = 300$ (J/K).

The solution of the simplified model (4.151) is

$$\theta(t) = \frac{1}{C_{n(TH)}} \int_0^t P_{C0(cell)}(x)dx + \theta_{AMB}. \tag{4.153}$$

Substituting Equation 4.154 into Equation 4.153 yields the cell temperature over time

$$\theta(t) = \frac{1}{C_{n(TH)}} \frac{1}{NM} R_{C0} P_{C0(DCH)} C_N \ln \frac{C_N U_{C0\,\max}^2}{C_N U_{C0\,\max}^2 - 2 P_{C0(DCH)} t} + \theta_{AMB}. \tag{4.154}$$

The cell maximum temperature is

$$\theta(t = T_P) = \frac{1}{C_{n(TH)}} \frac{1}{NM} R_{C0} P_{C0(DCH)} C_N \ln \frac{C_N U_{C0\,\max}^2}{C_N U_{C0\,\max}^2 - 2 P_{C0(DCH)} T_P} + \theta_{AMB}. \tag{4.155}$$

Substituting the data given into Equation 4.155 yields the cell maximum temperature $\theta_{n(max)} = \mathbf{40.735\,°C}$.

Please note that the results obtained by the "exact" and simplified method are almost the same. Once the discharging is finished, the ultra-capacitor module is recharged with 10% of the rated power. The temperature response can be computed from Equations 4.149 and 4.142 substituting $P_{C0(CH)} = 0.1 P_{C0(DCH)}$ and $\theta_0 = \theta_{n(max)}$. However, such a calculation is not necessary. As the recharge power is 10% of the discharge, the ultra-capacitor losses will be only 1% of that of the discharge. Hence, the module's temperature will decrease toward the ambient θ_{AMB}.

2. Once the ultra-capacitor module is discharged to the minimum voltage, discharging has stopped. Immediately, the ultra-capacitor module is recharged with rated power $P_{C0(CH)} = 400\,\text{kW}$. The temperature of the hot-spot cell can be computed using a simplified method (Equations 4.151–4.153). The charging losses are

$$P_{C0(\varsigma)(cell)}(t) = \frac{1}{NM} 2R_{C0} \frac{P_{C0(CH)}^2 C_N}{C_N U_{C0\,min}^2 + 2P_{C0(CH)}t}. \tag{4.156}$$

Substituting Equation 4.156 into Equation 4.153 yields the cell temperature during the ultra-capacitor charge

$$\theta(t) = \frac{1}{C_{n(TH)}} \frac{1}{NM} R_{C0} P_{C0(CH)} C_N \ln \frac{C_N U_{C0\,min}^2 + 2P_{C0(CH)}t}{C_N U_{C0\,min}^2} + \theta(T_P) + \theta_{AMB}, \tag{4.157}$$

where $\theta(T_P)$ is the cell initial temperature given by Equation 4.155. From the date given, the discharging maximum temperature, and Equation 4.157, we have the cell temperature at the end of discharge/charge cycle $\theta_{n(max)} = \mathbf{41.63\,°C}$.

The cell temperature development during the entire discharge/recharge cycle is

$$\theta(t) = \frac{1}{C_{n(TH)}} \frac{1}{NM} R_{C0} C_N$$

$$\times \left\{ \begin{array}{l} [h(t) - h(t - T_P)] \left[P_{C0(DCH)} \ln \dfrac{C_N U_{C0\,max}^2}{C_N U_{C0\,max}^2 - 2P_{C0(DCH)}t} \right] \\[4mm] + h(t - T_P) \left[P_{C0(CH)} \ln \dfrac{C_N U_{C0\,min}^2 + 2P_{C0(CH)}(t - T_P)}{C_N U_{C0\,min}^2} \right] \end{array} \right\} + \theta_{AMB}, \tag{4.158}$$

where $h(t)$ is the step function defined as

$$h(t) = \begin{cases} 0 & t \le 0 \\ 1 & t > 0 \end{cases}. \tag{4.159}$$

Figure 4.24 shows the module's hot-spot temperature development over time, when the ultra-capacitor module is discharged/recharged with nominal power.

Exercise 4.9

The ultra-capacitor module from Exercise 4.8 may operate in continuous discharge/recharge mode.

1. Roughly estimate the module hot-spot temperature after five consecutive discharge/recharge cycles. The ultra-capacitor is discharged and recharged with nominal power $P_{C0} = 400\,\text{kW}$.

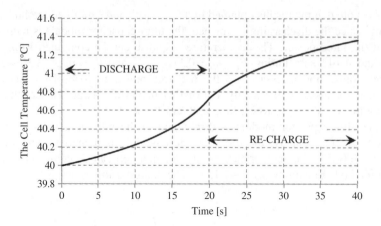

Figure 4.24 The ultra-capacitor hot-spot cell temperature over the discharge/recharge cycle

2. Estimate the maximum number of cycles if the hot-spot temperature is $\theta_{n(max)} = 65\,°\mathrm{C}$.
3. Calculate the hot-spot cell thermal resistance if the module is required to operate under the continuous full power discharge/recharge condition.

Solution

1. The module hot-spot temperature after five consecutive discharge/recharge cycles can be estimated assuming the adiabatic thermal process. The temperature rise after the first cycle is computed from Equation 4.158, $\Delta\theta_{n(1)} = 1.63\,°\mathrm{C}$. After five cycles, the temperature rise is $\Delta\theta_n \cong 5\Delta\theta_{n(1)} = 8.15\,°\mathrm{C}$. The cell absolute maximum temperature is $\theta_{n(max)} = 48.15\,°\mathrm{C}$.
2. As already mentioned in Exercise 4.8, the charge/recharge period is two orders of magnitude shorter than the module time constant. Hence, the ultra-capacitor losses can be averaged over one discharge/recharge cycle and the cell temperature computed from the average power losses.
   The cell can be described by the following differential equation

$$\frac{d\theta}{dt} + \frac{\theta}{\tau_n} = \frac{R_n}{\tau_n}P_{C0\varsigma(cell)(AV)} + \frac{\theta_{AMB}}{\tau_n}, \tag{4.160}$$

where $P_{C0\varsigma(cell)(AV)}$ are the cell average losses. The average losses are

$$P_{C0\varsigma(cell)(AV)} = \frac{1}{2T_P}\int_0^{2T_P}[P_{C0\varsigma(cell)(DCH)}(t) + P_{C0\varsigma(cell)(DCH)}(t)]dt$$

$$= \frac{R_{C0}C_N}{NM\,2T_P}\left[\begin{array}{l} P_{C0(DCH)}\ln\dfrac{C_N U_{C0\,max}^2}{C_N U_{C0\,max}^2 - 2P_{C0(DCH)}T_P} \\[2mm] +P_{C0(CH)}\ln\dfrac{C_N U_{C0\,min}^2 + 2P_{C0(CH)}T_P}{C_N U_{C0\,min}^2} \end{array}\right]. \tag{4.161}$$

Substituting the data given in Equation 4.161 yields the cell average losses $P_{C0\varsigma(cell)(AV)} = 12.25\,\text{W}$. The losses can also be computed from one cycle temperature rise computed in the previous exercise. From the cell temperature equation (Equation 4.153) we have

$$P_{C0\varsigma(cell)(AV)} = C_{n(TH)}\frac{\Delta\theta}{\Delta T} = C_{n(TH)}\frac{\Delta\theta}{2T_P}. \tag{4.162}$$

The cell temperature over time is a solution of Equation 4.160 given as

$$\theta(t) = R_n P_{C0\varsigma(cell)(AV)}\left(1 - \exp\left(-\frac{t}{\tau_n}\right)\right) + \theta_{AMB}. \tag{4.163}$$

The maximum load time T_{max} is computed from the condition that the hot-spot temperature is limited to the maximum $\theta_{n(max)} = 65°\text{C}$. From this condition and Equation 4.163 we have

$$T_{max} = -\tau_n \ln\left(1 - \frac{\theta_{n(max)} - \theta_{AMB}}{R_n P_{C0\varsigma(cell)(AV)}}\right). \tag{4.164}$$

The maximum time T_{max} corresponds to maximum number of cycles NC_{max}

$$N_{Cmax} = floor\left(\frac{T_{max}}{2T_P} - 1\right). \tag{4.165}$$

From the data given and Equations 4.164 and 4.165 we have $T_{max} = \textbf{747 seconds}$ and $NC_{max} = \textbf{18}$. After 18 cycles, the hot-spot temperature will be close to the maximum and hence the ultra-capacitor cannot be discharged/recharged any further.

3. If the ultra-capacitor module has to permanently operate under full discharge/charge conditions, the hot-spot maximum temperature is a steady state temperature

$$\theta_{n(max)} = R_n P_{C0\varsigma(cell)(AV)} + \theta_{AMB}. \tag{4.166}$$

The thermal resistance is

$$R_n \leq \frac{\theta_{n(max)} - \theta_{AMB}}{P_{C0\varsigma(cell)(AV)}}. \tag{4.167}$$

Substituting the data given into Equation 4.167 yields $R_n \leq \textbf{2.04 K/W}$. To achieve such low thermal resistance, the ultra-capacitor module must be air-forced-cooled. The air flow velocity has to be computed according to the cells' arrangement.

4.6 Ultra-Capacitor Module Testing

Testing and experimental verification of the ultra-capacitor module is one of the most crucial steps of the ultra-capacitor module design process. Once the initial mock-up prototype or final module has been designed and fabricated, the module has to be tested and experimentally verified. The test objective is to obtain relevant parameters such as capacitance, internal resistance, leakage current, self-discharge characteristics, and thermal impedances.

The test is a multi-step process that can be basically described as "the design of experiments (DoE)" process. The test steps can be summarized as:

1. Test set-up definition.
2. Set of input variables and their values are defined.
3. Set of output variables to be monitored are defined.
4. Analysis of the input and output variables and transfer function identification.

4.6.1 Capacitance and Internal Resistance

The ultra-capacitor's capacitance and internal resistance are measured from the ultra-capacitor current-to-voltage time response when the ultra-capacitor is discharged with a constant current step [9–12]. This characterization method is the most adopted method in industry [9]. Another possibility for measuring the ultra-capacitor impedance (capacitance and resistance) is impedance spectroscopy [13]. In the following sections, we will describe the current-to-voltage time response method in more detail.

4.6.1.1 Charge/Discharge Test Set-Up

Figure 4.25a shows a circuit diagram of a test set-up that is used to test the ultra-capacitor's charge/discharge performance. The ultra-capacitor, denoted as UUT (ultra-capacitor under test), is connected to a controlled bi-directional current source denoted as "Test Source." The current source is controlled and provides charge/discharge current i_{C0} according to the profile specified. The ultra-capacitor current i_{C0} is sensed with a shunt resistor R_S and the sensed voltage amplified with a voltage amplifier A_1. The ultra-capacitor voltage u_{C0} is measured with a voltage probe/amplifier A_2. The cells' temperatures are measured with appropriate probes attached to the cells. All values measured are recorded using a high resolution long buffer data logger.

Figure 4.25b shows some detail of a possible realization of the test set-up. The test source realized is a bi-directional double conversion dc–dc converter composed of two back-to-back connected dc–dc converters, a dc link capacitor C_{BUS}, and a buffer energy storage ultra-capacitor C_{02}. The dc–dc converters are two-level single-cell non-isolated converters. The first one (SW_{1A}/SW_{1B} and L_1) is connected to the UUT. The second (SW_{2A}/SW_{2B} and L_2) is connected to an internal ultra-capacitor energy storage C_{02}. The role of this ultra-capacitor is to store and restore energy when the test ultra-capacitor is discharged and charged respectively. Please note the additional auxiliary power supply connected to the converter dc bus. The role of this additional power supply is to cover the conversion losses and keep the dc bus voltage constant.

4.6.1.2 Measurement Procedure: General Terms

The UUT is charged and discharged with constant currents $I_{C0(CH)}$ and $I_{C0(DCH)}$ respectively. The voltage is cycled between the minimum and maximum voltage U_{C0min} and U_{C0max}. A typical current–voltage profile is illustrated in Figure 4.26. The capacitance and internal resistance are measured from the terminal voltage response when a constant current step discharge is applied.

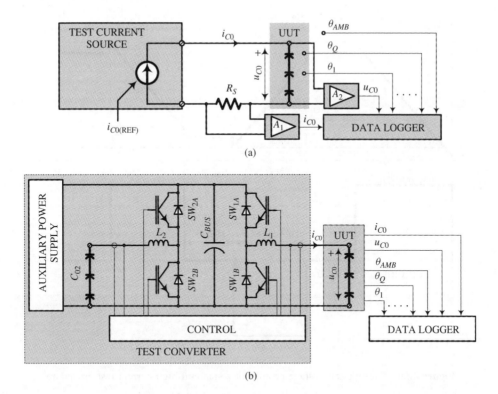

Figure 4.25 (a) Test set-up for high current testing and (b) details of possible realization

The ultra-capacitor is charged to the rated voltage U_{C0max} and held on that level for a period T_{HOLD}. At the instant t_0, the ultra-capacitor starts being discharged with constant current $I_{C0(DCH)}$. The hold method, hold time, and discharge current magnitude depend on the standard applied. Different standards will be addressed later on in the following section.

The ultra-capacitor internal resistance is measured from the voltage step when the ultra-capacitor switches from the hold mode to constant current discharge mode. Details are depicted in Figure 4.27a. The voltage step is defined as a cross-section point of the back-projection of the ultra-capacitor voltage and a vertical defined by the instant t_0 when the discharge current has been applied.

$$R_{C0} = \frac{\Delta u_{C0}}{I_{C0(DCH)}}, \tag{4.168}$$

where $I_{C0(DCH)}$ is the constant discharge current.

The capacitance is computed by the definition

$$C_N = I_{C0(DCH)} \frac{\Delta t}{\Delta u_{C0}} = I_{C0(DCH)} \frac{\Delta t}{U_{C01} - U_{C02}}, \tag{4.169}$$

where U_{C01} and U_{C02} are the voltages determined by the standard applied.

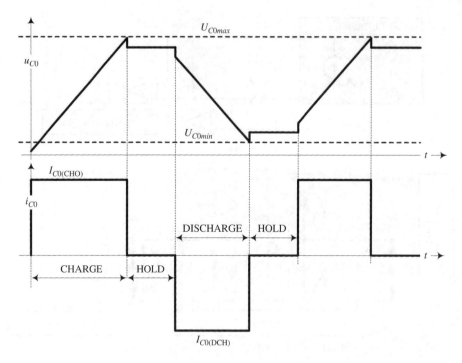

Figure 4.26 Current and voltage profiles of a standard high current test procedure

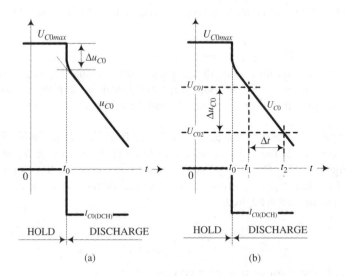

Figure 4.27 Measurement of the ultra-capacitor input impedance. (a) Internal resistance R_{C0} and (b) capacitance C_N

There are several standards and norms that define the parameter identification process. The three most used standards, IEC62391, IEC62576, and EUCAR, are briefly addressed below.

4.6.1.3 IEC62391 Method

The standard IEC62391 [19, 20] defines the hold method and time T_{HOLD}, charge/discharge current $I_{C0(CH)}$ $I_{C0(DCH)}$, and measurement voltage points U_{C01}, U_{C02}. Values are summarized in Table 4.4.

4.6.1.4 IEC62576 Method

The IEC62576 standard is very similar to the IEC62391 standard. The charge/discharge currents are specified for minimum of 95% efficiency. The charge/discharge current can be computed from Equation 4.48 as

$$I_{0(DCH)} \leq \frac{1 - 0.95}{R_{C0}6 \dfrac{\left(\dfrac{1 - k_0}{U_N}\left(U_{C\max}^2 - U_{C\min}^2\right) + k_0(U_{C\max} - U_{C\min})\right)}{4\dfrac{1 - k_0}{U_N}(U_{C\max}^3 - U_{C\min}^3) + 3k_0(U_{C\max}^2 - U_{C\min}^2)}}, \qquad (4.170)$$

where the resistance R_{C0} and coefficient k_0 are the ultra-capacitor target (expected) parameters. The coefficient k_0 is in range 0.7–0.8.

The ultra-capacitor internal resistance R_{C0} is measured in the same way as described for IEC62391 (Equation 4.168).

The capacitance is measured from the energy definition

$$C_N = \frac{2\displaystyle\int_{t1}^{t2} u_{C0}i_{C0}dt}{U_{C01}^2 - U_{C02}^2} = \frac{2E_{DCH}}{U_{C01}^2 - U_{C02}^2}, \qquad (4.171)$$

where E_{DCH} is the energy realized from the ultra-capacitor between t_1 and t_2.

The standard test parameters are summarized in Table 4.5.

4.6.1.5 EUCAR Method

The EUCAR test procedure is very similar to the IEC62391 [21] procedure. The resistance and capacitance are computed from Equations 4.168 and 4.169 respectively. The major

Table 4.4 Test parameters defined by the EC62391 standard

Hold method	Hold time	Charge current	Discharge current	Referent voltages	
Open circuit	T_{HOLD} 30 min	$I_{C0(CH)}$ 75 mA/F	$I_{C0(DCH)}$ 75 mA/F	U_{C01} 80% of U_{C0max}	U_{C02} 40% of U_{C0max}

Table 4.5 Test parameters defined by the IEC62576 standard

Hold method	Hold time	Charge current	Discharge current		
Constant voltage	T_{HOLD} 300 s	$I_{C0(CH)}$ Equation 4.170	$I_{C0(DCH)}$ Equation 4.170	U_{C01} 90% of U_{C0max}	U_{C02} 70% of U_{C0max}

Table 4.6 Test parameters defined by the EUCAR standard

Hold method	Hold time	Charge current	Discharge current		
Constant voltage	T_{HOLD} 30 s	$I_{C0(CH)}$ 50 mA/F	$I_{C0(DCH)}$ 5 mA/F	U_{C01} 60% of U_{C0max}	U_{C02} 30% of U_{C0max}

difference between the EUCAR and IEC62391 procedure is the discharge current. The EUCAR discharge current is very low compared to IEC62391, only 5 mA/F. Such a low discharge current may create an issue with measurement of the voltage step when the ultra-capacitor is discharged. As discussed in Section 4.2.1, the voltage step on the internal resistance depends on the capacitance C_N as the carbon load factor k_{LC}, (Equation 4.7 and Figure 4.6). If the carbon loading factor is $k_{LC} = 5$ mA/F, we can see that the voltage step is below 1% of the ultra-capacitor rated voltage. Measurement of such a low voltage step requires very careful selection of the test equipment and voltage measurement method (the standard EUCAR test parameters are summarized in Table 4.6).

4.6.2   Leakage Current and Self-Discharge

4.6.2.1   Test Set-Up

The test set-up circuit diagram is depicted in Figure 4.28. The UUT is connected to a high precision power supply denoted "TEST VOLTAGE SOURCE." The power supply has the capability to charge the ultra-capacitor to nominal voltage and regulate the terminal voltage at the reference over a long time. The UUT and power supply are connected via two switches SW_0 and SW_1 and a shunt resistor R_S connected in parallel with the switch SW_1. The ultra-capacitor current is sensed with the shunt R_S. The sensed voltage is amplified with a voltage amplifier A_1. The ultra-capacitor voltage is measured directly with a voltage probe A_2. If required, the ultra-capacitor temperatures are measured too. All values measured are recorded with a high resolution long buffer data logger.

4.6.2.2   Leakage Current Measurement

The ultra-capacitor leakage current is by definition the charging current required to maintain the ultra-capacitor at the specified voltage value. The longer the ultra-capacitor is held at voltage, the lower the leakage current of the ultra-capacitor. The measured result will be influenced by the temperature, the voltage at which the device is charged, the test history of the device, and the aging conditions.

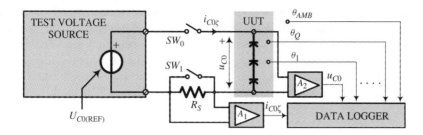

Figure 4.28   The leakage current and self-discharge characteristic measurement

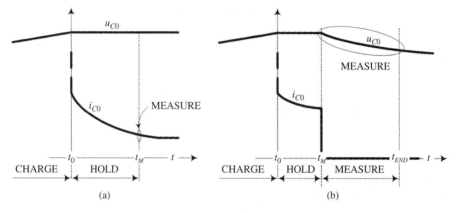

Figure 4.29   (a) Measurement of the ultra-capacitor leakage current and (b) self-discharge characteristics

Figure 4.29a shows detail of the test sequence. The ultra-capacitor is charged until the voltage has reached the nominal value at the instance t_0. The switch SW_1 is closed to prevent overload of the shunt resistor while the ultra-capacitor is being charged. Once the ultra-capacitor voltage is charged, the switch SW_1 can be opened. The voltage is regulated by a high precision power supply at the nominal voltage over the $T_{HOLD} = t_M - t_0$ period. Immediately at the end of the hold period, at the instance t_M, the ultra-capacitor current is measured and the measured value is taken as the leakage current. According to the manufacturer's specification, the hold period is $T_{HOLD} = 72$ hours [2].

4.6.2.3   Self-Discharge Characteristics

The self-discharge test is designed to observe the natural decay of the ultra-capacitor voltage over time after it is fully charged to a certain voltage. The measured result will be influenced by the temperature, the voltage at which the device is charged, the test history, and the aging condition.

Figure 4.29b shows detail of the test sequence. The ultra-capacitor is charged until the voltage has reached the nominal value at the instance t_0. The switch SW_1 is closed to prevent overload of the shunt resistor while the ultra-capacitor is being charged. The voltage is regulated by a high precision power supply at the nominal voltage over $T_{HOLD} = t_M - t_0$ period. Immediately at the end of the hold period, at the instance t_M, the switch SW_0 (Figure 4.28) is opened and the ultra-capacitor is left open circuited. The voltage is measured over the acquisition period $T_{ACQ} = t_{END} - t_M$. The measured value is taken as the self-discharge characteristic. According to the manufacturer's specification, the hold period is $T_{HOLD} = 1$ hour, while the acquisition period is $T_{ACQ} = 72$ hours [2].

4.7 Summary

The ultra-capacitor module is the core of the short-term energy storage system. Performances of the storage system, such as efficiency, life span, reliability, size, and cost strongly depend on the way the ultra-capacitor module is selected and designed. In this chapter, the ultra-capacitor module design is extensively addressed.

Three main parameters of ultra-capacitor module have to be selected according to the system specification:

1. The ultra-capacitor module voltage rating. Basically, there are four different operating voltage levels to be selected:
   (a) maximum operating voltage,
   (b) minimum operating voltage,
   (c) intermediate operating voltage, and
   (d) the module rated voltage.
2. The ultra-capacitor module's rated capacitance.
3. The ultra-capacitor module's internal resistance.

Selection of these parameters is an iterative process. The module voltage rating is selected according to the system operating voltage, the interface dc–dc converter topology, and the module's life expectancy. The module capacitance is selected according to the voltage rating, energy capability required, and conversion losses/efficiency specified.

Once the ultra-capacitor module is specified (voltage, capacitance, and resistance), the next step is the module design. The design steps can be summarized as follows:

1. Selection of the ultra-capacitor cell from the data sheet.
2. Selection of the number of series connected cells.
3. Selection of the number of parallel connected cells.
4. The module's thermal design.
5. The string voltage balancing circuit.

The final step in the entire design process is module testing and verification. The module capacitance and resistance is measured using the charge/discharge method. The module is charged and discharged with a constant current and the voltage response is measured. The module leakage current and self-discharge rate are also measured. The capacitance, resistance, leakage current, and self-discharge rate are measured according to the standards IEC62391-1, IEC62391-2, and EUCAR.

References

1. LS Mtron Ultra-Capacitors Data Sheet, http://www.ultracapacitor.co.kr (accessed 22 April 2013).
2. Maxwell Technologies Ultra-Capacitors http://www.maxwell.com/ultracapacitors (accessed 22 April 2013).
3. Enhui, Z., Q. Zhiping, and W. Tongzhen, (2010) Research on combination of series and parallel with super-capacitor module. 2nd IEEE International Symposium on Power Electronics for Distributed Generation Systems, PEDGS 2010, pp. 685–690.
4. Linzen, D., S. Buller, E. Karden, and R.W. De Doncker, (2003) Analysis and evaluation of charge balancing circuits on performance, reliability and lifetime of supercapacitor systems, Industry Applications Conference, October 12–16, 2003, Vol. 3, pp. 1589–1595.
5. Barrade, P., S. Pittet, and A. Rufer, (2000) Energy storage system using a series connection of supercapacitors, with an active device for equalizing the voltages. IPEC 2000, International Power Electronics Conference, Tokyo, Japan, April 3–7, 2000.
6. Grbović, P.J. (2009) Loss-free balancing circuit for series connection of electrolytic capacitors using an auxiliary switch mode power supply. *IEEE Transaction Power Electronics*, **24**, 1, 221–231.
7. Grbović, P.J. (2012) Art of control of advanced power semiconductors: from theory to practice, full day tutorial. IPEMC (ECCE Asia) 2012, IEEE Energy Conversion Congress and Exposition, Harbin, China, June 2–6, 2012.
8. Grbović, P.J., Delarue, P. and Le Moigne, P. (2011) A novel three-phase diode boost rectifier using hybrid half-DC-BUS-voltage rated boost converter. *IEEE Tansactions on Industrial Electronics*, **58**, 4, 1316–1329.
9. Maxwell Technologies (2009) Test Procedures for Capacitance, ESR, Leakage Current and Self-Discharge Characterizations of Ultracapacitors, Maxwell Technologies Application Note, July 2009.
10. IEC (2006) 62391-1. *Fixed Electric Double Layer Capacitors for Use in Electronic Equipment-Part I: Generic Specification*, IEC 40/1378/CD. International Electrotechnical Commission.
11. IEC 62391-2. *Fixed Electric Double Layer Capacitors for Use in Electronic Equipment-Part II: Sectional Specification Electric Double Layer Capacitors for Power Applications*, IEC 40/1378/CD. International Electrotechnical Commission.
12. EUCAR Traction Battery Working Group (2003) Specification of Test Procedure for Super-Capacitors Electric Vehicle Applications, EUCAR, April 2003.
13. Buller, S., Karden, E., Kok, D. and De Doncker, R.W. (2002) Modeling the dynamic behavior of super-capacitors using impedance spectroscopy. *IEEE Transaction Industry Applications*, **38**, 6, 1622–1626.
14. Conway, B.E. (1999) *Electrochemical Supercapacitors, Scientific Fundamentals and Technological Applications*, Kluwer Academic/Plenum Publisher, New York.
15. Miller, J.M. (2011) *Ultracapacitor Applications*, IET Power and Energy Series, Vol. **59**, The Institute of Engineering and Technology, Stevenage.
16. Grbović, P.J. (2012) Ultra-capacitors as energy storage for power conversion applications: from theory to practice. Half Day Tutorial, EnergyCon 2012, Florence, Italy, September 9–12, 2012.
17. Rafik, F., Gualous, H., Gallay, R. *et al.* (2007) Frequency, thermal and voltage supercapacitor characterization and modeling. *Journal of Power Sources*, **165**, 928–934.
18. Shiffer, J., Linzen, D. and Uwe Sauer, D. (2006) Heat generation in double layer capacitors. *Journal of Power Sources*, **160**, 1, 765–772.
19. Lee, D.H., Kima, U.S., Shin, C.B. *et al.* (2008) Modeling of the thermal behavior of an ultra-capacitor for a 42-V automotive electrical system. *Journal of Power Sources*, **175**, 664–668.
20. Gualous, H., Louahlia-Gualous, H., Gallay, R. and Miraoui, A. (2009) Super-capacitor thermal modeling and characterization in transient state for industrial applications. *IEEE Transaction Industry Applications*, **45**, 3, 1035–1044.
21. Xu, X., Sammakia, B.G., Murray, B.T., *et al.*, (2011) Thermal modeling and management of super-capacitor modules by high velocity impinging fan flow. International Mechanical Engineering Congress and Exposition, Denver, Colorado, IMECE 2011, November 11–17, 2011.
22. Hijazi, A., Kreczanik, P., Bideaux, E. *et al.* (2012) Thermal network model of super-capacitor stack. *IEEE Transaction Industrial Electronics*, **59**, 2, 979–987.

5

Interface DC–DC Converters

5.1 Introduction

In Chapter 3, we addressed and discussed different power conversion applications and the integration of ultra-capacitor energy storage into these applications. The basic system we discussed had a structure as depicted in Figure 5.1. The power supply and the load are interconnected via a rectifier (4) and an inverter (1). The ultra-capacitor energy storage device (2) is connected to the system dc bus via an interface converter (3). We also intensively discussed the design and selection of the ultra-capacitor bank. Here, in this chapter, we will address some aspects of the selection and design of the interface dc–dc converter.

In Section 3.2.4.3, we addressed the different methods of integration of the power conversion system and ultra-capacitor. As shown in Equation 3.2 and Figure 3.7, ultra-capacitor energy strongly depends on the terminal voltage. The ultra-capacitor can be directly connected to the system. However, such a system is not flexible and optimized. To achieve system flexibility and high efficiency, a bi-directional dc–dc power converter is mandatory as a link between the ultra-capacitor and the system dc bus, Figure 5.2. The interface converter topology and control scheme vary from case to case, depending on the application's requirement.

Here, in this chapter, we will discuss the interface dc–dc converter topologies. First, different concepts and state-of-the-art topologies will be discussed and compared. Then, two-level multi-cell non-isolated dc–dc converters as benchmark converters will be analyzed in detail.

5.2 Background and Classification of Interface DC–DC Converters

A dc–dc power converter is a static power converter that interconnects two dc systems, as illustrated in Figure 5.3. A dc–dc converter can be seen as a dc transformer that is the equivalent of an ac power transformer used in ac systems with a fixed frequency. The role of the dc–dc converter (dc transformer) is to convert the input dc quantity to

Ultra-Capacitors in Power Conversion Systems: Applications, Analysis and Design from Theory to Practice,
First Edition. Petar J. Grbović.
© 2014 John Wiley & Sons, Ltd. Published 2014 by John Wiley & Sons, Ltd.

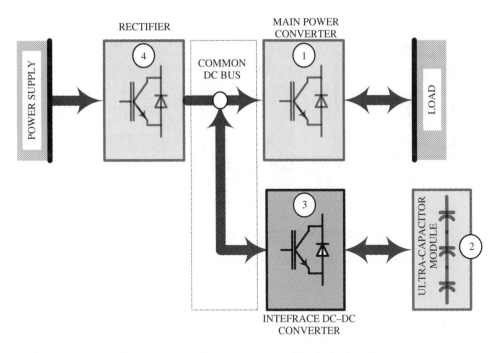

Figure 5.1 Power conversion system with ultra-capacitor energy storage

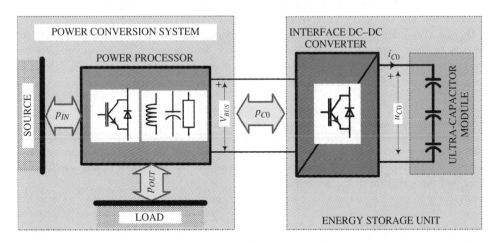

Figure 5.2 Interconnection between a power conversion system and ultra-capacitor energy storage

the output dc quantity. The dc quantities can be voltage or current of constant or varying magnitude.

The converter in general consists of an input filter, a switching matrix, and an output filter. The switching matrix is a network of two or more controlled switches such as insulated gate bipolar transistors (IGBTs), metal oxide semiconductor field effect transistors

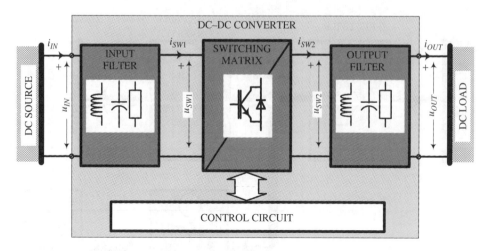

Figure 5.3   Generic block diagram of a dc–dc power converter

(MOSFETs), integrated gate commutated thyristorss (IGCTs), junction field effect transistors (JFETs), and switching diodes. The switches are periodically switched on and off, where the on/off time is defined and controlled by an external control circuit. By controlling the on/off time, it is possible to control power flow from one dc system to another. The input and output filters are passive low pass LC filters. The role of the input/output filters is to filter high frequency fluctuation of the input/output voltage/current caused by periodic and fast changes of the switches' status.

Dc–dc converters can be classified in different categories, depending on different aspects of their application. This could be power flow between the ultra-capacitor and the system, isolation between the converter input and output, the number of levels of the converter output voltage, the number of parallel connected devices, and so on.

5.2.1   Voltage and Current Source DC–DC Converters

The general dc–dc converter illustrated in Figure 5.3 can be of voltage source type or current source type. The converter type is determined by the nature of the input/output filter and the structure of the switch matrix.

5.2.1.1   Current Source Converters

If the converter input current i_{SW1} is continuous and the output current i_{SW2} is a train of pulses, we say the converter is of current source type. The input voltage u_{SW1} is a train of pulses while the output voltage u_{SW2} is continuous. A simplified circuit diagram of a current source dc–dc converter is depicted in Figure 5.4a. The input filter of a current source converter is an inductor L_{IN}, which is a current source filter. The output filter is a capacitor C_{OUT}, which is a voltage source filter. The switch matrix consists of voltage bi-directional and current uni-directional power semiconductor switches. Figure 5.5a shows a circuit diagram of a basic current source switch cell.

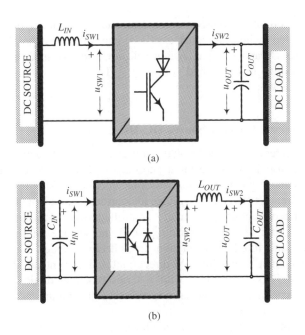

Figure 5.4   (a) Current source dc–dc converter and (b) voltage source dc–dc converter

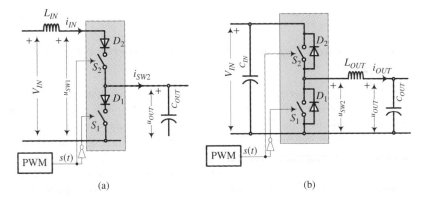

Figure 5.5   Basic switch cell as the building block of (a) current source converters and (b) voltage source converters

5.2.1.2   Voltage Source Converters

If the converter input voltage u_{SW1} is continuous and output voltage u_{SW2} is a train of pulses, we say the converter is of voltage source type. The converter input current i_{SW1} is a train of pulses while the output current i_{SW2} is continuous. A simplified circuit diagram of a voltage source dc–dc converter is depicted in Figure 5.4b. The input filter is a capacitor C_{IN}, which is a voltage source filter. The output filter is an inductor L_{OUT}, which is a current source filter. The switch matrix consists of current bi-directional and voltage uni-directional power semiconductor switches.

A basic switch cell of a voltage source converter is depicted in Figure 5.5b. The switch cell's so-called switch lag consists of two controlled switches S_1 and S_2 and two freewheeling diodes D_1 and D_2. The switch cell input voltage is a constant voltage V_{IN}, while the output voltage is a switched voltage u_{SW}. This basic switch cell is the building block of any kind of voltage source converter.

5.2.1.3 Voltage versus Current Source DC-DC Converters

There are a few differences between voltage source and current source dc–dc converters. The first one is the topology of the input and output filter. However, the total rating of the filter inductors and capacitors is almost same. There is no significant difference.

The second difference between the voltage and current source converter is the arrangement of power semiconductors, particularly controlled switches. The switches used in a voltage source converter are current bi-directional and voltage uni-directional switches, such as MOSFETs and IGBTs with parallel freewheeling diodes. The switches used in current source converters are voltage bi-directional and current uni-directional devices, such as IGBTs and IGBT/MOSFETs with a series blocking diode. Moreover, a basic bi-directional dc–dc converter requires two switch cells: Figure 5.5a. This is a major drawback of current source dc–dc converters.

In the majority of power conversion applications, dc–dc converters are of voltage source type. Current source converters are used mainly in super magnet energy storage (SMES) energy storage applications. Here, in this chapter, we will consider voltage source converters only.

5.2.2 Full Power and Fractional Power Rated Interface DC–DC Converters

If the interface dc–dc converter is connected in such a way that the entire power is processed by the converter, we say the converter is a full power rated converter. This case is illustrated in Figure 5.6a. The converter input voltage and current are the same as the system dc bus voltage end current. The converter output voltage and current correspond to the ultra-capacitor voltage and current.

If the interface dc–dc converter is connected in such a way that only a fraction of the power is processed by the converter, we say the converter is a fractional power rated converter. This case is illustrated in Figure 5.6b. The converter input voltage corresponds to the dc bus voltage, while the input current is a fraction of the dc bus current. The converter output voltage is the difference between the dc bus voltage and ultra-capacitor voltage, while the output current corresponds to the ultra-capacitor current.

In the majority of industrial applications, full power rated converters are used. Hence, in further discussion, we will consider this kind of converter only.

5.2.3 Isolated and Non-Isolated Interface DC–DC Converters

The interface dc–dc converter can be an isolated and non-isolated type. If the converter input is galvanically isolated from the output, we say the converter is of an isolated type, Figure 5.7a. Otherwise the converter is of a non-isolated type, Figure 5.7b.

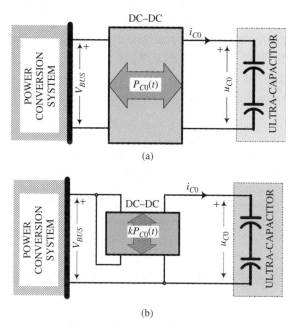

Figure 5.6 (a) Full power rating dc–dc converter and (b) fractional power rating dc–dc converter

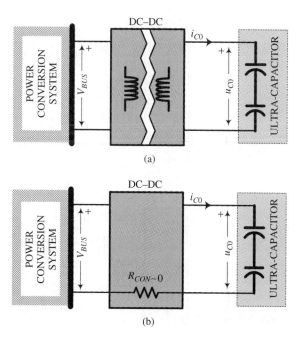

Figure 5.7 (a) Isolated dc–dc converter and (b) non-isolated dc–dc converter

Isolated interface converters are used in applications that require isolation due to safety or functional issues or in the case of large differences between the input and the output voltage. In all other cases, non-isolated interface converters are preferred because of their higher efficiency, smaller size, and lower cost.

5.2.4 Two-Level and Multi-Level Interface DC–DC Converters

In Section 5.2.1.2 we explained that the output voltage of the dc–dc converter switching cell is a train of voltage pulses. If the output voltage takes two discrete values, zero and full input dc bus voltage, the converter is a two-level voltage source converter, Figure 5.8a. The converter switch cell is the basic cell depicted in Figure 5.5b. The cell is composed of two bi-directional switches and one dc bus capacitor. If the output voltage can take more than two discrete values, the converter is a multi-level voltage source converter, Figure 5.8b. The number of switch cells and capacitors is $N_C = (N_L - 1)$, where N_L is the number of output voltage levels.

5.2.5 Single-Cell and Multi-Cell Interleaved Interface DC–DC Converters

In general, the interface dc–dc converter consists of one or more basic switch cells (Figure 5.5) and an input/output filter. If the converter consists of only one switch cell, the converter is a two-level single-cell converter, Figure 5.9a. If the dc–dc converter is

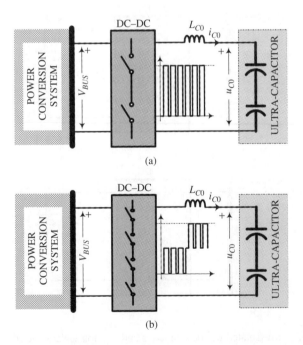

(a)

(b)

Figure 5.8 (a) Two-level dc–dc converter and (b) multi-level dc–dc converter

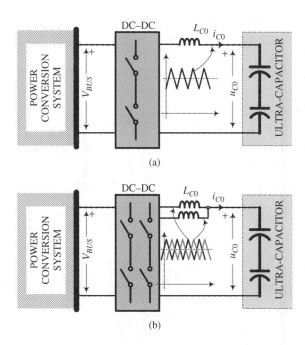

Figure 5.9 (a) Single-cell dc–dc converter and (b) multi-cell interleaved dc–dc converter

composed of N basic switch cells that are connected in parallel and driven by signals with a certain delay between each other, the converter is two-level N-cell interleaved converter, Figure 5.9b. Two-level single-cell and N-cell interleaved converters are very often used in high power applications. In this chapter we will mainly address this type of interface dc–dc converter. Advantages and drawbacks will be addressed in detail.

5.3 State-of-the-Art Interface DC–DC Converters

5.3.1 Two-Level DC–DC Converters

Most of the interface dc–dc converter topologies are based on an ordinary two-level single-cell topology, Figure 5.10a. The main drawback of these topologies is the fact that the switches are rated on the full dc bus voltage. As the dc bus voltage may go up to 800–1400 V, or even more, the switches are rated on 1200 or 1700 V. This becomes an issue if the converter switching frequency is quite high; let's say above 10–15 kHz. A two-level dc–dc converter with soft switching is an alternative [1, 2]. This solution offers lower switching losses. However, since the converter operates in discontinuous conduction mode (DCM), the peak current and the current ripple are greater than those one operates in the continuous conduction mode (CCM). This leads to an increase in the conversion losses, particularly the inductor core losses. Another issue is the output current ripple that has to be filtered by an additional output filter capacitor.

Paralleled power converters are used in power conversion applications that require high power capability. The load current is shared among the paralleled converters, and

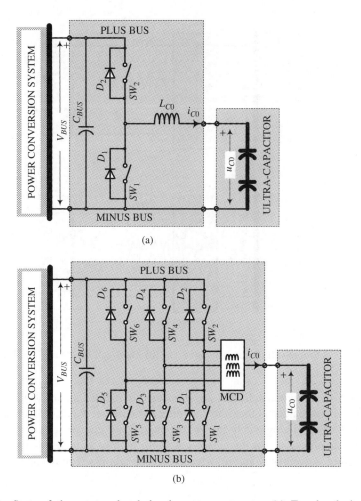

Figure 5.10 State of the art two-level dc–dc power converter. (a) Two-level single-cell and (b) two-level three-cell interleaved dc–dc converter

therefore power semiconductors with lower current capability can be used. To achieve proper current sharing between the individual cells, inter-phase inductors are used [3]. By shifting driving signals for $2\pi/N$ radians between the paralleled converters, so-called interleaving, additional improvements can be achieved. The benefits of using the interleaving technique include reduction of size, losses, and cost of the input and output filter, reduction of current stress of the dc bus capacitor, better utilization of the power semiconductors switches, and improved dynamic performances and controllability.

The interleaved dc–dc converter is an attractive solution widely used in various power conversion applications, such as voltage regulation modules (VRMs) [4–6], renewable applications [7], automotive and traction applications [8–11], power factor correctors (PFCs), and power supplies [12, 13].

5.3.2 Three-Level DC–DC Converters

Three-level converters are a well adopted solution in applications with a high input voltage and high switching frequency [14–16]. The switches are stressed on half of the total dc bus voltage. This allows us to use lower voltage rated switches that have better switching and conduction performance compared to the switches rated on the full blocking voltage. Therefore, the converter overall performance, including cost and efficiency, can be significantly better when compared to two-level converters, especially when the switching frequency is above 20 kHz or MOSFETs are used.

Figure 5.11 shows two the most commonly used topologies of a three-level bi-directional dc–dc converter. Figure 5.11a shows a floating-output three-level dc–dc converter [17], while Figure 5.11b shows a flying capacitor three-level dc–dc converter.

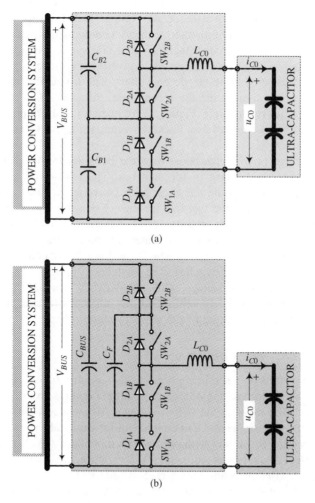

(a)

(b)

Figure 5.11 State of the art three-level interface dc–dc power converter. (a) Flying output and (b) flying capacitor and fixed output

There is no significant difference between the flying capacitor and floating output three-level dc–dc converter. The number and rating of active as well as passive components is the same. The major disadvantage of the floating output converter is the fact that the output voltage is floating. Therefore, the ultra-capacitor common mode voltage is high, being equal to half of the dc bus voltage. To avoid potential EMC problems, a large common mode output filter is required. In contrast to this, the flying capacitor converter has no such issue. However, the flying capacitor three-level converter has the issue of start-up and control of the flying capacitor voltage.

5.3.3 Boost-Buck and Buck-Boost DC–DC Converters

A common characteristic of the above discussed dc–dc converters is that the input to output voltage ratio is lower or equal to 1. Thus the ultra-capacitor voltage u_{C0} cannot be higher than the dc bus voltage V_{BUS}. In some applications, it is desired or even required to charge the ultra-capacitor above the dc bus voltage and discharge below the dc bus voltage. This is not possible using the above mentioned non-isolated converters.

Figure 5.12a,b shows the circuit diagrams of a single-cell cascade boost-buck and buck-boost converter. The boost-buck dc–dc converter (Figure 5.12a) is composed of two back-to-back connected basic dc–dc converters. The first converter (SW_3 and SW_4) operates as a boost converter, while the second one (SW_1 and SW_2) operates as a buck converter. The buck-boost converter (Figure 5.12b) operates in the opposite way. The first converter (SW_3 and SW_4) operates as a buck converter, while the second (SW_1 and SW_2) operates as a boost converter.

The main advantage of the buck-boost over the boost-buck converter is the number of passive components and rating of the switches. The buck-boost converter has one inductor and two capacitors, while the boost-buck converter has two inductors and one capacitor. Taking into account the fact that the energy density of a capacitor is an order of magnitude higher than energy density of an inductor; it is obvious that the buck-boost converter (Figure 5.12b) is smaller, cheaper, and more efficient than the boost-buck converter (Figure 5.12a). Another advantage of the buck-boost converter is the voltage rating of power semiconductor switches. The switches SW_1 and SW_2 are rated to the maximum of the ultra-capacitor voltage, while the switches SW_3 and SW_4 are rated to the maximum of the dc bus voltage. In contrast to this, all four switches of the boost-buck converter are rated to the maximum of the ultra-capacitor voltage.

5.3.4 Isolated DC–DC Converters

Isolated dc–dc converters are used in applications that require: (i) safety and/or functional isolation between the input and the output, (ii) multiple outputs from a single input, and (iii) a large ratio of the input to the output voltage. All isolated topologies can be split into four main groups: (i) single-end and double-end forward and fly-back, (ii) half-bridge and full-bridge forward converters with pulse width modulated (PWM) control, (iii) dual active bridge (DAB) with phase shift and PWM control, and (iv) series and parallel resonant converter (series resonant converter, SRC and PRC).

Single/double-end forward and fly-back converters are used mainly in low power applications, such as auxiliary power supplies. In medium and high power applications, DAB and SRC converters are dominant [18]. Figure 5.13 shows a simplified circuit diagram

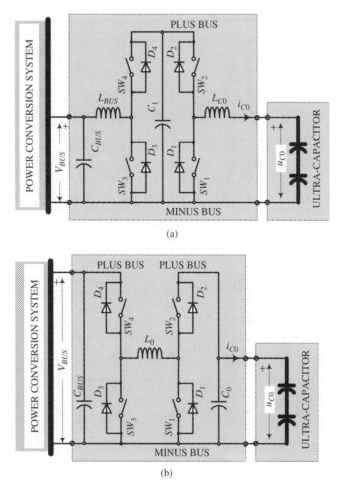

Figure 5.12 State of the art two-level interface buck-boost dc–dc power converter. (a) Common inductor and (b) common dc bus capacitor

of DAB and SRC. These topologies are an attractive solution when the ratio between the dc bus voltage and the ultra-capacitor voltage is high, greater than 2. If the ratio is lower than 2, the conversion efficiency is lower than that of a non-isolated topology.

5.3.5 Application Summary

All topologies of the interface dc–dc converters typically used in medium and high power applications are summarized in Table 5.1.

5.3.5.1 Non-Isolated Interface Converters

If the power conversion application does not specifically require galvanic isolation between the input and output, two-level and three/multi-level non-isolated dc–dc

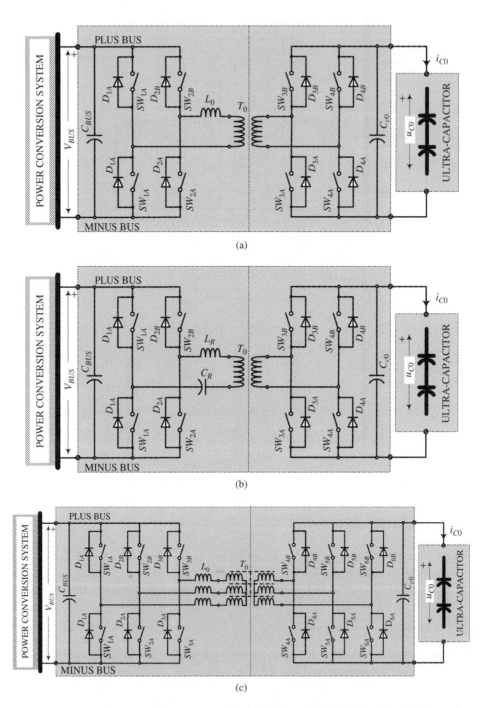

Figure 5.13 Isolated and bi-directional dc–dc converter topologies. (a) Single-phase dual active bridge (DAB), (b) single phase series resonant converter (SRC), and (c) three-phase dual active bridge (DAB)[18]

Table 5.1 Interface dc–dc converters

Non-isolated converters		Isolated converters	
Two-level	Three-level and multi-level	DAB	B-SRC
All the topologies can be single-cell as well as multi-cell interleaved			
Single-cell		Multi-cell interleaved	

converters are used as the interface between the ultra-capacitor and power conversion system.

Two-level single-cell and multi-cell interleaved converters are used in following cases:

- Medium and high power applications.
- Low voltage applications, the DC bus voltage $V_{BUS} < 1400\,V$:
  - Switching frequency $f_{SW} < 10\,kHz$,
  - 600, 1200, and 1700 V IGBTs and PiN diodes are used,
  - CollMOS used only in DCM.

Three-level and multi-level converters are used in the following cases:

- Low voltage applications, the DC bus voltage $V_{BUS} < 1400\,V$:
  - low and medium power,
  - high frequency and high power density applications,
  - switching frequency $f_{SW} > 10\,kHz$
  - 600 and 1200 V IGBTs and PiN diodes used
  - CollMOS used only in DCM.
- Medium voltage applications:
  - the dc bus voltage $V_{BUS} > 1400\,V$,
  - switching frequency $f_{SW} < 10\,kHz$,
  - 1200 and 1700 V IGBTs and PiN diodes used.

5.3.5.2 Isolated Interface Converters

If the power conversion application specifically requires galvanic isolation between the input and output, DAB converters and bi-directional series resonant converters (B-SRC) are used.

5.4 The Ultra-Capacitor's Current and Voltage Definition

Before going deeper into discussion and analysis of the interface dc–dc converter, we will briefly derive some equations of the ultra-capacitor current and voltage for different charge/discharge modes. These equations will be frequently used in the analysis in that follows in this chapter.

The ultra-capacitor current can be constant but can also change in time, depending on the ultra-capacitor charge/discharge mode. In the case of current control mode, the current is constant.

$$i_{C0} = I_{C0(DCH)} = Const,$$

$$i_{C0} = I_{C0(CH)} \quad = Const. \tag{5.1}$$

In the case of power control mode, the current changes significantly over time. This was discussed in Section 2.5.3. The charge and discharge current are given by Equations 2.68 and 2.79. The equations are repeated here

$$i_{C0} \cong -\frac{P_{C0(DCH)}}{U_{C(0)}} \sqrt{\frac{C_{C0}U_{C(0)}^2}{C_{C0}U_{C(0)}^2 - 2P_{C0(DCH)}t}}, \quad P_{C0(DCH)} > 0$$

$$i_{C0} \cong -\frac{P_{C0(CH)}}{U_{C(0)}} \sqrt{\frac{C_{C0}U_{C(0)}^2}{C_{C0}U_{C(0)}^2 - 2P_{C0(CH)}t}}, \quad P_{C0(CH)} < 0 \tag{5.2}$$

where

$P_{C0(DCH)}$ and $P_{C0(CH)}$ are the discharge and charge power and
$U_{C(0)}$ is the ultra-capacitor initial voltage.

According to the convention on Figure 2.19, the power source/drain is defined as

$$P_{C0} = -u_{C0}i_{C0}, \tag{5.3}$$

where the power is positive in the drain (discharging) mode and negative in the source (charging) mode.

The ultra-capacitor voltage, and therefore duty cycle, change significantly over time. The current charge/discharge mode was analyzed in Section 2.5.2.1. The voltage is given by Equations 2.44 and 2.53. For the sake of simplicity, we will assume the ultra-capacitor as a linear capacitor. Therefore, the voltage is

$$u_C(t) = U_{C(0)} - \frac{1}{C_{C0}}I_{0(DCH)}t,$$

$$u_C(t) = U_{C(0)} + \frac{1}{C_{C0}}I_{0(CH)}t, \tag{5.4}$$

where

$I_{C0(CH)}$ and $I_{C0(CH)}$ are the discharge and charge current and
$U_{C(0)}$ is the ultra-capacitor initial voltage.

In the case of power charge/discharge mode, the voltage is computed from Equation 5.2 and Equation 5.3

$$u_{C0}(t) = -\frac{P_{C0(DCH)}}{i_{C0}(t)} = \sqrt{U_{C(0)}^2 - \frac{2P_{C0(DCH)}t}{C_{C0}}}, \quad P_{C0(DCH)} > 0$$

$$u_{C0}(t) = -\frac{P_{C0(CH)}}{i_{C0}(t)} = \sqrt{U_{C(0)}^2 - \frac{2P_{C0(CH)}t}{C_{C0}}}, \quad P_{C0(CH)} < 0 \tag{5.5}$$

where

$P_{C0(CH)}$ and $P_{C0(CH)}$ are the discharge and charge power and
$U_{C(0)}$ is the ultra-capacitor initial voltage.

5.5   Multi-Cell Interleaved DC–DC Converters

5.5.1   Background of Interleaved DC–DC Converters

5.5.1.1   Direct Paralleled Converters

Paralleled power converters are used in power conversion applications that require high power capability. The load current is shared among the paralleled converters and therefore power semiconductors with a lower current capability can be used. Figure 5.14a shows a circuit diagram of a dc–dc power converter based on hard paralleling of power semiconductor switches. Current sharing between individual switches is critical. The switches have to be properly selected and gate driving signals must be precisely synchronized. A small dispersion in the switching may create a huge dynamic as well as static current unbalance. As a consequence, the switches have to be oversized, which is not a cost effective nor optimal solution.

High power inductors are made of N paralleled windings. The main reason for this is manufacturing simplicity. If a single wire with a large cross-section is used, it will be very difficult to wind the winding. Hence, to simplify manufacturing, the winding wire is split into N thinner wires that are connected into an input and an output node. Having multi-wire winding, we can disconnect the wires from the input node and connect each wire to the corresponding output of the converter cell, as is depicted in Figure 5.14b. The current sharing and balancing is now much easier since there is an inductance between the switch modules. A small deviation in the switch command will not produce any significant current imbalance.

5.5.1.2   Interleaved Paralleled Converters

By shifting driving signals for $2\pi/N$ radians between the paralleled converters, so-called interleaved driving, some additional improvements can be achieved. Benefits of using the interleaving technique include reduction of the size, losses, and cost of the input and output filter, reduction of the current stress of the dc bus capacitor, better utilization of the power semiconductors switches, and improved dynamic performance and controllability.

The interleaved dc–dc converter is an attractive solution widely used in various power conversion applications, such as VRM [4–6], renewable applications [7], automotive and traction applications [8–11], PFCs, and power supplies [12, 13].

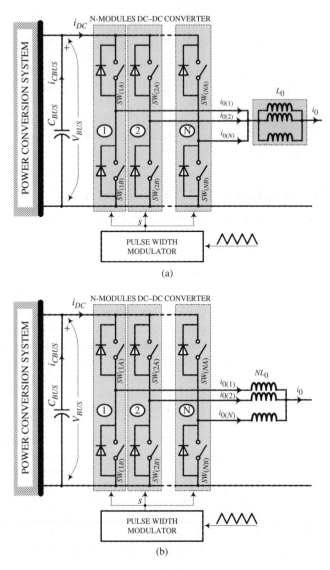

Figure 5.14 High power two-level dc–dc converter. (a) Direct hard paralleling without balancing inductors and (b) paralleling with individual inductors NL_0

Interleaved dc–dc converters are multi-cell converters composed of N paralleled dc–dc converter cells. The elementary cell can be a two-level or multi-level dc–dc converter. In this chapter, two-level dc–dc cells composed of two switches and two freewheeling diodes are considered only. The elementary dc–dc cells are driven by PWM pulses that are shifted for $2\pi/N$ radians. Figure 5.15 shows an example circuit diagram of an N-cell interleaved dc–dc converter. The left side of the converter is the so-called higher voltage

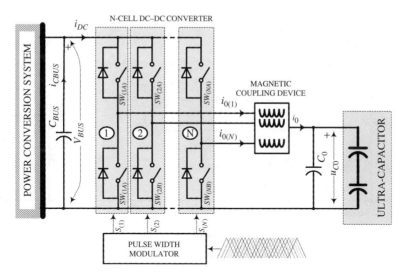

Figure 5.15 *N*-cell two-level interleaved dc–dc converter with magnetic coupling device

side and it will be denoted as the dc bus in future discussion. The right side of the circuit is the lower voltage side and it will be denoted as the output in future discussion.

Outputs of the switching cells are connected to the filter magnetic device denoted as the "magnetic coupling device." The structure of "the magnetic coupling device" depends on the interleaving concept. Two interleaving concepts are used. The first concept is so-called interleaving with individual inductors. The magnetic coupling device is a set of individual inductors with an inductance of L each. The second concept is so-called interleaving with inter-cell transformers (ICTs). The magnetic coupling device is a set of ICTs and an additional filter inductor [19–21]. The ICTs can have low leakage inductance or high leakage inductance. In the first case, an additional inductor L_0 is required. In the second case, the leakage inductance of the ICT is used as the output filter inductance [20, 21].

For the sake of simplicity, we will briefly describe a two-cell coupled interleaved dc–dc converter as an example. Later on we will extend and generalize the analysis to an *N*-cell coupled dc–dc converter.

5.5.2 *Analysis of a Two-Cell Interleaved Converter*

A circuit diagram of a two-cell coupled interleaved dc–dc converter is depicted in Figure 5.16a. The converter ICT is an ideal transformer with an additional filter inductor L_0. The converter consists of two cells composed of switches SW_{1A} to SW_{2B}, each switch having freewheeling diode D_{1A} to D_{2B}. The switches SW_{1A} and SW_{1B} are driven by the switching function s_1 and its complementary signal, while the switches SW_{2A} and SW_{2B} are driven from the switching function s_2 and its complementary signal. The signals s_1 and s_2 are generated by PWM1 and PWM2. The modulator's inputs are the duty cycles d_1 and d_2 and the triangular carriers v_{T1} and v_{T2}.

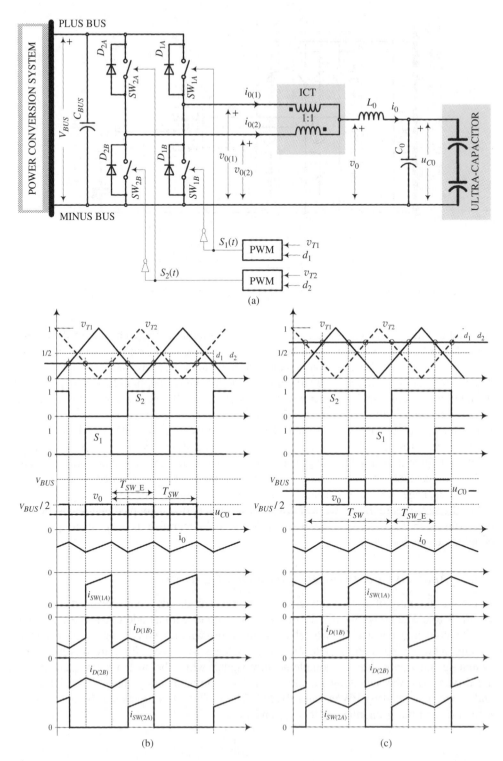

Figure 5.16 Two-cell two-level interleaved dc–dc converter with a magnetic coupling device ICT. (a) Circuit diagram, (b) waveforms when $d < 1/2$, and (c) waveforms when $d > 1/2$

The switching functions $s_1(t)$ are $s_2(t)$ are defined over one switching period as,

$$s_1(t) = \begin{cases} 1 & kT_{SW} \le t \le (k + d_1) T_{SW} \\ 0 & (k + d_1)T_{SW} \le t \le (k + 1)T_{SW} \end{cases},$$

$$s_2(t) = \begin{cases} 1 & \left(k - \frac{1}{2}\right) T_{SW} \le t \le \left(k - \frac{1}{2} + d_2\right) T_{SW} \\ 0 & \left(k - \frac{1}{2} + d_2\right) T_{SW} \le t \le \left(k - \frac{1}{2} + 1\right) T_{SW} \end{cases}, \qquad (5.6)$$

where
  k is an integer $k \in [, +\infty)$ and
  $d_1 \; d_2$ are the duty cycles.

The cells' output voltages are two-level voltages at switching frequency f_{SW}. The waveforms are phase shifted for π radians, as given in the following equation.

$$v_{0(1)}(t) = V_{BUS} s_1(t),$$

$$v_{0(2)}(t) = V_{BUS} s_2(t) = V_{BUS} s_1\left(t - \frac{T_{SW}}{2}\right) = v_{0(1)}\left(t - \frac{T_{SW}}{2}\right), \qquad (5.7)$$

where $s_1(t)$ and $s_2(t)$ are switching functions (Equation 5.6) Figure 5.16b and 5.16c show the waveforms.

Figure 5.17 shows an equivalent circuit diagram of the ICT. The ICT is a single-phase transformer with the transfer ratio 1:1. The transformer magnetizing inductance is $L\mu$. For the sake of simplicity, the magnetizing inductance $L\mu$ is split into two inductances $2L\mu$, one on each side of the transformer, as shown in Figure 5.17a.

The ICT input voltages are the cells' output voltages $v_{0(1)}(t)$ and $v_{0(2)}(t)$. From the transformer equations and circuit diagram of Figure 5.17a we compute the output voltage v_0 as

$$v_0(t) = \frac{v_{0(1)}(t) + v_{0(2)}}{2} = \frac{V_{BUS}}{2}(s_1(t) + s_2(t)). \qquad (5.8)$$

Substituting Equation 5.6 into Equation 5.8 and applying the averaging operator to it yields the average output voltage

$$\langle v_0 \rangle = \frac{V_{BUS}}{2} \frac{1}{T_{SW}} \int_0^{T_{SW}} (s_1(t) + s_2(t))dt = V_{BUS} d. \qquad (5.9)$$

The filter circuit is described by the following differential equation

$$L_0 \frac{di_0}{dt} = v_0(t) - u_{C0}. \qquad (5.10)$$

Applying the averaging operator to Equation 5.10 yields

$$\frac{1}{T_{SW}} \int_0^{T_{SW}} \left(L_0 \frac{di_0}{dt}\right) dt = \langle v_0 \rangle - \langle u_{C0} \rangle = 0 \Rightarrow \langle v_0 \rangle = \langle u_{C0} \rangle = u_{C0}. \qquad (5.11)$$

The ultra-capacitor terminal voltage u_{C0} is the voltage across the output filter capacitor C_0. As the capacitor is assumed to be large enough to filter the current ripple, the voltage is instantaneous and average values are almost the same. Thus, $\langle u_{C0} \rangle = u_{C0}$.

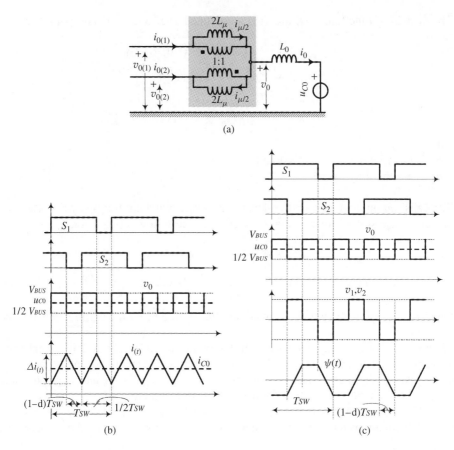

Figure 5.17 (a) Equivalent circuit diagram of the ICT. (b) The output current waveform, and (c) the voltages and flux waveforms

The output voltage v_0 is a combination of two switching functions. Since there are two switching functions with two states each, there will be four output voltage combinations. Depending on the duty cycle ($d < 1/2$ or $d > 1/2$), there will be different states of the output filter. A summary is given in Table 5.2. The states are indicated in Figure 5.16b,c.

State A

$$L_0 \frac{di_0}{dt} = 0 - u_{C0}. \tag{5.12}$$

The current is decreasing since $di_0/dt < 0$.

State B

$$L_0 \frac{di_0}{dt} = \frac{V_{BUS}}{2} - u_{C0}. \tag{5.13}$$

The current is increasing since $u_{C0} < V_{BUS}/2$ and therefore $di_0/dt > 0$.

State C

$$L_0 \frac{di_0}{dt} = \frac{V_{BUS}}{2} - u_{C0}. \tag{5.14}$$

Table 5.2 The switching function combinations

Switching function state s_1s_2	00	01	10	11
Output voltage v_0	0	$\frac{1}{2} V_{BUS}$	$\frac{1}{2} V_{BUS}$	V_{BUS}
The states at $d < 1/2$	A	B	B	X
The states at $d > 1/2$	X	C	C	D

The current is decreasing since $u_{C0} > V_{BUS}/2$ and therefore $di_0/dt < 0$.

State D

The current is increasing since $u_{C0} < V_{BUS}$ and therefore $di_0/dt > 0$.

State X

This state is not possible.

From Equation 5.10 and the switching states Equations 5.11–5.14 we can find the current i_0 in the form

$$i_0(t) = I_0 + \Delta i_0(t), \tag{5.15}$$

where

I_0 is the steady state current and
$\Delta i_0(t)$ is the current ripple.

The current ripple waveform is a triangular waveform with a peak-to-peak magnitude Δi_0 defined as

$$\Delta i_0(t) = \frac{\Delta i_0(d)}{2} \begin{cases} -1 + 2\dfrac{\Delta i_0(d)}{dT_{SW}}t & 0 \leq t \leq d\dfrac{T_{SW}}{2} \\ 1 - 2\dfrac{\Delta i_0(d)}{(1-d)T_{SW}}\left(t - d\dfrac{T_{SW}}{2}\right) & d\dfrac{T_{SW}}{2} \leq t \leq \dfrac{T_{SW}}{2} \end{cases}. \tag{5.16}$$

The ripple peak-to-peak magnitude Δi_0 is computed from Equations 5.6, 5.12–5.14 as

$$\Delta i_0(d) = \underbrace{\left[\frac{V_{BUS}}{16f_{SW}L_0}\right]}_{\Delta i_{0\,max}} 8 \begin{cases} (1 - 2d)d & 0 \leq d \leq \frac{1}{2} \\ (1 - d)(2d - 1) & \frac{1}{2} \leq d \leq 1 \end{cases}, \tag{5.17}$$

where Δi_{0max} is the maximum of the current ripple.

The ripple RMS value is

$$\Delta i_{0(RMS)} = \frac{2}{T_{SW}} \int_0^{T_{SW}/2} \Delta i_0^2(t)dt = \frac{\Delta i_0(d)}{2\sqrt{3}}. \tag{5.18}$$

The ICT terminal voltages $v_1(t)$ and $v_2(t)$ are

$$v_1(t) = v_2(t) = \frac{v_{0(1)}(t) - v_{0(2)}}{2} = \frac{V_{BUS}}{2}(s_1(t) - s_2(t)). \tag{5.19}$$

The ICT total flux is defined from the definition

$$\frac{d\psi}{dt} = v_1(t) = \frac{V_{BUS}}{2}(s_1(t) - s_2(t)). \tag{5.20}$$

Substituting the switching functions (Equation 5.6) into Equation 5.20 yields the total flux variation over one switching cycle

$$\Delta\psi = \frac{V_{BUS}}{2}\int_0^{T_{SW}/2}(s_1(t) - s_2(t))dt = \frac{V_{BUS}}{2}\frac{1}{f_{SW}}[d + (1 - 2d)floor(2d)]. \quad (5.21)$$

The flux density peak is computed from Equation 5.21 as

$$B_{peak} = \frac{\Delta\psi}{2nA_e} = \frac{V_{BUS}}{4nA_e}\frac{1}{f_{SW}}[d + (1 - 2d)floor(2d)] = B_{max}2[d + (1 - 2d)floor(2d)] \quad (5.22)$$

where
   n is the number of turns per a winding and
   A_e is the core effective cross-section.

Figure 5.18 shows the simulated waveform of the ICT flux density B versus duty cycle d. The flux density has a maximum at the duty cycle of $d = 0.5$, as predicted by Equation 5.22.
   The ICT instantaneous currents $i_{0(1)}$ and $i_{0(2)}$ are

$$i_{0(1)}(t) = i_{0(2)}(t) = \frac{i_0(t)}{2} = \frac{i_{C0}}{2} + \frac{\Delta i_0(t)}{2}, \quad (5.23)$$

where
   $\Delta i_0(t)$ is the output current ripple and
   i_{C0} is the ultra-capacitor current.

Equation 5.23 is an approximation that is correct if and only if the magnetizing inductance $L\mu$ is significantly larger than the filter equivalent inductance. In that case the magnetizing current i_μ can be neglected compared to the filter ripple current $\Delta i_0(t)$. If this is not the case, the magnetizing current has to be taken into account in Equation 5.23.

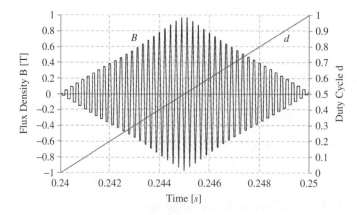

Figure 5.18   Simulated waveforms of the ICT flux density B versus duty cycle d. The waveform simulated at switching frequency $f_{SW} = 4.5\,kHz$

The RMS current can be computed by the definition

$$I_{0(1)(RMS)} = I_{0(2)(RMS)} = \sqrt{\frac{1}{T_{SW}} \int_0^{T_{SW}} \left(\frac{i_{C0}}{2} + \frac{\Delta i_0(t)}{2}\right)^2 dt}. \tag{5.24}$$

Applying Parseval's theorem [22] on Equation 5.24 yields an RMS current

$$I_{0(1)(RMS)} = I_{0(2)(RMS)} = \frac{1}{2}\sqrt{i_{C0}^2 + \Delta i_{0(RMS)}^2}, \tag{5.25}$$

where $\Delta i_{0(RMS)}$ is RMS value of the current ripple given by Equation 5.18.

The total output current RMS is

$$I_{0(RMS)} = \sqrt{i_{C0}^2 + \Delta i_{0(RMS)}^2} . \tag{5.26}$$

5.5.3 N-Cell General Case Analysis

Let's now analyze a general *N*-cell interleaved converter, Figure 5.15. A simplified circuit diagram of the converter is depicted in Figure 5.19a. The *N*-cell bridge is modeled by voltage sources $v_{0(1)} \ldots v_{0(N)}$. Figure 5.19b shows the equivalent circuit diagram of the converter. The circuit (model) consists of an equivalent voltage source v_0, an equivalent inductance L_0, the output capacitor C_{OUT}, and the ultra-capacitor impedance Z_{C0}.

The switch cells' output voltages are two-level voltages at the switching frequency f_{SW}. The voltages are shifted for $2\pi/N$ radians. The output voltage of the *k*th cell is

$$v_{0(k)}(t) = V_{BUS} s_k(t), \tag{5.27}$$

where

$s_l(t)$ is the *k*th cell switching function and

k is an integer $k = (1, 2 \ldots N)..$

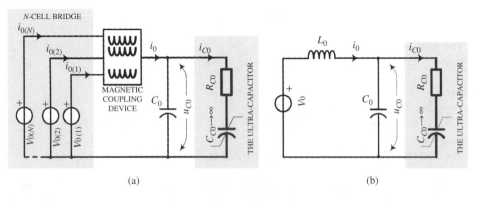

Figure 5.19 (a) Simplified circuit diagram of *N*-cell interleaved converter with magnetic coupling device ICT. (b) Equivalent circuit diagram

Assuming fully symmetrical ICT and that the switching cells are identical, the duty cycles of all the cells are equal,

$$d_1(t) = d_1(t) = .. = d_k(t) = .. = d_N(t).$$ (5.28)

From Equations 5.28 and 5.27 follows

$$v_{0(k)}(t) = V_{BUS} s_1\left(t - \frac{k-1}{N}T_{SW}\right) = v_{0(1)}\left(t - \frac{k-1}{N}\right).$$ (5.29)

The switching functions are defined as

$$s_k(t) = s_1\left(t - \frac{k-1}{N}T_{SW}\right),$$

$$s_1(t) = \begin{cases} 1 & kT_{SW} \le t \le (k+d_1)T_{SW} \\ 0 & (k+d_1)T_{SW} \le t \le (k+1)T_{SW} \end{cases}.$$ (5.30)

The equivalent output voltage v_0 is $N+1$-level voltage given by the following equation

$$v_0(t) = \sum_{k=1}^{N} \frac{v_{0(k)}(t)}{N} = \frac{V_{BUS}}{N}\sum_{k=1}^{N} s_k(t), \quad k = (1, 2, \dots N).$$ (5.31)

The frequency of the equivalent voltage v_0 is N times the basic switching frequency f_{SW}. It could be proven by expanding the switching functions $s_k(t)$ into a Fourier series and substituting the series into Equation 5.31. For more detail, please see Section 5.5.3.3.

Figure 5.20 shows the waveforms of the cells' switching functions s_1 to s_4 and the equivalent output voltage v_0 in the case of a four-cell interleaved dc–dc converter.

The output current of each individual cell is denoted as $i_{0(k)}$, while the total output current is i_0. From the first Kirchhoff law it follows that the total output current is the sum of the output currents of individual converters. As we have assumed fully symmetrical ICT and switching cells, the total output current is equally distributed among all N cells

$$i_0(t) = \sum_{k=1}^{N} i_{0(k)}(t) = N i_{0(cell)}(t).$$ (5.32)

The total output current is

$$i_0(t) = i_{C0} + \Delta i_0(t),$$ (5.33)

where
$\quad i_{C0}$ is the ultra-capacitor current and
$\quad \Delta i_0$ is the output current ripple.

The current of each individual switching cell is

$$i_{0(cell)}(t) = \frac{i_0(t)}{N} = \frac{i_{C0}}{N} + \frac{\Delta i_0(t)}{N},$$ (5.34)

where N is the number of interleaved cells.

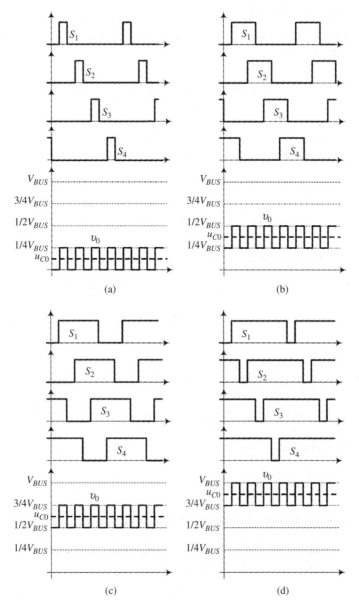

Figure 5.20   Sketched waveforms of the cell switching functions s_1 to s_4 and equivalent output voltage v_0. (a) $0 < d < 1/4$, (b) $1/4 < d < 1/2$, (c) $1/2 < d < 3/4$, and (d) $3/4 < d < 1$

5.5.3.1 Steady State Analysis

Let's calculate the average value of the output voltage v_0 (Equation 5.31). By the definition of average value we have

$$\langle v_0 \rangle = \frac{1}{T_{SW}} \int_0^{T_{SW}} v_0(t)dt = \frac{V_{BUS}}{N} \frac{1}{T_{SW}} \int_0^{T_{SW}} \sum_{k=1}^{N} s_k(t)dt$$

$$= \frac{V_{BUS}}{N} \sum_{k=1}^{N} \frac{1}{T_{SW}} \int_0^{T_{SW}} s_k(t)dt = \frac{V_{BUS}}{N} \sum_{k=1}^{N} \langle s_k \rangle .k = (1, 2, \ldots N). \qquad (5.35)$$

The average value of the switching function s_k is computed from Equation 5.30 as

$$\langle s_k \rangle = \frac{1}{T_{SW}} \int_0^{T_{SW}} s_k(t)dt = d_k, k = (1, 2, \ldots N). \qquad (5.36)$$

If the ICT is fully symmetrical and the switching cells are identical, the cell duty cycles are equal,

$$d_1 = \ldots \ldots = d_k = \ldots = d_N = d. \qquad (5.37)$$

Substituting Equations 5.36 and 5.37 into Equation 5.35 yields

$$\langle v_0 \rangle = V_{BUS} d. \qquad (5.38)$$

The equivalent circuit of Figure 5.19 can be described by a differential equation

$$L_0 \frac{di_0}{dt} = v_0 - u_{C0}. \qquad (5.39)$$

Applying the averaging operator to Equation 5.39 yields

$$\frac{1}{T_{SW}} \int_0^{T_{SW}} \left(L_0 \frac{di_0}{dt} \right) dt = \langle v_0 \rangle - \langle u_{C0} \rangle = 0 \Rightarrow \langle v_0 \rangle = \langle u_{C0} \rangle. \qquad (5.40)$$

The ultra-capacitor terminal voltage u_{C0} is the voltage across the output filter capacitor C_0. As the capacitor is assumed sufficient to filter the current ripple, the voltage is instantaneous and the average values are almost the same. Thus, $\langle u_{C0} \rangle = u_{C0}$.

Now we can determine the duty cycle from Equations 5.38 and 5.40 as

$$d = \frac{u_{C0}}{V_{BUS}}. \qquad (5.41)$$

Substituting Equation 5.4 into Equation 5.41 yields the duty cycle in the current charge/discharge mode

$$d(t) = \frac{U_{C(0)}}{V_{BUS}} - \frac{1}{C_{C0} V_{BUS}} I_{0(DCH)} t = D_0 - \frac{1}{C_{C0} V_{BUS}} I_{0(DCH)} t,$$

$$d(t) = \frac{U_{C(0)}}{V_{BUS}} + \frac{1}{C_{C0} V_{BUS}} I_{0(CH)} t = D_0 + \frac{1}{C_{C0} V_{BUS}} I_{0(CH)} t, \qquad (5.42)$$

where D_0 is the initial duty cycle. Substituting Equation 5.5 into Equation 5.41 yields the duty cycle in the power charge/discharge mode

$$d(t) = \sqrt{\frac{U_{C(0)}^2}{V_{BUS}^2} - \frac{2P_{C0(DCH)}t}{C_{C0}V_{BUS}^2}} = \sqrt{D_0^2 - \frac{2P_{C0(DCH)}t}{C_{C0}V_{BUS}^2}}, \quad P_{C0(DCH)} > 0.$$

$$d(t) = \sqrt{D_0^2 - \frac{2P_{C0(CH)}t}{C_{C0}V_{BUS}^2}}, \quad P_{C0(CH)} < 0. \tag{5.43}$$

5.5.3.2 The Output Current Ripple

In can be proved that the current ripple of an N-cell interleaved converter is a triangular periodic waveform defined as

$$\Delta i_0(t) = \frac{\Delta i_0(d)}{2} \begin{cases} -1 + N\dfrac{\Delta i_0(d)}{dT_{SW}}t & 0 \le t \le d\dfrac{T_{SW}}{N} \\ 1 - N\dfrac{\Delta i_0(d)}{(1-d)T_{SW}}\left(t - d\dfrac{T_{SW}}{N}\right) & d\dfrac{T_{SW}}{N} \le t \le \dfrac{T_{SW}}{N} \end{cases} . \tag{5.44}$$

The current ripple frequency is N times the basic switching frequency f_{SW}. The current peak-to-peak ripple $\Delta i_0(d)$ can be defined in a general form

$$\Delta i_0(d) = \left(\frac{V_{BUS}}{4f_{SW}L_0}\right)\frac{4}{N^2}[(Nd - floor(Nd)) - (Nd - floor(Nd))^2]$$

$$= \left(\frac{V_{BUS}}{4f_{SW}L_0}\right)K_{\Delta i}(d), \tag{5.45}$$

where d is the converter duty cycle and $T_{SW} = 1/f_{SW}$ is the basic switching period. The ripple current reaches zero when the duty cycle is equal to a multiple of $1/N$. The ripple current reaches the maximum $\Delta i_{0(max)}$ at duty cycle

$$d = \frac{1}{N}\left(k - 1 + \frac{1}{2}\right), k = (1, 2, \ldots N). \tag{5.46}$$

The peak-to-peak maximum current ripple $\Delta i_{0(max)}$ is

$$\Delta i_{0(max)} = \left(\frac{V_{BUS}}{4f_{SW}L_0}\right)\frac{1}{N^2}, \tag{5.47}$$

where L_0 is the filter equivalent inductance, V_{BUS} is the dc bus voltage, and f_{SW} is the basic switching frequency. The current ripple Equation 5.45 can be redefined in the form

$$\Delta i_0(d) = \left(\frac{V_{BUS}}{4f_{SW}L_0}\frac{1}{N^2}\right)N^2K_{\Delta i}(d) = \Delta i_{0\,max}N^2K_{\Delta i}(d). \tag{5.48}$$

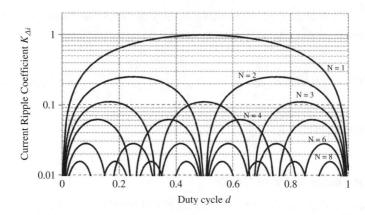

Figure 5.21 The output current ripple coefficient $K\Delta i_{C0}$ versus duty cycle d and number of interleaved cells N

The output current ripple coefficient $K_{\Delta i}$ versus the duty cycle d is depicted in Figure 5.21. The ripple shown in Figure 5.21 is the ripple normalized to the maximum ripple $\Delta i_{0(max)}$ of a single cell dc–dc converter ($N = 1$). Figure 5.22 shows the waveform of the output current i_0, current of one cell i_{01}, and duty cycle d of a four-cell coupled interleaved dc–dc converter.

The ripple RMS value computed over the T_{SW}/N period (the basic period of the current ripple) is

$$\Delta i_{0(RMS)} = \frac{N}{T_{SW}} \int_0^{T_{SW}/N} \Delta i_0^2(t) dt = \frac{\Delta i_0(d)}{2\sqrt{3}}. \qquad (5.49)$$

5.5.3.3 The DC BUS Current Analysis

The converter model is depicted in Figure 5.23. Each cell is modeled by an output controlled voltage source and an input controlled current source. The output voltage sources are controlled by the dc bus voltage v_{BUS} and switching function $s_{(k)}$, while the input current sources are controlled by the output current $i_{0(k)}$ and switching function $s_{(k)}$.

The dc side current i_{DC} is the sum of currents of all N converters, where each current is defined by the output current $i_{0(k)}$ and the switching function $s_{(k)}$,

$$i_{DC}(t) = \sum_{k=1}^{N} s_{(k)}(t) i_{0(k)}(t). \qquad (5.50)$$

Switching functions Equation 5.30 can be expended in Fourier series in a general form

$$s_{(k)}(t) = d + \frac{2}{\pi} \sum_{p=1}^{\infty} \frac{1}{p} \sin(pd\pi) \cos\left(p\omega_{SW} t + \frac{2\pi}{N}(k-1)\right), k = (1, 2, \ldots N) \quad (5.51)$$

where
    N is the number of interleaved converters and
    k is the index of the kth converter.

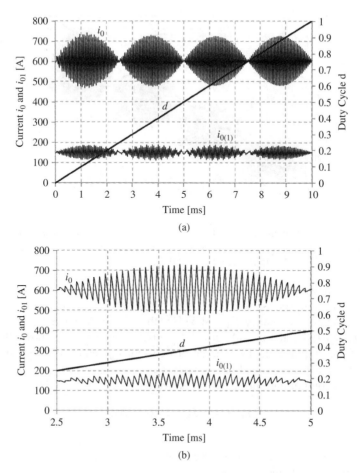

(a)

(b)

Figure 5.22 (a) Waveform of the output current i_0 and current of one cell $i_{0(1)}$. (b) Zoomed in waveforms. A four-cell coupled interleaved dc–dc converter, switching frequency $f_{SW} = 4.5$ kHz, the dc bus voltage $V_{BUS} = 700$ V, filter inductance $L_0 = 10\,\mu$H, and output current $I_{C0} = 600$ A

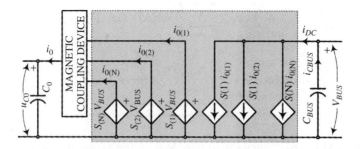

Figure 5.23 Model of an N-cell interleaved dc–dc converter

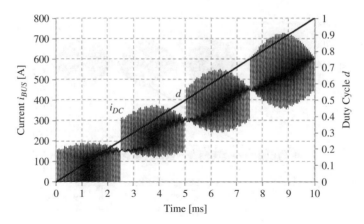

Figure 5.24 Simulated waveform of the dc bus current i_{DC} versus duty cycle d. A four-cell coupled interleaved dc–dc converter, switching frequency $f_{SW} = 4.5$ kHz, dc bus voltage $V_{BUS} = 700$ V, filter inductance $L_0 = 10$ μH, and output current $I_{C0} = 600$ A

Figure 5.24 shows the simulated waveform of the dc bus current i_{DC} versus duty cycle. A four-cell coupled interleaved dc–dc converter was simulated at the following conditions: switching frequency $f_{SW} = 4.5$ kHz, the dc bus voltage $V_{BUS} = 700$ V, filter inductance $L_0 = 10$ μH, and output current $I_{C0} = 600$ A. Figure 5.25 shows the simulated waveform of the output current i_0, the dc bus current i_{DC}, and a cell output voltage u_{01} of a four-cell coupled interleaved dc–dc converter.

In general, the output current of the kth converter is

$$i_{0(k)}(t) = I_{0(k)} + \Delta i_{0(k)}(t), \tag{5.52}$$

where

$I_{0(k)}$ is the average (dc) output current and
$\Delta i_{0(k)}$ is the current ripple.

Substituting Equations 5.51 and 5.52 into Equation 5.50 yields

$$
i_{DC}(t) = \sum_{k=1}^{N} d_{(k)} I_{0(k)} + \sum_{k=1}^{N} I_{0(k)} \frac{2}{\pi} \sum_{p=1}^{\infty} \frac{1}{p} \sin(pd\pi) \cos\left(p\omega_{SW}t + \frac{2\pi}{N}(k-1) \right)
$$

$$
+ \sum_{k=1}^{N} d_{(k)} \Delta i_{0(k)}(t) + \underbrace{ \sum_{k=1}^{N} \Delta i_{0(k)}(t) \frac{2}{\pi} \sum_{p=1}^{\infty} \frac{1}{p} \sin(pd\pi) }_{\cong 0} . \tag{5.53}
$$
$$\times \cos\left(p\omega_{SW}t + \frac{2\pi}{N}(k-1) \right)$$

The output RMS current ripple is usually low, in the range of 10–20% of the output rated current. Hence, the contribution of this current to the dc bus RMS current can be insignificant. However, under certain conditions, such as light load conditions or a high number of interleaved cells, the output current ripple may have a significant contribution to the dc bus current ripple. Hence, it has to be properly taken into account.

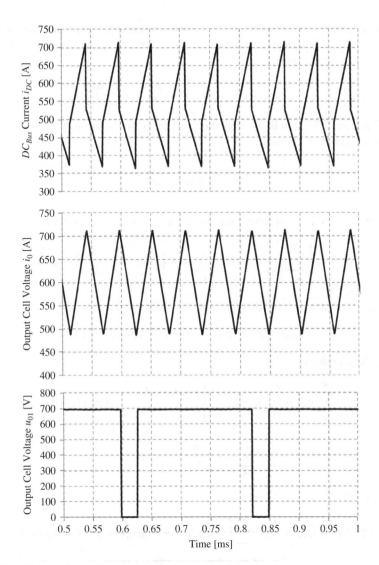

Figure 5.25 Simulated waveform of the output current i_0, the dc bus current i_{DC} and one-cell output voltage u_{01}. A four-cell coupled interleaved dc–dc converter, switching frequency $f_{SW} = 4.5\,\mathrm{kHz}$, duty cycle $d = 0.75$, the dc bus voltage $V_{BUS} = 700\,\mathrm{V}$, filter inductance $L_0 = 10\,\mu\mathrm{H}$, and output current $I_{C0} = 600\,\mathrm{A}$

For the sake of simplicity, one can assume that the converter cells are identical (identical inductors and switches). Thus, the average currents and duty cycles are equal,

$$d_{(1)} = d_{(2)} = \ldots = d_{(N)} = d$$

$$I_{0(1)} = I_{0(2)} = .. = I_{0(N)} = \frac{I_0}{N}. \tag{5.54}$$

The current ripple of each individual cell strongly depends on the structure of the output filter. In the case of interleaving with ICTs with significant magnetizing inductance, the individual cell current ripple is

$$\Delta i_{0(1)}(t) = \Delta i_{0(2)} = .. = \Delta i_{0(N)}(t) = \frac{\Delta i_0(t)}{N}. \tag{5.55}$$

From Equation 5.54 it follows that the third term of Equation 5.53 is the sum of all the individual currents' ripples

$$\sum_{k=1}^{N} d_{(k)} \Delta i_{0(k)}(t) = d \Delta i_0(t). \tag{5.56}$$

Taking into account all the above mentioned facts, one can define the dc bus current as

$$i_{DC}(t) \cong d I_0 + I_0 \frac{2}{\pi} \sum_{p=1}^{\infty} \left[\frac{1}{p} \sin(pd\pi) \sum_{k=1}^{N} \cos\left(p\omega_{sw} t + \frac{2\pi}{N}(k-1) \right) \right] + d \Delta i_0(t). \tag{5.57}$$

Applying the trigonometric identity Equations 5.57 to 5.58

$$\cos(\alpha + \beta) = \cos(\alpha)\cos(\beta) - \sin(\alpha)\sin(\beta), \tag{5.58}$$

and using the identities

$$\sum_{k=1}^{N} \cos p \frac{2\pi}{N}(k-1) = \begin{cases} 1 & p = iN \\ 0 & p \neq iN \end{cases} \quad i = (1, \infty),$$

$$\sum_{k=1}^{N} \sin p \frac{2\pi}{N}(k-1) = 0 \tag{5.59}$$

yields the dc bus current

$$i_{DC}(t) \cong d I_0 + \underbrace{\frac{I_0}{N} \frac{2}{\pi} \sum_{i=1}^{\infty} \left[\frac{1}{i} \sin(iNd\pi) \cos(iN\omega_{sw} t) \right] + d \Delta i_0(t)}_{i_{CBUS}(t)}. \tag{5.60}$$

From the dc bus current Equation 5.60, one can conclude that all the harmonics that are not multiples of the number of interleaved converters (N) are eliminated from the current spectra. The first harmonic that appears in the total dc current is the Nth harmonic of the basic switching frequency.

From Equation 5.60 one can compute the dc bus capacitor as

$$i_{CBUS}(t) \cong \frac{I_0}{N} \frac{2}{\pi} \sum_{i=1}^{\infty} \left[\frac{1}{i} \sin(iNd\pi) \cos(iN\omega_{sw} t) \right] + d \Delta i_0(t) = i_1(t) + i_2(t). \tag{5.61}$$

The dc bus capacitor current is composed of two components, namely $i_1(t)$ and $i_2(t)$. The first one, denoted $i_1(t)$, is the correlation of the average output current I_0 and the equivalent switching function. The second component $i_2(t)$ is the correlation of the average duty cycle m and the output current ripple $\Delta i_0(t)$.

Let's assume that the output filter inductor is an inductor with neglected resistance. Under this assumption it can be proven that the currents $i_1(t)$ and $i_2(t)$ are orthogonal functions and as such that they satisfy the condition

$$\int_t^{t+T} i_1(t)i_2(t)dt = 0. \tag{5.62}$$

This feature of the dc bus capacitor current will be used in the following section to compute the capacitor RMS current.

From Equations 5.60 and 5.62 the dc bus capacitor RMS current can be computed as

$$I_{CBUS(RMS)} = \sqrt{\frac{1}{T}\int_0^T i_1^2(t)dt + \frac{1}{T}\int_0^T i_2^2(t)dt + \underbrace{\frac{1}{T}\int_0^T i_1 i_2(t)dt}_{=0}}. \tag{5.63}$$

The RMS current Equation 5.63 can be computed in two ways. The first one is direct calculation using the time domain waveform of the i_{CBUS} current. This is the easy way if the converter is a single-cell converter. However, in the case of N-cell interleaved converters this method is not practical. The second method is to use the current spectrum Equation 5.60. Applying Parseval's theorem [22] to Equation 5.60 and using Equation 5.62 yields the capacitor RMS current

$$I_{CBUS(RMS)} = \sqrt{\left(\frac{I_0}{N}\frac{\sqrt{2}}{\pi}\right)^2 \sum_{i=1}^{\infty}\frac{1}{i^2}\sin^2(iNd\pi) + d^2\Delta i_{0(RMS)}^2}. \tag{5.64}$$

Using the identity

$$\sin^2\alpha = \frac{1}{2}(1 - \cos 2\alpha), \tag{5.65}$$

we have

$$\sum_{i=1}^{\infty}\frac{1}{i^2}\sin^2(iNd\pi) = \frac{1}{2}\left(\sum_{i=1}^{\infty}\frac{1}{i^2} - \sum_{i=1}^{\infty}\frac{\cos(iNd2\pi)}{i^2}\right). \tag{5.66}$$

The sums of Equation 5.66 are computed in close form as

$$\sum_{i=1}^{\infty}\frac{1}{i^2} = \frac{\pi^2}{6}, \tag{5.67}$$

and

$$\sum_{i=1}^{\infty}\frac{\cos(iNd2\pi)}{i^2} = \frac{\pi^2}{6}\frac{3(2Nd)^2 - 6(2Nd) + 2}{2}, \quad 0 \le 2Nm \le 2. \tag{5.68}$$

The argument of the function Equation 5.68 is limited in the range of 0–2. In reality the argument varies in a broad range, depending on N. Therefore Equation 5.68 cannot be directly used. As the *cos* function is a periodic function with a period of 2π radians, the argument can be redefined as follows

$$y = 2(Nd - floor(Nd)).\tag{5.69}$$

Substituting Equation 5.69 into Equations 5.68 and 5.67 yields

$$\sum_{i=1}^{\infty} \frac{1}{i^2}\sin^2(iNd\pi) = \frac{\pi^2}{2}[(Nd - floor(Nd)) - (Nd - floor(Nd))^2].\tag{5.70}$$

To compute the second term of Equation 5.64, let's first compute the output current RMS ripple. The output RMS current ripple can be computed from Equation 5.45 as

$$\Delta i_{0(RMS)} = \Delta i_{0(max)} \frac{2}{\sqrt{3}}[(Nd - floor(Nd)) - (Nd - floor(Nd))^2],\tag{5.71}$$

where d is the duty cycle and N is the cell number.

Constant Output Current

Substituting Equations 5.70 and 5.71 into Equation 5.64 yields the dc bus capacitor RMS current,

$$I_{CBUS(RMS)} = I_0 \sqrt{\begin{array}{l} \dfrac{1}{N^2}[(Nd - floor(Nd)) - (Nd - floor(Nd))^2] \\ + \left(\dfrac{\Delta i_{0(max)}}{I_0} d \dfrac{2}{\sqrt{3}}\right)^2 [(Nd - floor(Nd)) - (Nd - floor(Nd))^2]^2 \end{array}}$$

$$= I_0 K_{RMS(I)}.\tag{5.72}$$

The current ripple coefficient $K_{RMS(I)}$ versus the converter duty cycle d and number of interleave converters is depicted in Figure 5.26. Please note from the graph that the current ripple reaches zero at the duty cycle that corresponds to multiple of $1/N$, where N is the number of interleaved converters. This behavior is expected since all harmonics of the input as well as the output current are cancelled at the duty cycle that is equal to $1/N$. Please also note that the maximum RMS current at the duty cycle that corresponds to the extreme values given by Equation 5.46 is increasing as the duty cycle increases. This increase is the influence of the output current ripple to the dc bus capacitor RMS current. As we can see, the influence of the output ripple current becomes significant as the number of interleaved cells increases. The capacitor current is predominantly determined by the output current ripple when a high number of interleaved converters is used. The minimum RMS current that can be achieved with very high number of interleaved cells is

$$\lim_{N \to \infty} I_{CBUS(RMS)} = \Delta i_{0(max)} \frac{d}{2\sqrt{3}}.\tag{5.73}$$

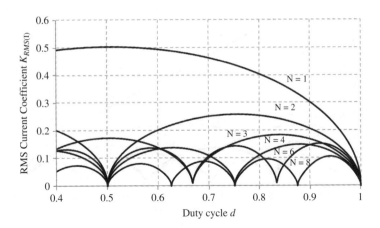

Figure 5.26 The dc bus capacitor RMS current coefficient versus duty cycle and number of interleaved cells. The ultra-capacitor current I_{C0} is constant as a parameter

Constant Power Operation

In the previous analysis, we assumed that the output current I_0 is constant. However, in most power conversion applications, the output current is not constant. The conversion power is constant, and therefore the output current increases as the duty cycle (output voltage) decreases. Typical applications are battery and ultra-capacitor dischargers for uninterruptible power supply (UPS) systems.

Let the conversion power P_{C0} and dc bus voltage V_{BUS} be constant. The output current I_0 is computed from Equation 5.41 as

$$I_0 = i_0(d) = \left(\frac{P_{C0}}{V_{BUS}}\right)\frac{1}{d}. \tag{5.74}$$

Substituting Equation 5.74 into Equation 5.72 yields the capacitor RMS current,

$$I_{CBUS(RMS)} = \frac{P_{C0}}{V_{BUS}}$$
$$\times \sqrt{\frac{[(Nd - floor(Nd)) - (Nd - floor(Nd))^2]}{(Nd)^2} + \left(\frac{V_{BUS}}{P_{C0}}\Delta i_{0(max)}d\frac{2}{\sqrt{3}}\right)^2 [(Nd - floor(Nd)) - (Nd - floor(Nd))^2]^2}.$$
$$= \frac{P_{C0}}{V_{BUS}}K_{RMS(P)} \tag{5.75}$$

The dc bus capacitor RMS current ripple versus duty cycle and number of interleaved cells is depicted in Figure 5.27. Please note that the capacitor current is increasing as the duty cycle is decreasing.

The minimum RMS current that can be achieved with a very high number of interleaved cells is the same as in the case of constant current Equation 5.72.

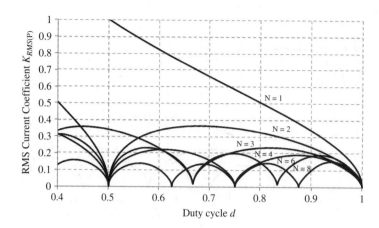

Figure 5.27 The dc bus capacitor RMS current coefficient versus duty cycle and number of interleaved cells. The ultra-capacitor power P_{C0} is constant as a parameter

5.5.3.4 DC Bus Capacitor Losses

The dc bus capacitor losses are discussed in detail in Section 5.8.4.1. The dc bus capacitor average losses (Equation 5.250) were derived in Section 5.8.4.1. In the case of an N-cell interleaved dc–dc converter, the dc bus capacitor losses are

$$P_{\xi(CBUS)} = \sum_{k=1}^{+\infty} ESR(kNf_{SW}) \cdot I^2_{C(RMS)(k)}, \tag{5.76}$$

where

$I_{C(RMS)(k)}$ is the RMS value of the kth harmonic of the dc bus capacitor current, (A),
N is the number of interleaved converters,
f_{SW} is the switching frequency, (Hz), and equivalent series resistance,
$ESR(kNf_{SW})$ is the capacitor resistance at the kth harmonic of N times switching frequency, (Ω).

However, since the capacitor resistance is constant at frequencies above few kilohertz, the losses are

$$P_{\xi(CBUS)} = ESR(Nf_{SW})I^2_{CBUS(RMS)}, \tag{5.77}$$

where the capacitor RMS current is given by Equation 5.72 or 5.75, depending on the converter load profile (constant current or constant power).

5.5.3.5 DC Bus Voltage Ripple

An exact calculation of the dc bus voltage peak-to-peak ripple can be made from the time domain current Equation 5.61 and the capacitor charge,

$$\Delta v_{BUS} = \frac{1}{C_{BUS}} \Delta q = \frac{1}{C_{BUS}} \int_{t1}^{t2} i_{CBUS}(t)dt. \tag{5.78}$$

However, for this we need to precisely determine the current waveform and the instance t_1 and t_2. It may become quite complicated if a high number of cells are interleaved. The analysis is even more complicated if the dc bus capacitor internal resistance is significant.

For a simplified fast calculation of the dc bus voltage ripple we can use the RMS sinusoidal approximation. Let's approximate the dc bus capacitor current with an equivalent sinusoidal current

$$i_{CBUS}(t) = \sqrt{2}I_{CBUS(RMS)} \sin(N\omega_{SW}t + \psi),\tag{5.79}$$

where

$I_{CBUS(RMS)}$	is the capacitor RMS current Equations 5.72 and 5.75,
N	is the number of interleaved cells,
ω_{SW}	is the switching angular frequency, and
ψ	is the equivalent current phase.

The dc bus voltage peak-to-peak ripple is computed from Equations 5.78 and 5.79 as

$$\Delta v_{BUS} = \frac{\sqrt{2}}{C_{BUS}N\pi f_{SW}}\sqrt{1 + (C_{BUS}N\omega_{SW}ESR)^2}I_{CBUS(RMS)}.\tag{5.80}$$

Figure 5.28 shows simulated waveforms of the dc bus voltage ripple Δv_{BUS} and current i_{DC}. The voltage ripple Δv_{BUS} is a real ripple and Δv_{BUS_sin} is the sinusoidal approximation Equation 5.80. A four-cell coupled interleaved dc–dc converter has been simulated under the following conditions: switching frequency $f_{SW} = 4.5\,\text{kHz}$, duty cycle $d = 0.75$, the dc bus voltage $V_{BUS} = 700\,\text{V}$, filter inductance $L_0 = 10\,\mu\text{H}$, the dc bus capacitor $C_{BUS} = 2350\,\text{uF}$, and constant output current $I_{C0} = 600\,\text{A}$.

Substituting Equations 5.72 and 5.75 into Equation 5.80 yields the dc bus voltage ripple under the conditions of constant output current I_{C0}

$$\Delta v_{BUS} = \frac{\sqrt{2}}{C_{BUS}\pi f_{SW}}\sqrt{1 + (C_{BUS}N\omega_{SW}ESR)^2}I_{C0}K_{\Delta V(i)},\tag{5.81}$$

or constant output power P_{C0}

$$\Delta v_{BUS} = \frac{\sqrt{2}}{C_{BUS}\pi f_{SW}}\sqrt{1 + (C_{BUS}N\omega_{SW}ESR)^2}\frac{P_{C0}}{V_{BUS}}K_{\Delta V(p)}.\tag{5.82}$$

The voltage ripple multipliers $K_{\Delta V(i)}$ and $K_{\Delta V(p)}$ are

$$K_{\Delta V(i)} = \frac{K_{RMS(i)}}{N},\quad K_{\Delta V(p)} = \frac{K_{RMS(p)}}{N}.\tag{5.83}$$

The RMS current multipliers $K_{RMS(i)}$ and $K_{RMS(p)}$ are defined by Equations 5.72 and 5.75.

The voltage ripple multipliers $K_{\Delta V(i)}$ and $K_{\Delta V(p)}$ versus duty cycle d and number of interleaved cells N are plotted in Figure 5.29.

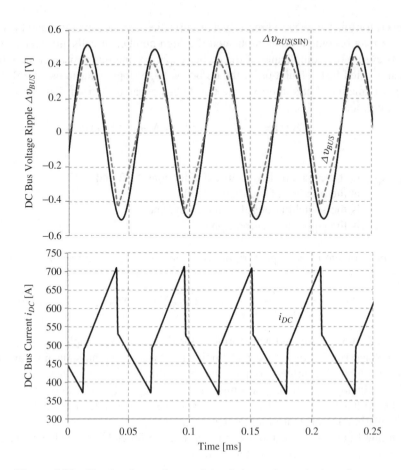

Figure 5.28   Simulated waveforms of the dc bus voltage ripple and current

5.6   Design of a Two-Level *N*-Cell Interleaved DC–DC Converter

In this section we will address some methods and techniques to design and select adequate passive, as well as active, components of an *N*-cell interleaved interface dc–dc converter. Because fully optimized design of magnetic components such as ICTs and inductors is a complex discipline, we will not go very deep into this matter. Some guidelines to select electrical and magnetic parameters of the ICT and filter inductor will be given. The dc bus capacitor design procedure will be also addressed. Finally, selection of power semiconductors (switches and diodes) will be given in the last part of the section.

5.6.1   ICT Design: A Two-Cell Example

5.6.1.1   Design Objective

In this section we will give a two-cell ICT design example. A typical high power two-cell ICT is depicted in Figure 5.30.

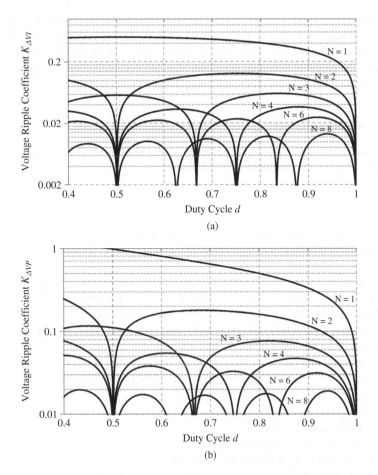

Figure 5.29 The dc bus voltage ripple multipliers versus duty cycle d and number of interleaved cells. (a) Constant output current (I_{C0}) and (b) constant output power (P_{C0})

Figure 5.30 High power two-cell ICT

Design of an ICT is an iterative multi-step process. The very first step is initial sizing of the ICT. The second step is the loss calculation. The third step is the ICT temperature calculation. If necessary, the selected core can be changed to a bigger or smaller one, depending on the hot-spot temperature. The process steps are described in detail in following subsections.

5.6.1.2 The ICT Initial Sizing

The initial selection of the ICT core is based on the area-product (AP) factor [21,23,24]. The AP factor of a magnetic core is defined as

$$AP = A_E A_W,$$ (5.84)

where
  A_E is the core effective cross-section area (m^2) and
  A_w is the core winding window area (m^2).

The AP factor is given as a parameter in the core data sheet [25]. Generally, the AP factor of a magnetic device depends on the device voltage and current, frequency, flux density, current density, the winding(s) arrangement, and so on.

To develop the AP factor, let's start from the core peak flux density Equation 5.22. From Equation 5.22 we can compute the core cross-section A_E as

$$A_E = \frac{V_{BUS}}{4n B_{peak} f_{SW}} [d + (1 - 2d) floor(2d)].$$ (5.85)

where
  n is the number of turns,
  f_{SW} is the switching frequency (kHz),
  B_{peak} is the core peak flux density (T), and
  d is the duty cycle.

The core winding window is

$$A_w = \frac{2n I_{RMS}}{J k_{FL}},$$ (5.86)

where
  I_{RMS} is the winding RMS current (A),
  J is the current density (A/m^2), and
  k_{FL} is the winding coefficient.

The factor k_{FL} depends on the wire profile and winding technique. If litz wire is used for the winding, the coefficient k_{FL} is very low, 0.2–0.4. If a flat wire or a foil is used for the ICT winding, the coefficient k_{FL} can be much higher, in the range 0.6–0.8.

Substituting Equations 5.25, 5.85, and 5.86 into Equation 5.84 yields the AP factor

$$AP = \frac{V_{BUS}}{4 B_{peak} f_{SW} J k_{FL}} \sqrt{i_{C0}^2 + \Delta i_{0(RMS)}^2} [d + (1 - 2d) floor(2d)].$$ (5.87)

The ultra-capacitor current i_{C0} is given by Equations 5.1 and 5.2. The current RMS ripple $\Delta i_{0(RMS)}$ is defined by Equation 5.49. Please note from Equations 5.1, 5.2, 5.49, and 5.87 the AP factor is a function of the duty cycle d,

$$AP = AP(d). \tag{5.88}$$

If the current ripple RMS value is reasonably low, let's say 10–20% of the ultra-capacitor RMS current, the ripple effect on the AP can be neglected. Under such an assumption, it could be proven that

$$\frac{\partial AP(d)}{\partial d} < 0. \tag{5.89}$$

Equation 5.89 basically means that the lower the duty cycle the higher the AP factor is. As we discussed in Chapters 2 and 4, the ultra-capacitor minimum voltage is usually limited to 50% of maximum voltage. Thus, the duty cycle is limited to $d_{min} = 0.5$. According this conclusion, we can define the worst case scenario AP factor

$$AP_{max} \cong \frac{V_{BUS} i_{C0(d=0.5)}}{8 B_{peak} f_{SW} J k_{FL}}. \tag{5.90}$$

Now we have almost defined the ICT AP factor. However, this is not the end of the design story! We still need to define the current density J and the flux peak density.

The current density J defines the copper losses and temperature rise. For the temperature rise, the current density depends on the winding size. The bigger the device, the lower the current density allowed. Detailed analysis can be found in [23]. It was proven in [23, 24] that the current density can be defined by

$$J = 4.5(AP)^{-0.125} 10^5, \tag{5.91}$$

where
   AP is the magnetic device area-product factor (m^4) and
   J is the current density (A/m^2).

As described in [23], the current density Equation 5.91 is given for a temperature increase of $\Delta\theta = 30$ K and natural air cooling. If a higher temperature rise is allowed or forced air cooling is used, the current density can be higher than Equation 5.91.

The core flux density has to be selected according to two criteria. The first one is the core saturation flux density. The second is the core losses at a given frequency and core material. A summary of the different frequencies and materials is given in Table 5.3.

The core losses p_ξ versus the flux density and frequency can be defined by the Steinmez equation

$$P_{\xi(core)} = m_{core} k_p B_{peak}^\alpha (f_{SW} 10^{-3})^\beta, \tag{5.92}$$

where
   B_{peak} is the flux density peak, (T),
   k_p, α, and β are the coefficients given by the core manufacturer,
   f_{SW} is the switching frequency, (Hz), and
   m_{core} is the core mass, (kg).

Table 5.3 The core peak flux density at different frequencies and core materials

Operating frequency	Low <1 kHz	Medium 1–15 kHz	High 15–100 kHz	Very high >100 kHz
The core material	—	Powder iron	Ferrites	Ferrites
B_{peak}	Losses $B_{peak} = B_{peak}(p_\xi)$	Amorphous Nano-crystalline Losses $B_{peak} = B_{peak}(p_\xi)$	Saturation $B_{peak} = B_{sat}$	Losses $B_{peak} = B_{peak}(p_\xi)$

The core losses are limited by the core temperature. If the core is naturally air-cooled, the relation between the losses and the core temperature is

$$P_{\xi(core)\,max} = A\Delta\theta^{1.1}10, \tag{5.93}$$

where

$P_{\xi(core)max}$ is the maximum loss (W),
A is the core surface (m^2), and
$\Delta\theta$ is the core temperature rise (K).

In the case of forced air-cooling, the core loss capability can be estimated as

$$P_{\xi(core)\,max} = A\frac{(5 + (v + 0.1)^{0.66})}{1.7}\Delta\theta^{1.1}, \tag{5.94}$$

where v is the cooling air velocity (m/s).

Is there any relation between the core mass m_{core} and surface A? Yes, there is a unique relation between the mass and the surface. Figure 5.31 shows an example of a C-shape amorphous core [25]. The core mass versus surface area is illustrated. A similar characteristic can be derived for any kind of core shape and material.

The function m_{core} versus A can be approximated by

$$m_{core} \cong k_m A, \tag{5.95}$$

where k_m is a coefficient (kg/m^2).

Substituting Equation 5.95 into Equations 5.92 and 5.94 yields peak flux density

$$B_{peak} = \left(\frac{\dfrac{(5 + (v + 0.1)^{0.66})}{1.7}\Delta\theta^{1.1}}{k_m k_p (f_{SW}\,10^{-3})^\beta}\right)^{\frac{1}{\alpha}}, \tag{5.96}$$

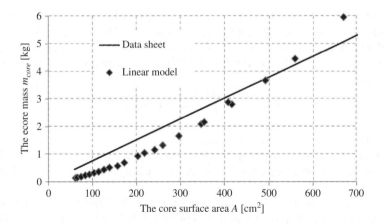

Figure 5.31 The core mass m_{core} versus surface area A of C-shaped cores [25]

Substituting Equations 5.91 and 5.96 into Equation 5.90 yields

$$AP_{max} \cong \left[\frac{V_{BUS} i_{C0(d=0.5)} (k_m k_p (f_{SW} 10^{-3})^\beta)^{\frac{1}{\alpha}}}{36 \left(\dfrac{(5 + (v + 0.1)^{0.66})}{1.7} \Delta\theta^{1.1} \right)^{\frac{1}{\alpha}} f_{SW} k_{FL}} 10^{-5} \right]^{1.43} , \qquad (5.97)$$

where the AP factor is given in (m^4).

Now we can pre-select the core from a list of available cores. From the core selected parameters and Equation 5.85 we can compute the number of turns

$$n = \frac{V_{BUS}}{2B_{peak} A_E f_{SW}}, \qquad (5.98)$$

and the cross-section of the winding wire,

$$A_{cu} = \frac{A_w k_{FL}}{2n}. \qquad (5.99)$$

5.6.1.3 The Core and Winding Losses

The core losses are

$$P_{\xi(core)} = m_{core} k_p B_{peak}^\alpha (f_{SW} 10^{-3})^\beta, \qquad (5.100)$$

where

B_{peak} is the flux density peak, (T),
k_p, α, and β are the coefficients given by the core manufacturer,
f_{SW} is the switching frequency, (Hz), and
m_{core} is the core mass, (kg).

The ICT winding losses are Joule's losses produced by the winding current $i_{0(1)}(t)$ flowing through a non-ideal winding having resistance R_0. The resistance R_0 is a frequency and therefore time-dependent resistance. This is similar to the case we discussed in Section 2.6 and Equation 2.97. The power dissipated on one winding can be expressed in a general form,

$$P_{\xi(cu)} = \frac{1}{T} \int_t^{t+T} i_{0(1)}^2(\tau) R_0(\tau) d\tau, \tag{5.101}$$

where T is the period of moving averaging.

The average losses of a winding can be computed using the same approach that we used to compute the ultra-capacitor frequency dependent losses, Section 2.6.1, Equations 2.102–3.109.

The ICT winding current is a periodic current defined by Equation 5.23. Expanding the current ripple $\Delta i_0(t)$ in Fourier series and substituting it into Equation 5.23 yields

$$i_{0(1)}(t) = i_{0(2)}(t) = \frac{i_{C0}}{2} + \frac{1}{2}\left[\sqrt{2}\sum_{k=1}^{\infty} I_{0(RMS)(k)} \sin(kN\omega_{SW}t + \phi_{(kN\omega_{SW})})\right], \tag{5.102}$$

where

    $I_{0(RMS)(k)}$ is the RMS value of kth harmonic of the current ripple $\Delta i_0(t)$ and
    i_{C0} is the ultra-capacitor current Equations 5.1 and 5.2.

Substituting Equation 5.102 into Equation 5.101 and using Parseval's theorem [22] yields the copper losses

$$P_{\xi(cu)} = \frac{i_{C0}^2}{4} R_{0(DC)} + \frac{1}{4}\sum_{k=1}^{n} I_{0(RMS)(k)}^2 R_{0(k)}, \tag{5.103}$$

where $R_{0(k)}$ is the ICT winding resistance that depends on the frequency.

The winding resistance is computed from the winding geometry and the number of turns [24]. The total ICT losses are

$$P_{\xi(ICT)} = P_{\xi(core)} + 2P_{\xi(cu)} = V_{core}\rho(k_p B_{peak}^\alpha f_{SW}^\beta) + \frac{i_{C0}^2}{2} R_{0(DC)} + \frac{1}{2}\sum_{k=1}^{n} I_{0(RMS)(k)}^2 R_{0(k)}. \tag{5.104}$$

The flux peak density Equation 5.22 and the winding resistance are functions of the number of turns for a given core and winding profile. Therefore, the total losses are also a function of the number of turns

$$P_{\xi(ICT)} = P_{\xi(core)}(n) + 2P_{\xi(cu)}(n). \tag{5.105}$$

Does the number of turns computed by Equation 5.98 give a minimum of total losses? Most likely it does not! Most likely we will need an optimization of the number of number of turns. Theoretically, this can be done from the condition

$$\frac{\partial P_{\xi(ICT)}(n)}{\partial(n)} = 0, \quad \rightarrow \quad \frac{\partial P_{\xi(core)}(n)}{\partial(n)} = -2\frac{\partial P_{\xi(cu)}(n)}{\partial(n)}. \tag{5.106}$$

Solution of Equation 5.106 yields the optimal number of turns

$$n = n_{opt}. \tag{5.107}$$

Equation 5.106 cannot be solved analytically. We can use a numerical tool such as Excel Solver to solve Equation 5.106. An even easier way to find the optimal number of turns is to plot the total losses versus the number of turns, Equation 5.105, and graphically find n_{opt} that gives minimum losses $P_{\xi(ICT)min}$.

Remark: The optimization objective can be different from minimization of total losses. It could be minimum core losses at a given maximum winding temperature at full load, or it could be the minimum losses at a given load profile, and so on.

5.6.1.4   The ICT Temperature Estimation

Once the ICT core and winding have been selected, the next design step is to confirm the temperature rise at different operating conditions. The ICT temperature is determined by the heat generated in the ICT and the capability of the ICT to evacuate the heat generated. The heat is generated by two mechanisms: the core and winding losses. The loss mechanisms have already been discussed in the previous section.

The ICT core and winding steady state temperatures can be estimated as

$$\theta_{core} = R_{cc} P_{\xi(core)} + R_{cw} P_{\xi(w)} + \theta_{amb},$$
$$\theta_w = R_{ww} P_{\xi(cw)} + R_{cw} P_{\xi(core)} + \theta_{amb}, \tag{5.108}$$

where
R_{cc} and R_{ww} are the ICT thermal self resistance, (K/W) and
R_{wc} is the mutual thermal resistance, (K/W).

The ICT thermal-self resistance determines the contribution of the core losses to the core temperature and the winding losses to the winding temperature. The mutual thermal resistance determines the contribution of the core losses to the winding temperature and the winding losses to the core temperature. The thermal resistances strongly depend on the ICT geometry core and winding geometry, cooling method, and temperature [24]. As this topic is beyond the scope of this section, it will not be discussed in detail. Here we will assume the thermal resistances R_{cc}, R_{ww}, and R_{wc} are known as design parameters. Thermal modeling of magnetic devices will be given in some more detail in Section 5.8.3.

5.6.1.5   The ICT Design Summary

The ICT design procedure is summarized in a flow chart depicted in Figure 5.32.

5.6.2   The Filter Inductor Design

5.6.2.1   The Inductor Initial Sizing

The initial selection of the filter inductor is based on the AP factor (Equation 5.84), [23, 24]. The process is basically the same as the design process for the ICT. To develop

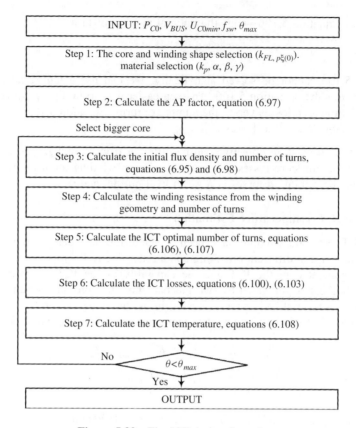

Figure 5.32   The ICT design flow chart

the AP factor of the inductor, we will start from the core peak flux density. The inductor flux density is

$$B_{peak} = \frac{L_0(i_{0(peak)})i_{0(peak)}}{nA_E},$$ (5.109)

where $i_{0(peak)}$ is the peak of the inductor current and $L_0(i_{0(peak)})$ is the inductance at the peak current. From Equation 5.110 we have the core cross-section A_E,

$$A_E = \frac{L_0(i_{0(peak)})i_{0(peak)}}{nB_{peak}},$$ (5.110)

where n is the number of turns. The winding window is

$$A_w = \frac{nI_{RMS}}{Jk_{FL}},$$ (5.111)

where

I_{RMS} is the winding RMS current,
J is the current density, and
k_{FL} is the winding coefficient.

The factor k_{FL} depends on the wire profile and winding technique. If litz wire is used for the winding, the coefficient k_{FL} is very low, 0.2–0.4. If a flat wire or a foil is used for the inductor winding, the coefficient k_{FL} can be much higher, in the range 0.6–0.8.

Substituting Equations 5.26, 5.110, and 5.111 into Equation 5.84 yields the inductor AP factor,

$$AP = \frac{L_0(i_{0(peak)})i_{0(peak)}}{B_{peak}} \frac{\sqrt{i_{C0}^2 + \Delta i_{0(RMS)}^2}}{Jk_{FL}}, \tag{5.112}$$

where i_{C0} is the ultra-capacitor current Equations 5.1 and 5.2, and $\Delta i_{0(RMS)}$ is the current RMS ripple Equation 5.18.

The discussion about the current ripple and its influence on the AP factor, which was conducted in Section 5.6.1.2 and Equations 5.88–5.90, applies also to the inductor design. Simply speaking, we can neglect the current ripple and assume the converter operates at a minimum duty cycle $d_{min} = 0.5$. Hence, the AP factor is

$$AP = \frac{L_0(i_{0(peak)})i_{0(peak)}}{B_{peak}} \frac{i_{C0(d=0.5)}}{Jk_{FL}}. \tag{5.113}$$

Now we have to select and compute an additional three parameters; peak flux density B_{peak}, the current density J, and the inductance L_0.

Let's start with the flux density. In contrast to the ICT, where the flux density is pure ac symmetrical flux density, the flux density of the inductor core is predominantly a dc component with a small ac component. Hence, the peak flux density is mainly determined by the saturation flux density, not by the core losses.

The current density depends on the core size. We can use the same approach used for the ICT design, and define the current density versus the AP factor as given in Equation 5.91.

The remaining issue is to compute the filter inductance at specified conditions. In previous analysis in Section 5.5.3.2 we assumed that the inductance L_0 is constant (current independent). This is a very rough approximation because the inductance may strongly vary with the inductor current. This is reflected in the current ripple, which is further reflected in the the filter design and performance. How we can deal with this issue? There are two approaches, namely a conservative one and an optimized one.

In the conservative approach, we can assume the inductance is constant but equal to minimum inductance at the nominal current

$$L_0 = L_0(I_{C0}) = Const. \tag{5.114}$$

Using this minimum value, we will slightly over-size filter inductor L_0.

In an optimized approach, we can compute the current ripple as a function of the duty cycle and the current-dependent inductance. Also, let the inductance be current-dependent inductance

$$L_0 = L_0(i_{C0}), \tag{5.115}$$

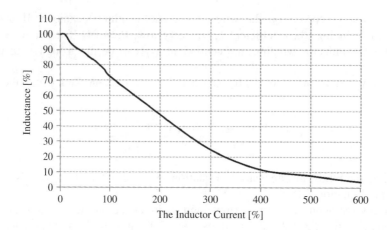

Figure 5.33 The inductance versus current of a typical iron-powder core inductor

where $L_0(i_{C0})$ is a strictly decreasing function of the ultra-capacitor current i_{C0}. Figure 5.33 shows the inductance versus current of a typical iron-powder core inductor.

The current-dependent inductance can be approximated by a linear equation

$$L_0 = L_{max}\left(1 - \frac{k}{I_{C0(n)}}|i_0|\right),\tag{5.116}$$

where L_{max} is the maximum inductance in the no-current condition, k is a coefficient, and $I_{C0(n)}$ is the nominal current. The coefficient is defined from a condition that the inductance at the nominal current is 70% of the no-load inductance. Thus, $k = 0.3$.

The ultra-capacitor nominal current is defined as the charge/discharge current at minimum voltage U_{C0min}.

$$I_{C0(n)} = \frac{P_{C0}}{u_{C0\,min}} = \frac{P_{C0}}{V_{BUS}\,d_{min}}.\tag{5.117}$$

Substituting Equation 5.116 into Equations 5.45 and 5.47 yields

$$\Delta i_0(d) = \left(\frac{V_{BUS}}{4f_{SW}L_{max}\left(1 - \frac{k}{I_{C0(n)}}|i_0|\right)}\right)\frac{1}{N^2}$$

$$\times 4[(Nd - floor(Nd)) - (Nd - floor(Nd))^2].\tag{5.118}$$

If the converter is loaded with constant current $i_{C0} = $ Const, it is obvious that maximum current ripple is achieved at maximum current and the critical duty cycle Equation 5.46. However, if the converter is loaded with constant power P_{C0}, it is not so obvious what the critical operating point is.

Substituting Equations 5.74 and 5.117 into Equation 5.118 yields the current ripple versus the conversion power P_{C0}, duty cycle d, and number of interleaved cells N,

$$\Delta i_0(d) = \left(\frac{V_{BUS}}{4f_{SW}L_{max}}\right)\frac{4}{N^2}\left(\frac{d}{d - k\dfrac{P_{C0}}{P_{C0(n)}}d_{min}}\right)$$

$$\times [(Nd - floor(Nd)) - (Nd - floor(Nd))^2]$$

$$= \left(\frac{V_{BUS}}{4f_{SW}L_{max}}\right)K_{\Delta i}(d), \tag{5.119}$$

where the ripple coefficient that takes into account the inductance variation is

$$K_{\Delta i}(d) = \frac{4}{N^2}\left(\frac{d}{d - k\dfrac{P_{C0}}{P_{C0(n)}}d_{min}}\right)[(Nd - floor(Nd)) - (Nd - floor(Nd))^2]. \tag{5.120}$$

Figure 5.34 shows the current ripple coefficient $K_{\Delta i}$ versus the duty cycle d and the number of interleaved converters N. The minimum duty cycle is $d_{min} = 0.5$ and the inductance coefficient is $k = 0.3$.

What is the critical duty cycle that gives the maximum current ripple? In the case of a linear inductor L_0, the maximum of the current ripple is given by Equation 5.47 and the critical duty cycle is defined by Equation 5.46. In the case of a nonlinear inductor, it is not evident what the critical case is.

From the condition

$$\frac{\partial \Delta i_0(d)}{\partial d} = \left(\frac{V_{BUS}}{4f_{SW}L_{max}}\right)\frac{\partial K_{\Delta i}(d)}{\partial d} = 0, \tag{5.121}$$

solution of Equation 5.121 yields the critical duty cycle d_{CR} that gives the maximum of the current ripple. However, solving Equation 5.121 can be difficult and impractical.

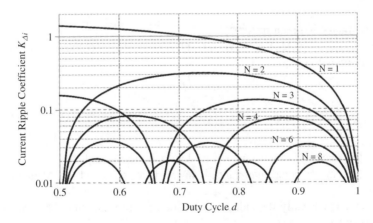

Figure 5.34 The output current ripple coefficient versus duty cycle and number of interleaved cells. The filter inductor is a nonlinear inductor. Minimum duty cycle $d_{min} = 0.5$ and inductance coefficient $k = 0.3$, Equation 5.116

It could be proven that the difference between the solutions of Equations 5.121 and 5.46 is not significant. Hence, we can assume an ideal inductor with constant inductance and compute the critical duty cycles from Equation 5.46

$$d_{CR} \cong \frac{1}{N} \left(floor \left(\frac{1}{d_{min}} \right) + \frac{1}{2} \right), \tag{5.122}$$

where d_{min} is the minimum operating duty cycle. Substituting the solution of Equation 5.122 into Equation 5.119 yields the maximum of the current ripple

$$\Delta i_{0\,max} = \left(\frac{V_{BUS}}{4 f_{SW} L_{max}} \right) K_{\Delta i}(d_{CR}). \tag{5.123}$$

Now we can compute the inductance L_{max}

$$L_{max} = \left(\frac{V_{BUS}}{4 f_{SW} \Delta i_{0\,max}} \right) K_{\Delta i}(d_{CR}). \tag{5.124}$$

However, how do we select the maximum current ripple? This is a question of system multi-objective optimization. Since this subject is outside the scope of the book, we will not discuss it here. From experience, we can say that the current ripple is usually in the range 20–40% of the rated current $I_{C0(n)}$.

Substituting Equations 5.91, 5.116, and 5.124 into Equation 5.113 yields the inductor AP factor

$$AP = 10^8 \left[\frac{\left(\frac{V_{BUS}}{4 f_{SW} \Delta i_{0\,max}} \right) K_{\Delta i}(d_{CR}) \left(1 - \frac{k}{I_{C0(n)}} i_{0(peak)} \right) i_{0(peak)} i_{C0(d=0.5)}}{B_{sat} k_{FL} 4.5} 10^{-6} \right]^{1.43}, \tag{5.125}$$

where the AP factor is given in (cm^4).

Now, having the core AP factor we can select the core from the catalog. The number of turns is computed as

$$n = \sqrt{\frac{L_{max}}{A_L}}, \tag{5.126}$$

where

L_{max} is the maximum inductance at no-load conditions and
A_L is the core selected inductance characteristic, so-called the Al factor (nH/n^2).

The A_L factor is basically the inductance of the one-turn inductor. The cross-section of the winding wire is

$$A_{cu} = \frac{A_w k_{FL}}{n}. \tag{5.127}$$

5.6.2.2 The Inductor Losses

The inductor total losses consist of the copper (winding) losses and the core losses. Both of them depend on the inductor instantaneous current that is given by

$$i_0(t) = I_{C0} + \frac{i_{C0}}{2} + \sqrt{2}\sum_{k=1}^{\infty} I_{0(RMS)(k)}\sin(kN\omega_{SW}t + \varphi_{(kN\omega_{SW})}),\tag{5.128}$$

where $\omega_{SW} = 2\pi f_{SW}$.

The converter switching frequency is f_{SW} and k is the harmonic order. The copper losses can be found in a general form

$$P_{\xi(cu)} = \sum_{k=0}^{n} I_{0RMS\,(k)}^2 R_{0(k)},\tag{5.129}$$

where $R_{0(k)}$ is the inductor winding resistance that depends on the frequency.

One can define the core losses in a similar way,

$$P_{\xi(core)} \cong \sum_{k=0}^{\infty} I_{0RMS\,(k)}^2 R_{C\,(k)},\tag{5.130}$$

where $R_{C(k)}$ is the core equivalent resistance that models the core losses as a function of the frequency [17, 42].

Here we have to highlight that the core losses model Equation 5.130 is an approximation of the real core losses.

The inductor losses model Equations 5.129 and 5.130 take into account the harmonics of the inductor current. To simplify computation, without neglecting the effect of higher harmonics, one can substitute the current ripple by an equivalent sinusoidal current having the same RMS value and frequency as the total current ripple. Such an approach is not completely correct, but it is sufficient for the simplified calculation we need.

$$i_0(t) = i_{C0} + \Delta i_0(t) \cong i_{C0} + \sqrt{2}\Delta I_{0(RMS)}\sin(N\omega_{SW}t + \psi).\tag{5.131}$$

The RMS value of the equivalent sinusoidal current ripple is given by Equation 5.49

$$\Delta I_{0(RMS)} = \frac{\Delta i_0}{2\sqrt{3}}.\tag{5.132}$$

Substituting Equation 5.119 into Equation 5.132 yields

$$\Delta I_{0(RMS)} = \left(\frac{V_{BUS}}{4f_{SW}L_{\max}}\right)\frac{1}{2\sqrt{3}}K_{\Delta i}(d).\tag{5.133}$$

The inductor average current (dc component) is defined by Equation 5.74. Substituting Equations 5.74, 5.131, and 5.133 into Equations 5.129 and 5.130 yields the inductor total losses,

$$P_{\xi(L0)} \cong R_{0(DC)}\left(\frac{P_{C0}}{V_{BUS}d}\right)^2 + (R_{0(Nf_{SW})} + R_{C(Nf_{SW})})\left(\frac{V_{BUS}}{4f_{SW}L_{\max}}\right)^2\frac{1}{12}[K_{\Delta i}(d)]^2\tag{5.134}$$

where

$R_{0(DC)}$ is the inductor resistance at low frequency and

$R_{0(NfSW)}$ and $R_{C(NfSW)}$ are the inductor winding resistance and the core equivalent
resistance at N times switching frequency.

For more details on the determination of the core equivalent resistance $R_{C(NfSW)}$, please
have a detailed look at [17, 42].

5.6.2.3   The Inductor Temperature Estimation

Once the inductor core and winding have been selected, the next design step is to confirm
the temperature rise at different operating conditions. The inductor temperature is determined by the heat generated in the inductor and the capability of the inductor to evacuate
the heat generated. The heat is generated by two mechanisms; the core and winding losses.
The loss mechanisms have been already discussed in the previous section.

The inductor core and winding steady state temperatures can be estimated as

$$\theta_{core} = R_{cc} P_{\xi(core)} + R_{cw} P_{\xi(w)} + \theta_{amb},$$
$$\theta_{w} = R_{ww} P_{\xi(cw)} + R_{cw} P_{\xi(core)} + \theta_{amb}, \qquad (5.135)$$

where

R_{cc} and R_{ww} are the inductor thermal self-resistance, (K/W) and
R_{wc} is the inductor mutual thermal resistance, (K/W).

The inductor thermal self-resistance determines the contribution of the core losses to the
core temperature and the winding losses to the winding temperature. The mutual thermal
resistance determines the contribution of the core losses to the winding temperature and
the winding losses to the core temperature. The thermal resistances strongly depend on
the inductor geometry core and winding geometry, cooling method, and temperature [24].
As this topic is beyond the scope of this section, it will not be discussed in detail. Here
we will assume the thermal resistances R_{cc}, R_{ww}, and R_{wc} are design parameters. Thermal
modeling of magnetic devices will be covered in some more detail in Section 5.8.3.

5.6.2.4   The Inductor Design Summary

The inductor design summary is given by the flow chart depicted in Figure 5.35.

5.6.3   DC Bus Capacitor Selection

The dc bus capacitor is selected based on three criteria: (i) the capacitor RMS current
and life span, (ii) the dc bus voltage ripple, and (iii) the dc bus voltage controllability.

5.6.3.1   RMS Current and Life Span

The dc bus capacitor RMS current was defined by Equations 5.72 and 5.75. Having
computed the RMS current, we can select the capacitor with sufficient current capability.

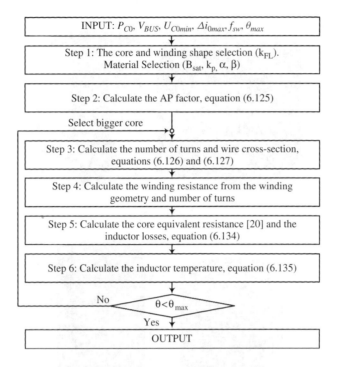

INPUT: P_{C0}, V_{BUS}, U_{C0min}, Δi_{0max}, f_{sw}, θ_{max}

Step 1: The core and winding shape selection (k_{FL}).
Material Selection (B_{sat}, k_p, α, β)

Step 2: Calculate the AP factor, equation (6.125)

Select bigger core

Step 3: Calculate the number of turns and wire cross-section,
equations (6.126) and (6.127)

Step 4: Calculate the winding resistance from the winding
geometry and number of turns

Step 5: Calculate the core equivalent resistance [20] and the
inductor losses, equation (6.134)

Step 6: Calculate the inductor temperature, equation (6.135)

No $\theta < \theta_{max}$

Yes

OUTPUT

Figure 5.35   The inductor design flow chart

The selection process depends on the capacitor technology. Currently, two technologies of capacitors are used in power conversion applications: film capacitors and electrolytic capacitors (Figure 5.36).

Film capacitors are characterized by high current capability and a long operating time. The selection criterion is $I_{CBUS(RMS)} < I_N$, where I_N is the capacitor rated current given by the capacitor manufacturer [27]. If the current rating of selected capacitor is not sufficient, M_{BUS} capacitors have to be parallel connected, where M_{BUS} is

$$M_{BUS} = 1 + floor\left(\frac{I_{CBUS(RMS)}}{I_N}\right). \tag{5.136}$$

The situation with electrolytic capacitors is little more complicated. In fact, electrolytic capacitors are characterized by an operating life span that strongly depends on the rated to operating RMS current and operating temperature [28]. Hence, the capacitor rated current has to be selected according to the life expectancy. The selection process is described step by step.

Step 1: The capacitor RMS current

Let's assume that the dc bus capacitor current is a periodic current defined as

$$i_{CBUS(t)} = \sum_{n=1}^{\infty} \sqrt{2} I_n \sin(n\omega t + \psi_n). \tag{5.137}$$

(a)

(b)

Figure 5.36 (a) DC bus high current film capacitors and (b) high power electrolytic capacitors [27, 28]. Copyright EPCOS AG 2013, with permission

The capacitor equivalent RMS current is computed as

$$I_{CBUS(RMS)} = \sqrt{\sum_{n=1}^{\infty} \left(\frac{I_n}{k_n}\right)^2},$$ (5.138)

where

I_n is the nth harmonic RMS current and
k_n is a coefficient given by the capacitor manufacturer [28].

Please note from Figure 5.37 that the frequency scaling coefficient is almost constant at frequencies above a few kilohertz. Therefore, we can assume that the dc bus capacitor current is a single harmonic at N times switching frequency and RMS computed by Equations 5.72 and 5.75. Substituting Equation 5.72 or 5.75 into Equation 5.138 yields

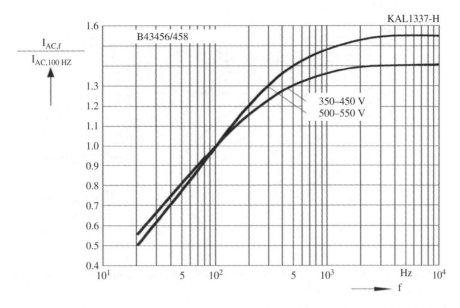

Figure 5.37 Scaling factor of large EPCOS electrolytic capacitors [28]. Copyright EPCOS AG 2013, with permission

$$I_{CBUS(RMS)} = I_0 \frac{K_{RMS(I)}}{k_{(Nf_{SW})}} \quad \text{or} \quad I_{CBUS(RMS)} = \frac{P_{C0}}{V_{BUS}} \frac{K_{RMS(P)}}{k_{(Nf_{SW})}}. \quad (5.139)$$

Step 2: The capacitor life span and loading factor

The dc bus capacitor life span is defined by three parameters, namely the ambient temperature, the capacitor losses, and the applied voltage [28]. Expected life can be defined as

$$\lambda_T = \lambda_{T0} 2^{\frac{T_0 - T_a}{10K}} A^{(1-K_L^2)\frac{\Delta T_0}{10K}} \left(\frac{U_C}{U_N} \right)^{-n}, \quad (5.140)$$

where λ_{T0} is the life span at referent conditions, T_0 is the maximum rated temperature, T_a is the capacitor ambient temperature, ΔT_0 is the capacitor core temperature rise, and A is a coefficient. The coefficient K_L is the capacitor loading factor

$$K_L = \frac{I_{CBUS(RMS)}}{I_N}, \quad (5.141)$$

where

$I_{CBUS(RMS)}$ is the capacitor RMS current scaled to the referent frequency Equation 5.139 and

I_N is the capacitor rated current.

The capacitor life span characteristic is usually given as a plot of the loading coefficient versus the capacitor temperature and life expectancy. An example is given in

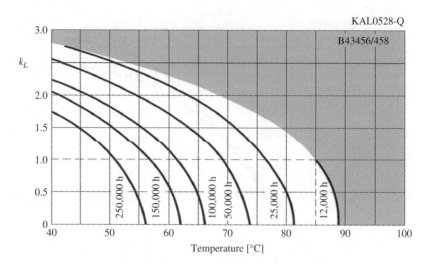

Figure 5.38 Life span versus current load factor and operating temperature [28]. Copyright EPCOS AG 2013, with permission

Figure 5.38. The vertical axis is the loading coefficient K_L. Two values are indicated. The first one denoted as $(I_{AC}/I_{AC,R})$ is the coefficient when the capacitor is air-cooled. The second one $(I_{AC}/I_{AC,R(B)})$ is the coefficient when forced base-cooling is applied.

Let the capacitor life expectancy be given as the design criterion. The capacitor temperature and cooling are also defined. From the plot we can determine the coefficient K_L.

Step 3: The capacitor cell selection and parallel/series combination

Let's select the capacitor cell, where the cell is characterized by the rated current I_N and voltage U_N. Now, having computed the loading factor and having selected the capacitor cell, we can compute the number of parallel connected cells

$$M_{BUS} = 1 + floor\left(\frac{I_{CBUS(RMS)}}{I_N K_L}\right). \tag{5.142}$$

In the case that the dc bus voltage is higher than the cell rated voltage, N_{BUS} cells have to be series connected to achieve the required voltage rating. The number of series connected cells is

$$N_{BUS} = 1 + floor\left(\frac{V_{BUS}}{U_N}\right). \tag{5.143}$$

Let the capacitance of the selected cell be $C_{BUS(cell)}$. The total dc bus capacitance C_{BUS} is

$$C_{BUS} = C_{BUS(cell)}\frac{M_{BUS}}{N_{BUS}}. \tag{5.144}$$

5.6.3.2 The DC Bus Voltage Ripple

Let the converter be loaded by constant power P_{C0}. The dc bus voltage ripple has already been defined by Equation 5.82. From the condition

$$\frac{\partial \Delta v_{BUS}(d)}{\partial d} = \left(\frac{\sqrt{2}}{C_{BUS}\pi f_{SW}} \frac{P_{C0}}{V_{BUS}} \right) \frac{\partial K_{\Delta V(P)}(d)}{\partial d} = 0, \tag{5.145}$$

we can compute the critical duty cycle d_{CR} that gives the maximum of the voltage ripple. However, Equation 5.145 cannot be solved analytically in closed form. Hence, an approximation is necessary. We can assume that the critical duty cycles are defined by Equation 5.46. Hence, the critical duty cycle that gives the absolute maximum of the voltage ripple is

$$d_{CR} \cong \frac{1}{N} \left(floor \left(\frac{1}{d_{min}} \right) + \frac{1}{2} \right), \tag{5.146}$$

where d_{min} is the minimum operating duty cycle. Substituting Equation 5.146 into Equation 5.82 yields

$$\Delta v_{BUS\,max} = \frac{\sqrt{2}}{C_{BUS}\pi f_{SW}} \frac{P_{C0}}{V_{BUS}} K_{\Delta V(p)}(d_{CR}). \tag{5.147}$$

The ripple coefficient $K_{\Delta V(P)}$ is

$$K_{\Delta V(P)}(d) = \frac{K_{RMS(p)}(d)}{N}. \tag{5.148}$$

where $K_{RMS(p)}$ is the RMS current coefficient defined by Equation 5.75.
Now we can select the dc bus capacitance

$$C_{BUS} = \frac{\sqrt{2}}{\Delta v_{BUS\,max}\pi f_{SW}} \frac{P_{C0}}{V_{BUS}} K_{\Delta V(P)}(d_{CR}). \tag{5.149}$$

The number of parallel connected cells is

$$M_{BUS} = 1 + floor \left(N_{BUS} \frac{C_{BUS}}{C_{BUS\,(cell)}} \right). \tag{5.150}$$

Usually, the RMS currant criterion and the dc bus voltage ripple criterion will give a different number of parallel connected capacitors, Equation $5.142 \neq 5.150$. To select the dc bus capacitor that satisfies both conditions, the number of paralleled capacitors has to be computed as

$$M_{BUS} = \max \left\{ \left[1 + floor \left(N_{BUS} \frac{C_{BUS}}{C_{BUS\,(cell)}} \right) \right], \left[1 + floor \left(\frac{I_{CBUS\,(RMS)}}{I_N K_L} \right) \right] \right\}. \tag{5.151}$$

5.6.3.3 The dc Bus Capacitor Losses and Temperature

The dc bus capacitor losses have been discussed in detail in Section 5.5.3.4. Equation 5.152 is the dc bus capacitor losses at the constant current and constant power condition.

$$P_{\xi(CBUS)} = ESR(f_B)\left(I_0 \frac{K_{RMS(i)}}{k_{(Nf_{SW})}}\right)^2,$$

$$P_{\xi(CBUS)} = ESR(f_B)\left(\frac{P_{C0}}{V_{BUS}} \frac{K_{RMS(p)}}{k_{(Nf_{SW})}}\right)^2. \tag{5.152}$$

The current ripple coefficients $K_{RMS(i)}$ and $K_{RMS(p)}$ are given by Equations 5.72 and 5.75. The current scaling factor is $k_{(Nf_{SW})}$. An example of the current scaling factor is given in Figure 5.37. For more details of the current scaling factor, please see Section 5.8.4.1 at the end of this chapter.

Let the dc bus capacitor be composed of N_{BUS} series connected and M_{BUS} parallel connected cells. The dc bus capacitor resistance is

$$ESR = ESR_{(cell)} \frac{N_{BUS}}{M_{BUS}}, \tag{5.153}$$

where $ESR_{(cell)}$ is the resistance of a cell.

The capacitor core and case temperature is computed by substituting Equation 5.152 into Equation 5.259.

$$\theta_{case} = ESR(f_B)\left(\frac{P_{C0}}{V_{BUS}} \frac{K_{RMS(p)}}{k_{(Nf_{SW})}}\right)^2 R_{ca\_amb} + \theta_{amb},$$

$$\theta_{core} = ESR(f_B)\left(\frac{P_{C0}}{V_{BUS}} \frac{K_{RMS(p)}}{k_{(Nf_{SW})}}\right)^2 (R_{co\_ca} + R_{ca\_amb}) + \theta_{amb}, \tag{5.154}$$

where R_{ca_amb} is the case to ambient thermal resistance, Equation 5.257.

Thermal modeling of electrolytic capacitors is fully addressed in Section 5.8.4.

5.6.3.4 The dc Bus Voltage Controllability

The dc bus voltage controllability is another important criterion for selecting the dc bus capacitor.

5.6.3.5 The Capacitor Selection Summary

The dc bus capacitor selection process can be summarized as in Figure 5.39.

5.6.4 Output Filter Capacitor Selection

What is the role of the output filter capacitor C_0? Is it desirable to have it or is it a must? The role of the filter capacitor is to shunt a part of the current ripple Δi_0 and

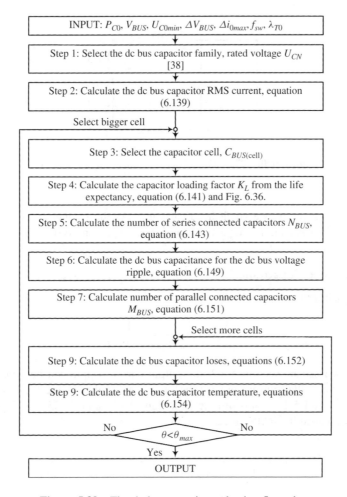

Figure 5.39 The dc bus capacitor selection flow chart

prevent a large current ripple flowing into the ultra-capacitor. The filter capacitor is not strictly necessary. However, since the ultra-capacitor resistance R_{C0} can be significant, the additional losses and heating of the ultra-capacitor may become an issue. To reduce the ultra-capacitor current ripple, a filter capacitor C_0 is connected in parallel with the converter output.

Figure 5.40 shows a circuit diagram of the ultra-capacitor and output filter. The inductance $L_{0\zeta}$ is the ultra-capacitor connection stray inductance. The ultra-capacitor resistance is R_{C0}. As already discussed in Chapter 2, the ultra-capacitor resistance strongly depends on the frequency. The frequency of interest is N times the switching frequency. Hence, the ultra-capacitor resistance is the resistance at that frequency.

What about the capacitance of the ultra-capacitor? It is indicated that the capacitance is $C_{C0} \to \infty$. Is this assumption correct? It depends on the ultra-capacitor size. It may be

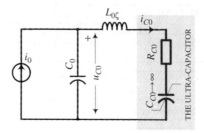

Figure 5.40   The output filter equivalent circuit diagram

that the application requires a small ultra-capacitor. As the frequency of interest is very high from the ultra-capacitor perspective, the capacitance can be quite low.

5.6.4.1   The Filter Capacitance

The ultra-capacitor current ripple is

$$\Delta i_{C0}(s) = \Delta i_0(s)\frac{1}{s^2 L_{0\xi} + s R_{C0} C_0 + 1}. \tag{5.155}$$

As already discussed in Section 5.6.2, the output current ripple can be approximated by an equivalent sinusoidal current having the same RMS value as the total current ripple, Equations 5.131 and 5.133. Substituting Equations 5.131 and 5.133 into Equation 5.155 yields the ultra-capacitor RMS current ripple

$$\Delta I_{C0(RMS)} = \Delta I_{0(RMS)}\frac{1}{\sqrt{(1 - (N\omega_{SW})^2 L_{0\xi} C_0)^2 + (N\omega_{SW} R_{C0} C_0)^2}} = \Delta I_{0(RMS)} A_t. \tag{5.156}$$

Let the filter attenuation A_t be given as a design requirement. The capacitance of the filter capacitor is computed from Equation 5.162. We can distinguish two extreme cases. The first one is an ideal case with neglected stray inductance $L_{0\zeta} \cong 0$. Under such a condition, we compute the filter capacitance as

$$C_0 = \sqrt{\left(\frac{1}{A_t^2} - 1\right)}\frac{1}{N\omega_{SW} R_{C0}}, \tag{5.157}$$

where
   N is the number of interleaved converters and
   ω_{SW} is the single cell switching frequency.

The second extreme case is the filter parallel resonance that occurs when the stray inductance is

$$L_{0\xi} = \frac{1}{(N\omega_{SW})^2 C_0}. \tag{5.158}$$

In this case the capacitance C_0 required to properly attenuate the current ripple is

$$C_0 = \frac{1}{A_t} \frac{1}{N\omega_{SW} R_{C0}}. \tag{5.159}$$

Substituting Equation 5.159 into Equation 5.158 yields the critical inductance

$$L_{0\xi(CR)} = \frac{A_t R_{C0}}{N\omega_{SW}}. \tag{5.160}$$

If the capacitance Equations 5.157 and 5.159 are out of range of the available capacitors, M_{C0} capacitors have to be connected in parallel to achieve the required capacitance Equation 5.157 or 5.159. The number of parallel connected capacitors is

$$M_{C0} = 1 + floor\left(\frac{C_0}{C_{0(cell)}}\right), \tag{5.161}$$

where $C_{0(cell)}$ is the capacitance of the selected capacitor cell.

5.6.4.2 RMS Current Ripple

The capacitor RMS current is computed from Figure 5.40

$$\Delta I_{C(RMS)} = \Delta I_{0(RMS)} \frac{N\omega_{SW} R_{C0} C_0}{\sqrt{1 + (N\omega_{SW} R_{C0} C_0)^2}} = \Delta I_{0(RMS)}\sqrt{1 - (A_t(\omega))^2}, \tag{5.162}$$

where the output current RMS ripple is given by Equations 5.119 and 5.133. Now we can select the capacitor with current capability according to Equation 5.162. If the selected capacitor has no required current capability, M_{C0} capacitors have to be connected in parallel to achieve the required current capability. The number of parallel connected capacitors is

$$M_{C0} = 1 + floor\left(\frac{\Delta I_{C(RMS)}}{I_N}\right), \tag{5.163}$$

where I_N is the capacitor rated current capability.

Finally, we have to select the number of paralleled capacitors from the worst case scenario of Equations 5.161 and 5.163

$$M_{C0} = max\left\{\left[1 + floor\left(\frac{C_0}{C_{0(cell)}}\right)\right], \left[1 + floor\left(\frac{\Delta I_{C(RMS)}}{I_N}\right)\right]\right\}. \tag{5.164}$$

5.6.5 Power Semiconductor Selection

5.6.5.1 The Design and Selection Objectives

Once we have selected the passive components of the converter (ICT, inductor, and capacitors), the remaining design step is the selection of power semiconductors. The objective of this design step is to properly select power semiconductors that fit into the converter specification. This includes selection of the semiconductor technology and family, voltage and current rating, and design of the cooling system. The power semiconductor of interest is the single switch/diode in the low power range and dual switch/diode module in the medium and high power range. Figure 5.41 shows state of the art single (low power) and dual (high power) IGBT/FWD devices.

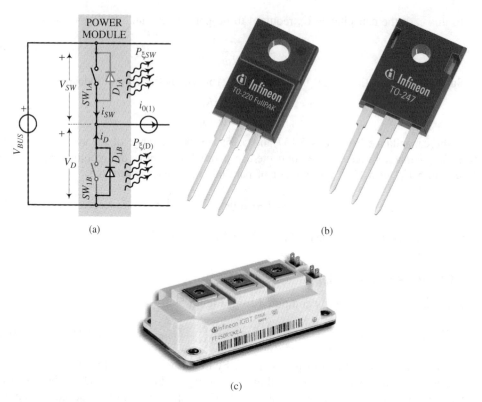

Figure 5.41 (a) Dual switch cell as the building block of interface dc–dc converters. (b) Single IGBT/FWD for low power applications [29]. (c) Dual IGBT/FWD module for high power applications [29]. Copyright Infineon Technologies AG, with permission

5.6.5.2 A Few Words about Advanced Semiconductor Switches and Diodes

Figure 5.42 compares the most often used semiconductor switches. The conduction losses (on-state voltage V_{CON}) and switching losses of three types of active switches are compared; Si IGBT, Si MOSFET, and SiC JFET are compared at two different voltage ratings: 600 and 1200 V. Regarding overall performance, switching, and conduction, in both voltage ranges the SiC JFET is superior. The 600 V rated MOSFET and IGBT have similar conduction performances, while the MOSFET is superior with regarding to switching performance. In the 1200 V range, the MOSFET has superior switching performance in comparison to the IGBT. The situation for conduction performance is the total opposite: the IGBT is superior. Here, we have to highlight that the technology of 600 and 1200 V MOSFETs is different. In the 600 V range, super-junction technology is predominant [30]. This technology offers a significant improvement of the switch conduction performance. However, in the 1200 V range there is no possibility for implementing super-junction technology.

Two types of switching diodes, Si PiN and SiC SBD, are compared at the 600 and 1200 V rating. Note that the switching losses of SiC SBD are lower than those of the Si

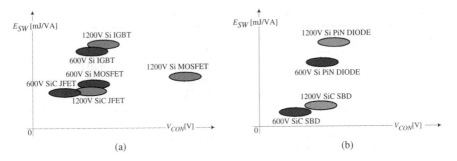

Figure 5.42   Advanced power semiconductor switch performances versus voltage rating. (a) Active switches and (b) fast diodes

PiN diode. The difference in on-state voltage is less significant. The SiC PiN diode is not considered because such a type of diode is applicable in very high voltage, rather than in low voltage, applications.

5.6.5.3   The Switch Voltage Rating and Technology

The very first step in the power semiconductor selection is the determination of the device voltage rating. Once the voltage rating has been determined, the next step is selection of the device type and technology according to the voltage rating determined previously.

The switches' and diodes' voltage rating is defined by the switch transient blocking voltage

$$V_{SW(\max)} = V_{D(\max)} = V_{BUS(\max)} + \Delta v, \tag{5.165}$$

where ΔV is the commutation over-voltage [30],

$$\Delta v = k_R L_\xi \frac{di_{SW}}{dt} + V_{FDY} \cong k_R L_\xi \frac{0.8 I_{SW}}{t_F} + V_{FDY}. \tag{5.166}$$

A factor k_R takes into account the resonance of the converter dc bus. This coefficient can be in the range 1–1.5, depending on the dc bus structure. V_{FDY} is the forward recovery voltage of the freewheeling diode. It can be as high as 50 V [30]. L_ξ is the commutation inductance and t_F is the switch current fall time [30].

In the low voltage range, up to 250 V, SuperJunction MOSFET is the switch of choice. The diode is the intrinsic body diode.

In the voltage range from 250 to 450 V, two switch technologies could be used. It could be the 600–700 V rated SuperJunction MOSFET or the 600–650 V rated IGBT. The first one offers better switching and similar conduction performance compared to the IGBT [30]. However, the intrinsic FWD of the SuperJunction MOSFET has inferior characteristics in comparison to the PiN diode used in the IGBT module. This issue could be resolved by the use of some complex topology, but it would cause additional expenses and losses. Thus, 600–650 V IGBT is the switch of choice. The diode is a fast soft recovery PiN diode (Table 5.4).

In the voltage range above 450 V, IGBT is definitely the switch of choice. The diode is a fast soft recovery PiN diode.

Table 5.4 The semiconductor's technology depending on the dc bus voltage

DC bus voltage	Up to 250 V	250–450 V	450–900 V	900–1400 V
Switch	MOSFET	600–650 V IGBT 600–650 V Super-Junction MOSFET[a]	1200 V IGBT	1700 V IGBT
Diode	Body diode	600–650 V 600–650 V[a]	1200 V	1700 V

[a]Limited performances due to intrinsic body diode.

5.6.5.4 Current Rating

The device current rating is determined by the ultra-capacitor maximum current and the current ripple. From Equation 5.34 we have the switch/diode current maximum

$$i_{SW\,max}(d) = i_{D\,max}(d) = \frac{i_{C0}(d)}{N} + \frac{1}{2}\frac{\Delta i_0(d)}{N},\tag{5.167}$$

where

N is the number of interleaved cells and

$i_{C0}(d)$ and $\Delta i_0(d)$ are the ultra-capacitor current and the output current ripple as functions of the duty cycle.

Substituting Equations 5.74 and 5.119 into Equation 5.167 yields

$$i_{SW\,max}(d) = i_{D\,max}(d) = \frac{1}{N}\left[\frac{P_{C0}}{V_{BUS}d} + \frac{1}{2}\left(\frac{V_{BUS}}{4f_{SW}L_{max}}\right)K_{\Delta i}(d)\right]$$

$$= \frac{1}{N}I_{C0(n)}\left[\frac{d_{min}}{d} + \frac{1}{2}\frac{\Delta i_{0\,max(0)}}{I_{C0(n)}}N^2K_{\Delta i}(d)\right],\tag{5.168}$$

where

$K_{\Delta i}$ is the current ripple coefficient defined in Equation 5.120,

$I_{C0(n)}$ is the ultra-capacitor nominal current Equation 5.117, and

$\Delta i_{0max(0)}$ is the maximum peak-to-peak current achieved at maximum inductance L_{max}.

$$\Delta i_{0\,max(0)} = \frac{V_{BUS}}{4f_{SW}L_{max}}\frac{1}{N^2}.\tag{5.169}$$

From the function extreme condition

$$\frac{\partial i_{SW\,max}(d)}{\partial d} = 0,\tag{5.170}$$

we have

$$\frac{2d_{min}I_{C0(n)}}{\Delta i_{0\,max(0)}N^2} = d^2\frac{\partial K_{\Delta i}(d)}{\partial d}.\tag{5.171}$$

The solution of Equation 5.171 is the critical duty cycle d_{CR} that gives the maximum of the device current. Substituting the solution of Equation 5.171 into Equation 5.168 yields the device's maximum current.

$$i_{SW\,max} = \frac{1}{N}I_{C0(n)}\left[\frac{d_{min}}{d_{CR}} + \frac{1}{2}\frac{\Delta i_{0\,max(0)}}{I_{C0(n)}}N^2 K_{\Delta i}(d_{CR})\right]. \tag{5.172}$$

Now we have defined the device's (the switch and diode) maximum current rating.

However, is this the end of the selection process? Is it enough to select the device current rating in a straightforward manner from Equation 5.172? In fact, no, it is not enough. From Equation 5.172 we cannot determine the exact current rating of the device. What we can do is an initial selection of the device. The main selection criterion for the device current rating is the device junction temperature under the worst case scenario. Based on experience, the initial selection of device current rating is

$$I_{SW(n)} = 2i_{SW\,max},$$
$$I_{D(n)} = 2i_{D\,max}. \tag{5.173}$$

Let's assume the cell output current $i_{0(1)}$ is positive as defined in Figure 5.41a. Also, for the sake of simplicity, let's assume the output current is smooth and ripple-free and defined from Equation 5.34 as

$$i_{0(1)}(t) = \frac{i_0(t)}{N} = \frac{i_{C0}(t)}{N} + \overbrace{\frac{\Delta i_0(t)}{N}}^{\cong 0} \cong \frac{i_{C0}(t)}{N}, \tag{5.174}$$

where N is the number of interleaved cells. The current ripple Δi_0 is neglected in the analysis. The dc bus voltage V_{BUS} is assumed as a constant or slowly changing voltage. The top switch SW_{1A} and bottom diode D_{1B} are periodically conducting the output current $i_{0(1)}$. The switch and diode currents are defined by the output current $i_{0(1)}$ and the switching function $s(t)$ as

$$i_{SW1A}(t) = i_{0(1)}(t) \cdot s(t), \quad i_{SW1B} = 0$$
$$i_{D1A} = 0, \quad i_{D1B}(t) = i_{0(1)}(t) \cdot (1 - s_1(t)). \tag{5.175}$$

5.6.5.5 Conversion Losses

The losses' origin and a methodology to calculate the losses under different conditions are intensively discussed in Section 5.7. In the following discussion we will call on equations derived in Section 5.7.

From Equations 5.6 and 5.224 we can compute the switch and diode moving average and RMS current as

$$I_{SW(AV)} = \frac{1}{T_{SW}}\int_t^{t+T_{SW}} i_{SW1A}(t)dt = i_{0(1)}(t) \cdot d(t),$$

$$I_{SW(RMS)} = \sqrt{\frac{1}{T_{SW}}\int_t^{t+T_{SW}} i_{SW1A}^2(t)dt} = i_{0(1)}(t) \cdot \sqrt{d(t)}, \tag{5.176}$$

$$I_{D(AV)} = \frac{1}{T_{SW}} \int_t^{t+T_{SW}} i_{D1A}(t)dt = i_{0(1)}(t) \cdot (1 - d(t)).$$

$$I_{D(RMS)} = \sqrt{\frac{1}{T_{SW}} \int_t^{t+T_{SW}} i_{D1A}^2(t)dt} = i_{0(1)}(t) \cdot \sqrt{1 - d(t)}. \tag{5.177}$$

Substituting Equations 5.176 and 5.177 into Equation 5.217 yields the switch and diode conduction losses

$$P_{SW(CON)} = \left[V_{sw(0)} \frac{i_{C0}(t)}{N} + r_{sw} \frac{i_{C0}^2(t)}{N^2} \right] \cdot d(t),$$

$$P_{D(CON)} = \left[V_{D(0)} \frac{i_{C0}(t)}{N} + r_D \frac{i_{C0}^2(t)}{N^2} \right] \cdot (1 - d(t)). \tag{5.178}$$

The cell current i_{C0} is defined in Equations 5.1 and 5.2, while the duty cycle is defined in Equations 5.42 and 5.43.

The switches' switching losses are

$$P_{SW(SW)} \cong f_{SW} \frac{(E_{ON} + E_{OFF})}{V_N I_N} V_{BUS} \frac{i_{C0}(t)}{N}, \tag{5.179}$$

where E_{ON} and E_{OFF} are commutation energy at rated voltage U_N and current I_N. The current $i_{C0}(t)$ is defined by Equations 5.1 and 5.2.

The diode commutation losses are

$$P_{D(SW)} \cong f_{SW} \frac{E_Q}{V_N I_N} V_{BUS} i_{C0}(t), \tag{5.180}$$

where E_Q is commutation energy at rated voltage U_N and current I_N.

5.6.5.6 Thermal Model and Temperature Rise

Thermal modeling and temperature management of power semiconductors are intensively discussed in Section 5.8.2. In the following discussion we will call on the equations derived in Section 5.8.2 and calculate the device junction temperature.

Before calculating the junction temperature, let's briefly describe the power semiconductor device and thermal model. The device we are dealing with is a basic switch cell, with a circuit diagram as depicted in Figure 5.43a. It could be a cell composed of two discrete switches/diodes or an integrated dual power module. The objective is to calculate the device junction temperature that is determined by the heat generated by device losses and the capability of the device to evacuate the heat generated. The losses have already been addressed in the previous section.

A simplified thermal model of a basic switch cell is depicted in Figure 5.43b. The devices losses are denoted by current source $P_{\xi(swA)}$ to $P_{\xi(DA)}$, while the junction-to-case, the case-to-heat sink, and the heat-sink-to-ambient thermal impedances are $Z_{j\_c(sw)}$, $Z_{j\_c(D)}$, Z_{c_hs}, and Z_{c_hs} respectively. For more detail, please see Section 5.8.2.

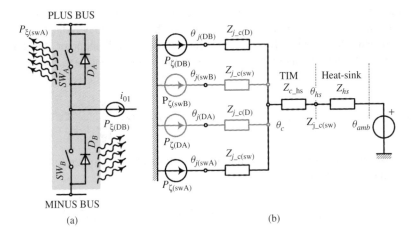

Figure 5.43 (a) Power semiconductor dual module and (b) simplified thermal model

The device junction-to-case temperature is

$$\Delta\theta_{j\_c}(s) = P_\xi(s)Z_{j\_c}(s) = P_\xi(s)\sum_{k=1}^{\infty}\frac{R_{j\_c(k)}}{1 + s\tau_{(k)}}, \tag{5.181}$$

where

P_ξ is the device loss, (W) and
Z_{j_c} is the device junction-to-case thermal impedance, (K/W).

In the case that the loss P_ξ is a step function (the losses are constant in time), the junction-to-case temperature response is a series of exponential functions

$$\Delta\theta_{j\_c}(t) = P_\xi\sum_{k=1}^{\infty}R_{j\_c(k)}\left(1 - e^{-\frac{t}{\tau_{(k)}}}\right). \tag{5.182}$$

If the losses are constant during a period that is significantly longer than the device time constant, the junction-to-case temperature is a steady state temperature, defined by the losses and total junction-to-case thermal resistance

$$\Delta\theta_{j\_c} = P_\xi\sum_{k=1}^{\infty}R_{j\_c(k)} = P_\xi R_{j\_c}. \tag{5.183}$$

The maximum allowed losses of the switch and diode can be computed from Equation 5.183 as

$$P_{\xi(sw)\,max} = \frac{\Delta\theta_{j\_c(sw)\,max}}{R_{j\_c(sw)}} \quad\text{and}\quad P_{\xi(D)\,max} = \frac{\Delta\theta_{j\_c(D)\,max}}{R_{j\_c(D)}}, \tag{5.184}$$

where

$\Delta\theta_{j\_c(sw)max}$ and
$\Delta\theta_{j\_c(D)max}$ are the switch and diode's maximum allowed junction-to-case temperature as defined by the device cycling capability.

If the maximum allowed device losses are lower than the actual, Equations 5.178–5.180, a new switch and diode has to be selected or the switching frequency reduced. The second option is undesirable since it will lead to the redesign of the ICT, inductor, and capacitors.

The case-to-heat-sink temperature is defined by the module losses and the case-to-heat-sink thermal impedance

$$\Delta\theta_{c\_hs}(s) = P_{\xi(MOD)}(s)Z_{c\_hs}(s) = \left(P_{\xi(sw)}(s) + P_{\xi(D)}(s)\right)\sum_{k=1}^{\infty}\frac{R_{c\_hs(k)}}{1 + s\tau_{(k)}}. \tag{5.185}$$

In the case that the losses are constant, the case-to-heat-sink temperature response is a series of exponential functions

$$\Delta\theta_{c\_hs}(t) = \left(P_{\xi(sw)} + P_{\xi(D)}\right)\sum_{k=1}^{\infty}R_{c\_hs(k)}\left(1 - e^{-\frac{t}{\tau_{c\_hs(k)}}}\right). \tag{5.186}$$

Steady state case-to-heat-sink temperature is determined by the module losses and the total case-to-heat-sink thermal resistance

$$\Delta\theta_{c\_hs} = \left(P_{\xi(sw)} + P_{\xi(D)}\right)\sum_{k=1}^{\infty}R_{c\_hs(k)} = \left(P_{\xi(sw)} + P_{\xi(D)}\right)R_{c\_hs}. \tag{5.187}$$

The heat-sink-to-ambient temperature is defined by the total losses and the heat-sink-to-ambient thermal impedance

$$\Delta\theta_{hs\_amb}(s) = N_M\left(P_{\xi(sw)}(s) + P_{\xi(D)}(s)\right)\sum_{k=1}^{\infty}\frac{R_{hs\_amb(k)}}{1 + s\tau_{hs\_amb(k)}}, \tag{5.188}$$

where

N_M is the number of modules attached to the same heat-sink,
R_{hs_amb} is the heat-sink thermal resistance, (K/W), and
τ_{hs_amb} is the heat-sink thermal time constant, (s).

High power heat-sinks can be approximated by second order thermal impedance. Time constants are usually in the range $\tau_1 = 5$–10 seconds and $\tau_2 = 50$–100 seconds.

In the case that the losses are constant, the heat-sink-to-ambient temperature response is a series of exponential functions

$$\Delta\theta_{hs\_amb}(t) = N_M\left(P_{\xi(sw)} + P_{\xi(D)}\right)\sum_{k=1}^{2}R_{hs\_amb(k)}\left(1 - e^{-\frac{t}{\tau_{hs\_amb(k)}}}\right). \tag{5.189}$$

If the losses are constant over a time that is significantly longer than the device time constant, the junction-to-case temperature is steady state temperature, defined by the losses and total junction-to-case thermal resistance

$$\Delta\theta_{hs\_amb} = N_M\left(P_{\xi(sw)} + P_{\xi(D)}\right)R_{hs\_amb}. \tag{5.190}$$

The switch and diode junction absolute temperature is computed from Equations 5.183, 5.187, and 5.190 as

$$\theta_{j(sw)} = \Delta\theta_{j\_c(sw)} + \Delta\theta_{c\_hs} + \Delta\theta_{hs\_amb} + \theta_{amb}$$
$$= P_{\xi(sw)}R_{j\_c(sw)} + (P_{\xi(sw)} + P_{\xi(D)})(R_{c\_hs} + N_M R_{hs\_amb}) + \theta_{amb}. \quad (5.191)$$

and

$$\theta_{j(D)} = \Delta\theta_{j\_c(D)} + \Delta\theta_{c\_hs} + \Delta\theta_{hs\_amb} + \theta_{amb}$$
$$= P_{\xi(D)}R_{j\_c(D)} + (P_{\xi(sw)} + P_{\xi(D)})(R_{c\_hs} + N_M R_{hs\_amb}) + \theta_{amb}. \quad (5.192)$$

Finally we can find the heat-sink thermal resistance from Equations 5.191 and 5.192 as

$$R_{hs\_amb} = \min \left\{ \begin{array}{l} \left(\dfrac{\theta_{j(sw)\,max} - \theta_{amb} - P_{\xi(sw)}R_{j\_c(sw)}}{(P_{\xi(sw)} + P_{\xi(D)})} - R_{c\_hs} \right)\dfrac{1}{N_M} \\[4mm] \left(\dfrac{\theta_{j(D)\,max} - \theta_{amb} - P_{\xi(D)}R_{j\_c(D)}}{(P_{\xi(sw)} + P_{\xi(D)})} - R_{c\_hs} \right)\dfrac{1}{N_M} \end{array} \right. . \quad (5.193)$$

Thermal design is usually finished with the heat-sink selection. If the heat-sink thermal resistance is too low and the heat-sink design is not realistic, new power semiconductors have to be selected or the switching frequency reduced. The second option is undesirable since it will lead to redesign of the ICT, inductor, and capacitors.

5.6.5.7 Design and Selection Summary

The design and selection process of power semiconductors is summarized in the flow chart depicted in Figure 5.44.

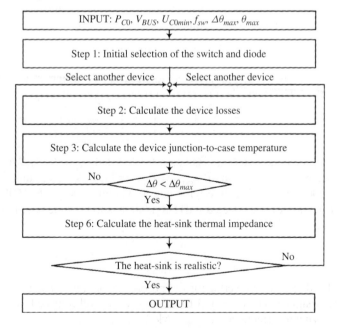

Figure 5.44 The switch/diode selection and thermal design flow chart

5.6.6 Exercises

Exercise 5.1

A power conversion system is equipped with ultra-capacitor energy storage. The system's parameters are as follows: rated power $P_{C0} = 100\,\text{kW}$, rated dc bus voltage $V_{BUS} = 750\,\text{V}$, maximum dc bus voltage $V_{BUSmax} = 800\,\text{V}$, maximum ultra-capacitor voltage $U_{COmax} = 750\,\text{V}$, minimum operating voltage $U_{COmin} = 400\,\text{V}$.

1. Calculate the current rating of the interface dc–dc power converter.
2. Calculate the voltage rating and the initial current rating of the power semiconductor switches if the dc–dc converter is a two-level four-cell interleaved converter. The output current ripple is 30% of the rated average current. Assume the turn-off over-voltage is $\Delta v \leq 200\,\text{V}$.

Solution

1. The interface dc–dc converter current rating is defined by the conversion power and the minimum operating ultra-capacitor voltage, Chapter 4, Section 4.2.3 and Equation 4.12. Hence, the converter current rating is

$$I_{C0(n)} = \frac{P_{C0}}{U_{CO\,min}}. \tag{5.194}$$

Substituting the data given into Equation 5.194 yields

$$\boldsymbol{I_{C0(n)} = 250\,A}.$$

2. The converter switches and diodes are exposed to full dc bus voltage. On top of the dc bus voltage, the devices have to sustain switching over-voltage Δv, in this case $\Delta v \leq 200\,\text{V}$. Thus, the maximum device voltage is

$$\boldsymbol{V_{SW} = 1000\,V}.$$

The device voltage rating is selected to be 1200 V, which is a standard device voltage rating. The device's maximum current is approximately

$$i_{SW\,max} \cong \frac{1}{N} I_{C0(n)} \left[1 + \frac{1}{2} \frac{\Delta i_{0\,max(0)}}{I_{C0(n)}} \right]. \tag{5.195}$$

Substituting the data given into Equation 5.195 yields $i_{swmax} = 71.9\,\text{A}$. The device's initially selected current rating is $I_{sw(N)} = I_{D(N)} = 143.8\,\text{A}$. According to this, we can select a 1200 V/150 A dual IGBT module in 34 mm package, FF150R12RT4, [29]. Final verification of the selected IGBT module has to be done based on the junction temperature.

Exercise 5.2

The interface dc–dc power converter for the power conversion system from the exercise above has to be designed. The converter topology is two-level four-cell interleaved dc–dc converter. The switching frequency is $f_{SW} = 6\,\text{kHz}$.

1. Calculate the worst case scenario dc bus capacitor RMS current.
2. Design the dc bus capacitor for the following objectives: life expectancy $\lambda_{BUS} = 100\,000\,\mathrm{h}$, dc bus voltage ripple $\Delta V_{BUS} = 1\,\mathrm{V}$. The capacitors are air-cooled with natural convection. The ambient temperature is $\theta_{amb} = 60\,°C$.
3. Calculate the dc bus capacitor losses and the capacitor cell surface temperature.

Solution

1. The dc bus capacitor RMS current is,

$$I_{CBUS(RMS)} = \frac{P_{C0}}{V_{BUS}} K_{RMS(p)}, \qquad (5.196)$$

where $k_{RMS(p)}$ is the RMS current ripple coefficient as a function of the duty cycle, the number of interleaved converters, and the output current ripple, Equation 5.75. The output current ripple is given as a design parameter

$$\Delta i_{0\,max(0)} = 0.3 I_{C0(n)}. \qquad (5.197)$$

Substituting the data given into Equation 5.197 yields

$$\Delta i_{0\,max} = 75\,\mathrm{A}.$$

To find the worst case scenario RMS current, we have to find the maximum of the ripple coefficient $k_{RMS(p)}$ from Equation 5.75. This can be done analytically. However, this is not practical. It will be easier to plot the coefficient $k_{RMS(p)}$ versus duty cycle and sort out the maximum point. The coefficient $k_{RMS(p)}$ is computed and plotted in Figure 5.45. From the plot we can note the maximum of the coefficient is approximately

$$k_{RMS(p)} = 0.225.$$

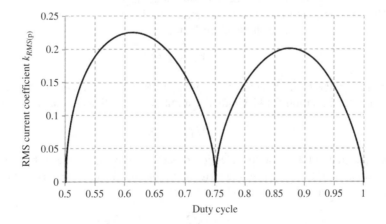

Figure 5.45 The dc bus capacitor RMS current ripple coefficients

Substituting the coefficient into Equation 5.196 yields the dc bus capacitor maximum RMS current

$$I_{CBUS(RMS)} = 30\,\text{A}.$$

The first harmonic of the dc bus capacitor current is N times the switching frequency, where N is the number of interleaved cells. In this case $N=4$, $f_{SW}=5\,\text{kHz}$ and therefore the dominant frequency of the dc bus capacitor current is 20 kHz.

2. Let's first select the voltage rating of the dc bus capacitor cell. Since the maximum dc bus voltage is 800 V, it is obvious that several capacitors have to be series connected. Standard voltage ratings of high power electrolytic capacitors are 350, 400, 450, 500, and 550 V. Let's initially select a 400 V rated capacitor. From Equation 5.143 we find the number of series connected capacitors is

$$N_{BUS} = 2.$$

However, the selected capacitors are at the limit of the voltage rating. There is no design margin. Hence, the selection is not appropriate. We need to select 450 V rated capacitors. In this case, there is a margin of 50 V per capacitor, which is 12.5% of the operating voltage. From the capacitor data sheet we will select large screw terminal electrolytic capacitors 1500 uF/450 V [28]. The selected capacitor has a rated current $I_N = 6.5\,\text{A}$.

Now we have to compute the capacitor equivalent current scaled to the capacitor base frequency. The capacitor scaling factor of the selected capacitors is depicted in Figure 5.37. From this graph we can find that the scaling factor is

$$K = 1.4$$

at frequency $Nf_{SW}=20\,\text{kHz}$. Substituting the scaling factor into Equation 5.139 yields the dc bus capacitor equivalent RMS current

$$I_{CBUS(RMS)} = 21.43\,\text{A}.$$

The next step is to compute the capacitor loading factor from the life expectancy, cooling method, and the ambient temperature. From the capacitor life span graph, Figure 5.46, we can find the loading factor

$$K_L = 1.15.$$

Substituting the loading factor, the equivalent RMS current, and the capacitor rated current into Equation 5.142 yields the number of paralleled capacitors

$$M_{BUS} = 3.$$

The final step in the capacitor selection process is to compute the dc bus capacitance according to the dc bus voltage ripple requirement. The dc bus capacitance is computed from Equations 5.148 and 5.149 as

$$C_{BUS} = \frac{\sqrt{2}}{\Delta v_{BUS\,max}\pi f_{SW}} \frac{P_{C0}}{V_{BUS}} \frac{K_{RMS(p)}}{N}. \tag{5.198}$$

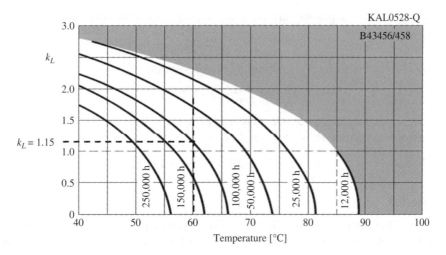

Figure 5.46 The dc bus capacitor RMS current ripple coefficients. Copyright EPCOS AG 2013, with permission

The worst case scenario RMS current ripple coefficient $k_{RMS(p)}$ has been sorted out in the first part of this exercise. Substituting the data given into Equation 5.198 yields the dc bus capacitance

$$C_{BUS} = 673 \ \mu F.$$

The number of paralleled capacitors according to the dc bus voltage ripple criterion is

$$M_{BUS} = 1.$$

Since the current RMS ripple and life span is more critical for the capacitor selection, the final selection of the dc bus capacitors is as follows.

Two series connected ($N_{BUS} = 2$) and three parallel connected ($M_{BUS} = 3$) 1500 μF/450 V EPCOS electrolytic capacitors from family B43456, B43458 are selected for the dc bus capacitor.

3. The ESR of the selected capacitors is ESR $= 120 \ m\Omega$ at 100 Hz. The equivalent resistance of the dc bus is from Equation 5.153

$$ESR = 80 \ m\Omega.$$

Total dc bus capacitor losses and losses per capacitor cell are

$$P_{\xi(BUS)} = 38.37 \ W, \quad P_{\xi(cell)} = 6.39 \, W.$$

To calculate the capacitor surface temperature we need to calculate the thermal resistance from Equation 5.257. Since the capacitors are naturally air cooled, the thermal resistance is

$$R_{ca\_amb} = \frac{1}{17.13A}, \tag{5.199}$$

where A is the capacitor surface. From the capacitor dimensions we have

$$A = 176 \, \text{cm}^2.$$

This gives the surface to ambient thermal resistance

$$R_{ca\_amb} = 3.31 \ \text{K/W}.$$

The capacitor surface temperature is given by Equation 5.259. Substituting the losses and thermal resistance into Equation 5.259 yields the case temperature

$$\theta_{case} = 21.19 + 60 = 81.19 \ ^\circ\text{C}.$$

Exercise 5.3

The output filter for the interface dc–dc power converter from the exercise above has to be designed. The converter topology is a two-level four-cell interleaved dc–dc converter. The switching frequency is $f_{SW} = 6 \, \text{kHz}$.

1. Calculate the filter inductance if the maximum peak-to-peak current ripple is 30% of the ultra-capacitor's maximum current. Assume the inductor is a nonlinear inductor with the coefficient $k = 0.25$.
2. Select the filter capacitor if the maximum RMS current ripple of the ultra-capacitor is $\Delta i_{C0(RMS)} < 2 \, \text{A}$. The ultra-capacitor resistance is $R_{C0} = 0.1 \, \Omega$. The connection inductance can be neglected.
3. Select the filter capacitance if the connection inductance is $L_{0\xi} \geq 1 \, \mu\text{H}$.
4. If the capacitor is selected according to condition 3 above, is there a risk of resonance if the connection inductance is much higher than $1 \, \mu\text{H}$?

Solution

1. The very first step in the output filter design is selection of the filter inductance. The inductance is given by

$$L_{max} = \left(\frac{V_{BUS}}{4 f_{SW} \, \Delta i_{0 \, max}} \right) K_{\Delta i}(d_{CR}). \tag{5.200}$$

where the ripple function $k_{\Delta i}(d_{CR})$ is given by Equation 5.120, while the critical duty cycle d_{CR} is given by Equation 5.122. From the number of interleaved cells $N = 4$ and the inductor characteristic $k = 0.25$ we compute

$$d_{CR} = 0.625 \quad \text{and} \quad k_{\Delta i}(d_{CR}) = 0.078.$$

The maximum peak-to-peak current ripple is computed from Equation 5.197 and the system data

$$\Delta i_{0 \, max} = 75 \, \text{A}.$$

Substituting the above computed parameters into Equation 5.200 yields the zero-load inductance

$$L_{max} = 39 \ \mu\text{H}.$$

2. Let's first calculate the filter gain (attenuation). The filter gain A_T is

$$A_T = \frac{\Delta i_{C0(RMS)}}{\Delta i_{0(RMS)}} = \frac{\Delta i_{C0(RMS)}}{\dfrac{\Delta i_{0\,max}}{2\sqrt{3}}}, \qquad (5.201)$$

where

Δi_{0max} is the maximum peak-to-peak ripple of the inductor current and
$\Delta i_{C0(RMS)}$ is the ultra-capacitor RMS current ripple.

From the data given we compute the filter gain

$$A_T = 0.0922.$$

From Equation 5.157 and the filter gain A_T we find the filter capacitance

$$C_0 = 868\,\mu F.$$

3. In the example above we selected the filter capacitor assuming the connection inductance is zero. However, the ultra-capacitor and the dc–dc interface converter are connected via a pair of cables. If the connection cables are not low inductance twisted cables, the inductance can be significant. On top of this, we have to consider the ultra-capacitor module inductance, which may also be significant. In total, we can assume the connection inductance is more than $1\,\mu H$.

   The interconnection inductance may have a significant influence on the filter performance. From Equation 5.156 we can compute the capacitance versus the connection inductance. The graph is depicted in Figure 5.47.

   From Figure 5.47 we can find the output capacitance is

$$C_0 = 400 \ \mu F.$$

   Please note that the required capacitance is more than twice as low as the capacitance of an ideal case with neglected connection inductance.

4. Let the filter capacitance be selected from the above calculation. From Equation 5.158 we find the critical inductance is

$$L_0 \xi(CR) = 0.158\,\mu H.$$

   Since the critical inductance is much lower than the minimum expected stray inductance, there is no risk of resonance between the filter capacitor and the connection stray inductor.

Exercise 5.4

The interface dc–dc power converter from the Exercise 5.1 has to be designed. The converter topology is a two-level four-cell interleaved dc–dc converter. The switching frequency is $f_{SW} = 6\,kHz$. The ultra-capacitor $C_{C0} = 10\,F$ is charged from the minimum voltage $U_{C0min} = 400\,V$ to maximum voltage $U_{C0max} = 750\,V$ with constant power $P_{C0} = 100\,kW$.

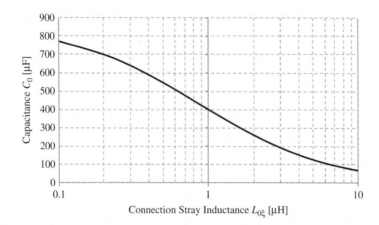

Figure 5.47 The output filter capacitance versus connection stray inductance. $A_T = 0.0922$, $Nf_{SW} = 20\,\text{kHz}$, and $R_{C0} = 0.1\,\Omega$

1. Calculate the conduction and switching losses of the IGBT and FWD. The selected IGBT/FWD have the following parameters: $V_{CE0} = 0.75\,\text{V}$, $r_{CE} = 8\,\text{m}\Omega$, $(E_{ON} + E_{OFF}) = 30\,\text{mJ}$ at $V_N = 600\,\text{V}$ and $I_N = 150\,\text{A}$, $V_{D0} = 0.9\,\text{V}$, $r_D = 5\,\text{m}\Omega$, $E_Q = 10\,\text{mJ}$ at $V_N = 600\,\text{V}$, and $I_N = 150\,\text{A}$.
2. Calculate the converter efficiency when the ultra-capacitor is charged from minimum voltage U_{C0min} to maximum voltage U_{C0max}.
3. Calculate the junction-to-case and case-to-heat-sink temperature of the IGBT and FWD. The IGBT and FWD thermal resistances are $R_{j\_case(IGBT)} = 0.19\,\text{K/W}$, $R_{j\_case(D)} = 0.31\,\text{K/W}$, and $R_{case\_hs} = 0.1\,\text{K/W}$.
4. Calculate the heat-sink maximum temperature if the junction maximum temperature is $\theta_{jmax} = 150\,°\text{C}$.

Solutions

1. Before calculating the IGBT and FWD loses, we need to determine the current and duty cycle profiles. The ultra-capacitor current, which is the converter output current, is defined in Section 5.3.5.2, Equations 5.1 and 5.2,

$$i_{C0} \cong I_{C0\,max}\sqrt{\frac{a}{a + bt}}. \tag{5.202}$$

The ultra-capacitor maximum current and coefficients a and b are

$$I_{C0\,max} = \frac{P_{C0}}{U_{C0\,min}}, \quad a = C_{C0}U_{C0\,min}^2, \quad b = 2P_{C0}, \tag{5.203}$$

where
    P_{C0} is the charging power, (W)
    U_{C0min} is the ultra-capacitor minimum voltage, (V), and
    C_{C0} is the ultra-capacitor capacitance, (F).

The converter duty cycle profile is defined in Section 5.5.3.1, Equations 5.42 and 5.43,

$$d(t) = D_{0\min}\sqrt{\frac{a+bt}{a}},$$ (5.204)

where the coefficients a and b are defined in Equation 5.203. The minimum duty cycle is

$$D_{0\min} = \frac{U_{C0\min}}{V_{BUS}}$$ (5.205)

Now we can calculate the IGBT conduction and switching losses from Equations 5.178 and 5.179,

$$P_{IGBT} = \frac{P_{C0}}{NV_{BUS}}V_{CE0} + \frac{P_{C0}}{NU_{C0\min}}\left(f_{SW}\frac{(E_{ON}+E_{OFF})}{V_N I_N}V_{BUS} + r_{CE}\frac{P_{C0}}{NV_{BUS}}\right)$$

$$\times \sqrt{\frac{a}{a+bt}}.$$ (5.206)

Substituting Equations 5.202–5.205 and the system parameters into Equation 5.206 yields

$$P_{IGBT} = 25 + 110.42\sqrt{\frac{a}{a+bt}}.$$ (5.207)

The FWD losses are defined in Equations 5.178 and 5.180

$$P_{WFD} = -\frac{P_{C0}}{NV_{BUS}}V_{D0} + \frac{P_{C0}}{NU_{C0\min}}$$

$$\times \left(V_{D0} + f_{SW}\frac{E_Q}{V_N I_N}V_{BUS} - r_D\frac{P_{C0}}{NV_{BUS}}\right)\sqrt{\frac{a}{a+bt}}$$

$$+ r_D\left(\frac{P_{C0}}{NU_{C0\min}}\right)^2\frac{a}{a+bt}.$$ (5.208)

Substituting Equations 5.202–5.205 and the system parameters into Equation 5.208 yields

$$P_{FWD} = -30 + 77.1\sqrt{\frac{a}{a+bt}} + 19.5\frac{a}{a+bt}.$$ (5.209)

The IGBT, FWD, and module losses are plotted in Figure 5.48.
2. Module losses are the sum of the IGBT and FWD losses,

$$P_{MOD} = -5 + 177.52\sqrt{\frac{a}{a+bt}}7 + 19.5\frac{a}{a+bt}.$$ (5.210)

The converter (silicon only) is

$$\eta = \frac{P_{C0}}{P_{C0} + NP_{MOD}} = \frac{P_{C0}}{P_{C0} - 20 + 710.08\sqrt{\frac{a}{a+bt}}7 + 78\frac{a}{a+bt}}.$$ (5.211)

The conversion efficiency versus time is plotted in Figure 5.49.

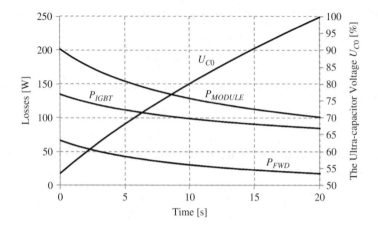

Figure 5.48 The IGBT, FWD, and IGBT module losses versus time

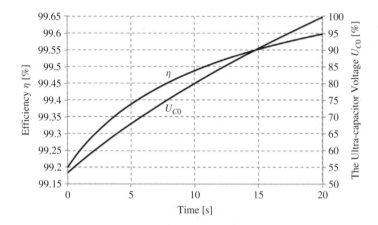

Figure 5.49 The converter excluding ICT and capacitors (silicon only) efficiency versus time

3. The IGBT and FWD junction-to-case steady state temperature has been defined in Equation 5.183. Substituting the device's thermal resistances and losses Equations 5.207 and 5.209 into Equation 5.183 yields

$$\Delta\theta_{IGBT} = 4.75 + 20.97\sqrt{\frac{a}{a+bt}},$$

$$\Delta\theta_{FWD} = -9.3 + 23.9\sqrt{\frac{a}{a+bt}} + 6.05\frac{a}{a+bt}. \tag{5.212}$$

The module-to-heat-sink temperature is given by Equation 5.187. Substituting the thermal resistance and the module losses Equation 5.210 into Equation 5.187 yields

$$\Delta\theta_{MOD} = -0.5 + 17.7\sqrt{\frac{a}{a+bt}}7 + 1.95\frac{a}{a+bt}. \tag{5.213}$$

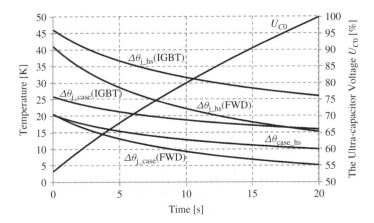

Figure 5.50 The IGBT and FWD junction-to-case temperature and the module case–to-heat-sink temperature versus time

The IGBT and FWD junction temperatures and the module case temperature are plotted in Figure 5.50.

4. The IGBT and FWD junction temperatures are

$$\theta_{j(IGBT)} = \Delta\theta_{MOD} + \Delta\theta_{IGBT} + \theta_{hs} \geq \theta_{max},$$
$$\theta_{j(FWD)} = \Delta\theta_{MOD} + \Delta\theta_{FWD} + \theta_{hs} \geq \theta_{max}. \qquad (5.214)$$

The heat-sink temperature can be computed from Equation 5.214

$$\theta_{hs} = \min[(\theta_{max} - (\Delta\theta_{MOD} + \Delta\theta_{IGBT})), (\theta_{max} - (\Delta\theta_{MOD} + \Delta\theta_{FWD}))]. \qquad (5.215)$$

The worst case scenario is $t = 0$, which gives

$$\boldsymbol{\theta_{hs\,(max)} = 104\,^\circ C.}$$

In order to protect the IGBT and FWD from thermal damage and destruction, the heat-sink temperature must be limited on the maximum temperature computed above.

5.7 Conversion Power Losses: A General Case Analysis

5.7.1 The Origin of the Losses

A power semiconductor device (switch and diode) is not an ideal switch. It has non-zero on-state voltage, non-zero off-state current, and finite transition time from on-to-off and from off-to-on state. All these non-ideals produce losses on the device whenever the device is loaded by current and exposed to voltage. Figure 5.51b shows the sketched waveforms of the device's terminal voltage and current over one switching cycle T_{SW}. We can distinguish four sectors; (1) turn-on, (2) conduction state, (3) turn-off, and (4) blocking state.

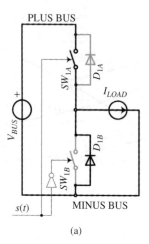

PLUS BUS

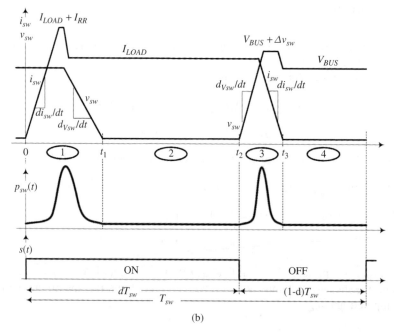

(a)

(b)

Figure 5.51 A power semiconductor's switch terminal voltage, current, and losses over one switching cycle

Power dissipated on the switch chip over one switching cycle T_{SW} can be computed from the definition and referring to the switch terminal voltage/current waveforms Figure 5.51b.

$$P_{SWITCH} = \frac{1}{T_{SW}} \left(\underbrace{\int_0^{t1} v_{sw} i_{sw}\, dt}_{1} + \underbrace{\int_{t1}^{t2} v_{sw} i_{sw}\, dt}_{2} + \underbrace{\int_{t2}^{t3} v_{sw} i_{sw}\, dt}_{3} + \underbrace{\int_{t3}^{T_{SW}} v_{sw} i_{sw}\, dt}_{4} \right),$$

(5.216)

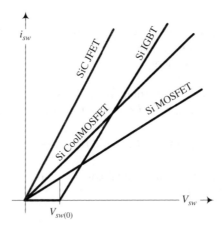

Figure 5.52   Output voltage-to-current characteristics of state of the art power semiconductors

As we will see in the following discussion, the switch's switching losses are mainly determined by the switch terminal voltage/current slope (di_{sw}/dt and dv_{sw}/dt).

5.7.2   Conduction Losses

The switch conduction losses are losses that correspond to the non-zero on-state voltage of the switch that conducts load current $i_{0(1)}$.

$$P_{CON} = \frac{1}{T_{SW}} \int_{t1}^{t2} i_{sw} v_{sw} \, dt = \frac{1}{T} \int_{0}^{dT_{SW}} i_{sw} v_{sw} \, dt \cong V_{sw(0)} I_{sw(AV)} + r_{sw} I_{sw(RMS)}^{2}, \quad (5.217)$$

where
  d is the duty cycle,
  $V_{sw(0)}$ is the switch on-state voltage knee, and
  r_{sw} is the on-state dynamic resistance (Figure 5.52).

5.7.3   Switching Losses

5.7.3.1   Turn-On Losses

Turn-on losses can be computed as

$$P_{ON} = \frac{1}{T_{SW}} \int_{0}^{t1} i_{sw} v_{sw} \, dt$$

$$\cong f_{SW} \frac{1}{2} \underbrace{\left(\frac{(I_{LOAD} + I_{RR})^{2}}{\left(\dfrac{di_{sw}}{dt} \right)_{ON}} V_{BUS} + \frac{V_{BUS}^{2}}{\left(\dfrac{dv_{sw}}{dt} \right)_{ON}} (I_{LOAD} + I_{RR}) \right)}_{E_{ON}} = f_{SW} E_{ON}, \quad (5.218)$$

where

I_{RR} is the FWD reverse recovery current,

$f_{SW} = 1/T_{SW}$ is the switching frequency, and

E_{ON} is the turn-on energy.

Generally, turn-on energy is a function of the dc bus voltage and the load current. However, the losses function can be approximated by a first-order function,

$$P_{ON} = f_{SW} E_{ON}(V_{BUS}, I_{LOAD}) \cong f_{SW} \frac{E_{ON(N)}}{U_N I_N} V_{BUS} I_{LOAD}, \tag{5.219}$$

where $E_{ON(N)}$ is the turn-on energy at nominal conditions U_N and I_N given as a parameter in the switch data sheet [30, 31].

5.7.3.2 Turn-Off Losses

Turn-off losses can be computed as

$$P_{OFF} = \frac{1}{T_{SW}} \int_{t2}^{t3} i_{sw} v_{sw} dt$$

$$\cong f_{SW} \frac{1}{2} \underbrace{\left(\frac{(V_{BUS} + \Delta v_{sw})^2}{\left(\frac{di_{sw}}{dt} \right)_{OFF}} I_{LOAD} + \frac{I_{LOAD}^2}{\left(\frac{dv_{sw}}{dt} \right)_{OFF}} (V_{BUS} + \Delta v_{SW}) \right)}_{E_{OFF}}, \tag{5.220}$$

where

Δv_{sw} is the switch over-voltage induced across the commutation equivalent inductance and

E_{OFF} is the turn-off energy.

Turn-off losses can also be approximated by a first-order equation

$$P_{OFF} = f_{SW} E_{OFF}(V_{BUS}, I_{LOAD}) \cong f_{SW} \frac{E_{OFF(N)}}{U_N I_N} V_{BUS} I_{LOAD}, \tag{5.221}$$

where $E_{OFF(N)}$ is the turn-off energy at nominal conditions U_N and I_N given as parameters in the switch data sheet [30, 31].

5.7.3.3 FWD Reverse Recovery Losses

Turn-off losses of freewheeling diode FWD can also be approximated by a first-order equation

$$P_Q = f_{SW} E_Q(V_{BUS}, I_{LOAD}) \cong f_{SW} \frac{E_{Q(N)}}{U_N I_N} V_{BUS} I_{LOAD}, \tag{5.222}$$

where $E_{Q(N)}$ is the turn-off energy at nominal conditions U_N and I_N given as parameters in the switch data sheet [30, 31].

5.7.4 Blocking Losses

Blocking losses correspond to the non-zero off-state current of the switch that blocks the dc bus voltage V_{BUS}

$$P_{BL} = \frac{1}{T_{SW}} \int_{t3}^{T_{SW}} i_{sw} v_{sw} \, dt \cong \frac{1}{T} \int_{dT}^{T_{SW}} i_{sw} v_{sw} \, dt = (1-d) V_{BUS} I_{sw(\varsigma)}, \qquad (5.223)$$

where $I_{sw(\varsigma)}$ is the switch leakage current.

Blocking losses are significantly lower than conduction and switching losses. However, in some cases, such as high voltage and high temperature applications, the blocking losses may become dominant. If the switch cooling system is not properly designed to take into account the blocking losses, the device may fail due to the thermal run away phenomena [32, 30].

5.7.5 Definition of the Moving Average and RMS Value

The moving average and RMS value of a variable $x(t)$ is defined as the local average and RMS value computed over the switching period T_{sw} that moves as a window over time [33],

$$X_{(AV)}(t) = \frac{1}{T_{SW}} \int_{t}^{t+T_{SW}} x(\tau) \, d\tau,$$

$$X_{(RMS)}(t) = \sqrt{\frac{1}{T_{SW}} \int_{t}^{t+T_{SW}} x^2(\tau) \, d\tau}. \qquad (5.224)$$

5.7.5.1 Remark on the Current and Duty Cycle

As we can see from Equations 5.2, 5.42, 5.43, the ultra-capacitor current and duty cycle are not constant. Does this have influence on the average and RMS moving value Equation 5.224? Yes, it generally does have an influence. And it has to be considered when moving average variables are computed (Equation 5.224). However, the current $i_{C0}(t)$ changes at a rate that is very low in comparison to the switching period. Therefore, we can assume the current is constant over a switching period T_{sw}. The duty cycle also changes at a very low rate. Moreover, if we take into account the nature of a regularly sampled PWM modulator, we will see that the switch duty cycle cannot be changed during a switching cycle. Simple speaking, the duty cycle is sampled and held during a switching period.

5.8 Power Converter Thermal Management: A General Case Analysis

5.8.1 Why is Thermal Management Important?

We clearly understand that thermal management of power conversion systems and components is very important for proper functionality of the system. But, why is thermal management so important? Is it good to have or an absolute must?

As we already discussed in Section 5.7, power semiconductors are not ideal devices. Whenever a switch or a diode is loaded by current and exposed to voltage, a certain amount of energy is realized as heat on the device chip. The heat has to be evacuated to the ambient via a cooling path. Depending on how effectively the heat is evacuated, the chip (junction) temperature rises to certain level. If the junction temperature exceeds the limit, the device may lose blocking capability and fail catastrophically. Thus, to avoid destruction, the device junction temperature must be kept below the maximum allowed temperature. To keep the junction temperature below the limit, the device cooling system must be properly designed. More detail on thermal management of power semiconductors is given in Section 5.8.2.

The situation with passive devices, such as transformers, inductors, and capacitors, is very similar. As discussed in Sections 5.5.3.4, 5.6.1.3, and 5.6.2.2, the current and voltage applied on these devices produce a certain amount of heat on them. The heat is evacuated to the ambient. The device temperature rises to certain level, where the temperature level strongly depends on the device's cooling efficiency. If the temperature exceeds the limit, the device life span may be dramatically shortened or the device may even be damaged or totally destroyed. To prevent this scenario, the device must be properly cooled. More details on thermal management of passive devices is given in Sections 5.8.2.3 and 5.8.4.

5.8.2 Thermal Model of Power Semiconductors

A power semiconductor device is a multi-layer device. A simplified structure of a power semiconductor device attached to a heat-sink is illustrated in Figure 5.53. The device consists of a chip, substrate, case base plate, thermal interface, and heat-sink. Power losses P_ξ are injected into the device chip. The losses are evacuated to the ambient through all the layers. Because the layers are thermally non-ideal, there will be a temperature gradient from the chip temperature to the ambient temperature. The temperature distribution of each layer is indicated on the right side of Figure 5.53.

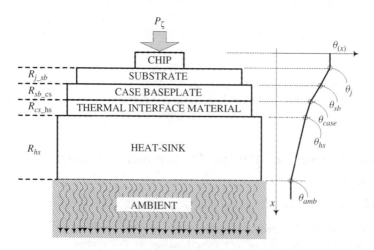

Figure 5.53   Overview of a power semiconductor cross-section with critical temperature points.

The heat transfer from the chip to the ambient, or at least to the heat-sink, is pure conduction. Hence, the thermal system can be described by a partial differential equation

$$\nabla^2\theta + \frac{p_\xi}{\sigma_{th}} = \frac{\rho}{\sigma_{th}}c_{th}\frac{\partial\theta}{\partial t},\qquad(5.225)$$

where

$\quad\nabla^2\quad$ is the Laplacian operator,
$\quad p_\xi\quad$ is the volumetric density of the losses, (W/ m^3)
$\quad\sigma_{th}\quad$ is the thermal conductivity, (W/mK)
$\quad\rho_{th}\quad$ is the material density, (kg/m^3), and
$\quad c_{th}\quad$ is the specific thermal capacitance, (J/kgK).

5.8.2.1   Model of a Single Power Semiconductor Attached to the Heat-Sink

Using an analogy between the heat field and the electromagnetic field, we can derive an equivalent circuit diagram of the power semiconductor module, Figure 5.53. The analogy between the thermal circuit and equivalent electric circuit is summarized in Table 5.5.

Figure 5.54a shows a thermal model of a single power semiconductor attached to a heat-sink. The acronym TIM stands for Thermal Interface Material.

The thermal impedances are defined as follows:

Z_{j_sb} is the device junction to substrate thermal impedance,
Z_{sb_c} is the device substrate to the case base-plate thermal impedance,
Z_{c_hs} is the device case to the heat-sink impedance, and
Z_{hs} is the heat-sink impedance.

Since high power modules consist of four groups of devices, two switches and two freewheeling diodes, all four groups of devices have to be included in the thermal model. Such a model is depicted in Figure 5.54b. However, the question is: how are the losses distributed on all four devices? Let the module current be positive according to the circuit diagram. In an ideal case, the losses will only be distributed to actively conducting devices, in this case the switch SW_A and diode D_B. The complementary devices SW_B and D_A should

Table 5.5   Analogy between thermal and electric circuits

Thermal system	Electric system
Heat, power (W)	Current (A)
Temperature (°C)	Voltage (V)
Thermal resistance (K/W)	Resistance (Ω)
Thermal capacitance (J/K)	Capacitance (F)

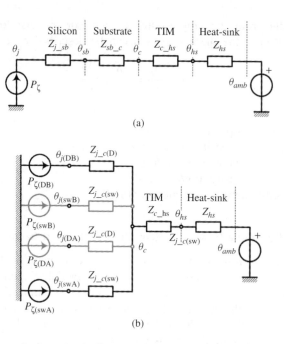

(a)

(b)

Figure 5.54 (a) An equivalent circuit diagram of the thermal model of a single power semiconductor device. (b) Thermal model of a dual power semiconductor module

not generate any losses, and therefore can be neglected by the model. This is indicated with light lines in the model, Figure 5.54b.

Is this assumption correct and, if so, under what conditions? Yes, the above assumption is correct, but only if the blocking losses of the non-active complementary devices (SW_B and diode D_A) can be neglected. However, if the power converter is designed in such a way that the conduction and switching losses are very low, the blocking losses of the complementary devices may become significant. If not taken properly into account, thermal run away may happen and the devices will be quickly destroyed [32, 30].

The device junction-to-case thermal impedance is

$$Z_{j\_c}(s) = \sum_{k=1}^{\infty} \frac{R_{j\_c(k)}}{1 + s\tau_{j\_c(k)}},$$ (5.226)

where

R_{j_c} is the junction-to-case thermal resistance, (K/W)
τ_{j_c} is the junction-to-case thermal time constant, (s).

The junction-to-case thermal impedance of a 1200 V 300 A dual IGBT/FWD is depicted in Figure 5.55. The impedance is approximated by a fourth order RC network.

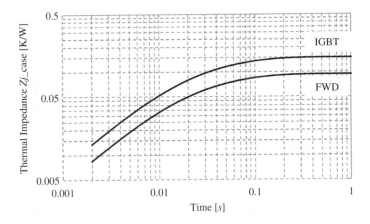

Figure 5.55 The junction-to-case thermal impedance of 1200 V 300 A IGBT/FWD

The case-to-heat-sink impedance is usually modeled by first or second order impedance,

$$Z_{c\_hs}(s) = \sum_{k=1}^{2} \frac{R_{c\_hs(k)}}{1 + s\tau_{c\_hs(k)}}.$$ (5.227)

The thermal resistance

$$R_{c\_hs} = R_{c\_hs(1)} + R_{c\_hs(2)},$$ (5.228)

is the module-to-heat-sink contact resistance, mainly determined by the thermal grease resistance.

5.8.2.2 Model of a Multi-Module Attached to a Massive Heat-Sink

In the previous section we derived a model of a dual power semiconductor module that is attached to a small heat-sink. Very often, in high power applications, we have several power semiconductor modules attached to a massive heat-sink. In that case the analysis and model from the previous section is accurate enough. The difficulties arise from the fact that each individual module injects losses into the heat-sink. Due to the non-ideal distribution of the heat through the heat-sink bulk, there is thermal interaction between the modules. If the module power density is high, in the range of 5–10 W/cm^2, there will be significant temperature variation between the modules. In order to properly design the converter cooling system, this effect should be considered.

Figure 5.56 shows a thermal model of a massive heat-sink and Q power modules attached to the heat-sink. The heat-sink bulk (average) to ambient impedance is modeled by Z_{hs}. The impedance between each module and the heat-sink bulk is Z_{11} to Z_{NMNM} respectively. These impedances are the so-called driving point thermal impedances. Thermal coupling between the modules is modeled by impedances Z_{nk} where $n \in [1, N_M)$ and $k \in [1, N_M)$. The coupling impedances have no physical meaning; they are only a mathematical representation of the coupling phenomena.

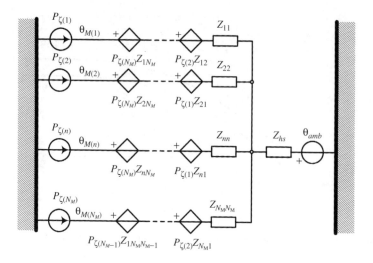

Figure 5.56 Model of a massive heat-sink equipped with N_M IGBT modules

The thermal model depicted in Figure 5.56 is described by the following matrix equation

$$
\begin{bmatrix} \theta_{M(1)} \\ \theta_{M(2)} \\ \cdot \\ \theta_{M(n)} \\ \cdot \\ \theta_{M(N_M)} \end{bmatrix} = \begin{bmatrix} Z_{11} & Z_{12} & \cdot & Z_{1n} & \cdot & Z_{1N_M} \\ Z_{21} & Z_{22} & \cdot & \cdot & \cdot & Z_{2N_M} \\ \cdot & \cdot & \cdot & \cdot & \cdot & \cdot \\ Z_{n1} & \cdot & \cdot & Z_{nn} & \cdot & Z_{nN_M} \\ \cdot & \cdot & \cdot & \cdot & \cdot & \cdot \\ Z_{N_M1} & Z_{N_M2} & \cdot & Z_{N_Mn} & \cdot & Z_{N_MN_M} \end{bmatrix} \begin{bmatrix} P_{\xi(1)} \\ P_{\xi(2)} \\ \cdot \\ P_{\xi(n)} \\ \cdot \\ P_{\xi(N_M)} \end{bmatrix} + \begin{bmatrix} 1 \\ 1 \\ \cdot \\ 1 \\ \cdot \\ 1 \end{bmatrix} \theta_{amb}. \qquad (5.229)
$$

The driving point thermal impedance of a large heat-sink is depicted in Figure 5.57. The impedance is measured from the temperature step response and the model identified using the Excel Solver tool box. The model shown in Figure 5.57 is a second order model with parameters $R_{hs(1)} = 0.035$ K/W, $\tau_{(1)} = 59$ seconds, $R_{hs(1)} = 0.034$ K/W, $\tau_{(1)} = 9$ seconds.

5.8.2.3 Temperature Calculation

The junction-to-heat-sink temperature is

and
$$
\theta_{j(sw)} = P_{\xi(sw)} Z_{j\_c(sw)}(s) + (P_{\xi(sw)} + P_{\xi(D)}) Z_{c\_hs}(s) + \theta_{hs}(s), \qquad (5.230)
$$
$$
\theta_{j(D)} = P_{\xi(D)} Z_{j\_c(D)}(s) + (P_{\xi(sw)} + P_{\xi(D)}) Z_{c\_hs}(s) + \theta_{hs}(s). \qquad (5.231)
$$

The heat-sink temperature θ_{hs} is not a constant parameter. It is a function of the heat-sink geometry, the number of power semiconductors modules attached to the heat-sink, and the losses of each module.

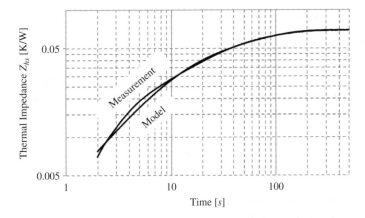

Figure 5.57 Self-driving thermal impedance of a massive heat-sink used in a 200 kW interface dc–dc converter

From the heat-sink thermal model Equation 5.229 we can define the temperature of the heat-sink below the nth power semiconductor module as

$$\theta_{hs(n)} = \sum_{k=1}^{N_M} P_{\xi(k)} Z_{nk}(s) + \theta_{amb}. \tag{5.232}$$

where

$P_{\xi(k)}$ is loss of the kth module, (W)
Z_{nk} is the heat-sink impedance of index n and k, (K/W).

If the device's losses are constant (step functions), the junction temperature versus time can be computed by substituting Equations 5.226 and 5.227 into Equations 5.230 and 5.231

$$\theta_{j(sw)}(t) = P_{\xi(sw)} \sum_{k=1}^{\infty} R_{j\_c(sw)(k)} \left(1 - e^{-\frac{t}{\tau_{j\_c(sw)(k)}}} \right)$$

$$+ (P_{\xi(sw)} + P_{\xi(D)}) \sum_{k=1}^{2} R_{c\_hs(k)} \left(1 - e^{-\frac{t}{\tau_{c\_hs(k)}}} \right) + \theta_{hs}(t), \tag{5.233}$$

and

$$\theta_{j(D)}(t) = P_{\xi(D)} \sum_{k=1}^{\infty} R_{j\_c(D)(k)} \left(1 - e^{-\frac{t}{\tau_{j\_c(D)(k)}}} \right)$$

$$+ (P_{\xi(sw)} + P_{\xi(D)}) \sum_{k=1}^{2} R_{c\_hs(k)} \left(1 - e^{-\frac{t}{\tau_{c\_hs(k)}}} \right) + \theta_{hs}(t). \tag{5.234}$$

The temperature of the heat-sink below the nth power semiconductor module is

$$\theta_{hs(n)} = \sum_{k=1}^{N_M} P_{\xi(k)} \sum_{i=1}^{2} R_{nk(i)} \left(1 - e^{-\frac{t}{\tau_{nk(i)}}} \right) + \theta_{amb}. \tag{5.235}$$

5.8.3   Thermal Model of Magnetic Devices

5.8.3.1   Losses of a Magnetic Device

A magnetic device consists of a ferromagnetic core and one or more windings wound around the core. The core magnetic field is generally a function of time. Simply speaking, it changes over time at a certain frequency between minimum and maximum values. The magnetic field variation produces losses in the core material and the core heats up. The core losses are generally

$$P_{\xi(core)} = m_{core} P_{m(core)}(f, B(t)) = V_{core} P_{v(core)}(f, B(t)), \qquad (5.236)$$

where

$\quad m_{core}$ is the mass of the core, (kg),
$\quad V_{core}$ is the volume of the core, (m^3),
$\quad P_{\xi m(core)}$ is the core specific losses, (W/kg)
$\quad P_{\xi v(core)}$ is the core losses density, (W/m^3), and
$\quad B(t)$ is the core flux density as a function of time, (T).

If the core flux density B is pure sinusoidal, the core losses can be analytically defined by the Steinmez equation [22, 24]

$$P_{\xi(core)} = m_{core} P_m B_{peak}^{\alpha} f^{\beta}, \qquad (5.237)$$

where

$\quad p_m$ is the loss coefficient, (W/kg)
$\quad B_{peak}$ is the peak of the core flux density, (T)
$\quad f$ is the frequency, (Hz), and
$\quad \alpha$ and β are the material coefficients.

In power electronics applications, the flux density is usually not pure sinusoidal. It is rather a dc component with a superimposed ripple. In such a case, the Steinmenz equation (Equation 5.237) has to be modified.

$$P_{\xi(core)} = m_{core} c_c(d) \left[p_m \left(\frac{\Delta B}{2} \right)^{\alpha} f^{\beta} \right], \qquad (5.238)$$

where

$\quad \Delta B$ is the flux density peak-to-peak ripple, (T) and
$\quad c_c(d)$ is a correction factor that takes into account the actual waveform of the flux density.

The winding losses are basically the Joule's losses. The winding is instantaneous and the average losses computed over a period T are

$$P_{\xi(w)} = i_w^2(t) R_{(w)}(t), \qquad (5.239)$$

$$P_{\xi(w)} = \frac{1}{T} \int_t^{t+T} i_w^2(\tau) R_{(w)}(\tau) d\tau, \qquad (5.240)$$

where

$i_w(t)$ is the winding instantaneous current, (A)
$R_w(t)$ is the winding time dependent resistance, (Ω), and
T is the period of moving averaging, (s).

Here we need to clarify the term "time dependent winding resistance." What does time dependent resistance mean? Is it really a resistance that changes over time? In fact the winding resistance is constant, being defined by the material, temperature, and wire size. However, due to the Eddy current effect, the wire current is not distributed equally to the entire cross-section of the wire. The current distribution strongly depends on the material property and frequency of the current [24, 29]. Therefore, equivalent cross-section and wire resistance depends on the current frequency,

$$R_{(w)} = R_{(w)}(\omega). \tag{5.241}$$

where ω is the angular frequency.

According to Fourier transformation, there is a unique relation between a variable in the frequency domain and its mirror in the time domain. Therefore, the frequency dependent resistance of the winding is also time-dependent resistance,

$$R_{(w)}(t) = \frac{1}{2\pi} \int_{-\infty}^{\infty} R_{(w)}(\omega) e^{j\omega t} d\omega. \tag{5.242}$$

Generally speaking, in power electronics applications, the magnetic device current i_w is a periodic current

$$i_w(t) = \sqrt{2} \sum_{k=0}^{\infty} I_{w(RMS)(k)} \sin(k(2\pi f_0)t + \psi_{(k)}), \tag{5.243}$$

where f_0 is the fundamental frequency, (Hz).

Substituting Equations 5.243 into 5.240 yields average losses in the form

$$P_{\xi(w)} = \sum_{k=0}^{+\infty} R_{(w)}(k f_0) \cdot I_{w(RMS)(k)}^2, \tag{5.244}$$

where $R_{(w)}(k f_0)$ is the winding frequency dependent resistance, (Ω).

A detailed derivation of Equation 5.244 from Equations 5.243, 5.242 and 5.240 has already been given in Section 2.6 and Equations 2.97–2.109.

5.8.3.2   Equivalent Thermal Circuit

The model shown in Figure 5.58 is a simplified model. The model has been developed under the following assumptions:

1. The core losses are uniformly distributed within the core.
2. The core and windings are uniformly thermally coupled over the entire core surface.

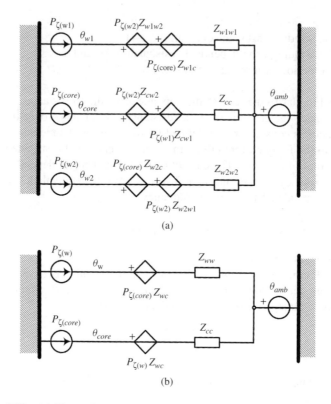

Figure 5.58 (a) Thermal model of an ICT and (b) thermal model of an inductor

Hence, the core losses can be represented by a lumped source $P_{\xi(core)}$ and there will not be hot-spots on the core.

If the ICT is properly designed, the above assumptions will be correct and the model from Figure 5.58 will give reasonably accurate results.

Temperatures of the ICT and inductor can be computed from the following equations

$$\begin{bmatrix} \theta_{w1} \\ \theta_{core} \\ \theta_{w2} \end{bmatrix} = \begin{bmatrix} Z_{w1w1} & Z_{w1\_cc} & Z_{w1w2} \\ Z_{c\_w1} & Z_{cc} & Z_{cw2} \\ Z_{w2w1} & Z_{w2c} & Z_{w2w2} \end{bmatrix} \begin{bmatrix} P_{\xi(w1)} \\ P_{\xi(core)} \\ P_{\xi(w2)} \end{bmatrix} + \begin{bmatrix} 1 \\ 1 \\ 1 \end{bmatrix} \theta_{amb}, \tag{5.245}$$

and

$$\begin{bmatrix} \theta_w \\ \theta_{core} \end{bmatrix} = \begin{bmatrix} Z_{ww} & Z_{wc} \\ Z_{cw} & Z_{cc} \end{bmatrix} \begin{bmatrix} P_{\xi(w)} \\ P_{\xi(core)} \end{bmatrix} + \begin{bmatrix} 1 \\ 1 \end{bmatrix} \theta_{amb}, \tag{5.246}$$

where

Z_{w1w1}, Z_{cc}, Z_{w2w2}, Z_{ww} are the driving point thermal impedances, (K/W) and Z_{w1c}, Z_{w1w2}, Z_{w2c}, Z_{w2w1}, Z_{cw1}, Z_{cw2}, Z_{wc} are the mutual thermal impedances, (K/W).

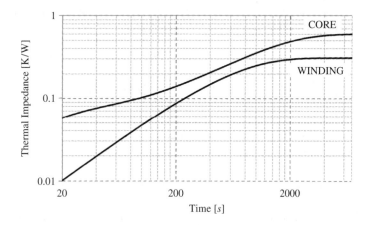

Figure 5.59 Experimentally measured thermal impedances of a high power two-cell ICT

The driving point thermal impedance of a high power two-cell ICT is depicted in Figure 5.59. The impedance is measured from the temperature step response and the model identified using the Excel Solver tool box. The model shown in Figure 5.59 is a second order model with parameters as follows. The core: $R_{cc(1)} = 0.54$ K/, $\tau_{cc(1)} = 1280$ seconds, $R_{cc(1)} = 0.060$ K/W, $\tau_{cc(1)} = 12$ seconds. The windings: $R_{ww(1)} = 0.028$ K/W, $\tau_{cc(1)} = 660$ seconds, $R_{cc(1)} = 0.28$ K/W, $\tau_{cc(1)} = 606$ seconds.

5.8.4   Thermal Model of Power Electrolytic Capacitors

5.8.4.1   The Capacitor Loss Mechanism

The dc bus capacitor is not an ideal capacitor. Whenever current is flowing through the capacitor some amount of energy is dissipated and realized as heat on the capacitor (Figure 5.60). The heat is realized on the capacitor internal resistance, the so-called ESR. Without losing generality of the analysis, we can assume the ESR is a frequency-dependent resistance. Having in mind frequency to time transformation, known best as Fourier transformation, we can write the ESR as time-dependent resistance.

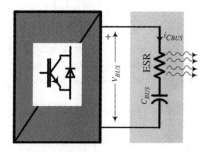

Figure 5.60 The dc bus capacitor with equivalent series resistance as a source of heat

Let the ESR(t) be time-dependent resistance and $i_{CBUS}(t)$ be the dc bus capacitor current. The capacitor's instantaneous and average losses computed over a period T are

$$p_{\xi(BUS)} = i_{CBUS}^2(t)ESR(t), \tag{5.247}$$

$$P_{\xi(BUS)} = \frac{1}{T}\int_t^{t+T} i_{CBUS}^2(\tau)ESR(\tau)d\tau, \tag{5.248}$$

where

    T is the period of moving averaging, (s) and

    $ESR(t)$ is the time-dependent internal resistance of the dc bus capacitor, (Ω).

The meaning of "time-dependent resistance" was given in the previous section, Equations 5.241 and 5.242.

In general power conversion applications, the dc bus capacitor current $i_{CBUS}(t)$ is a periodic function without a dc component,

$$i_{CBUS}(t) = \sqrt{2}\sum_{k=1}^{\infty} I_{C(RMS)(k)} \sin(k(2\pi f_0)t + \psi_{(k)}), \tag{5.249}$$

Substituting Equation 5.249 into Equation 5.248 yields average losses in the form

$$P_{\xi(CBUS)} = \sum_{k=1}^{+\infty} ESR(k f_0) \cdot I_{C(RMS)(k)}^2, \tag{5.250}$$

where f_0 is the fundamental frequency (Hz) and $ESR(kf_0)$ is the capacitor resistance at the kth harmonic of the fundamental frequency, (Ω).

Detailed derivation of Equation 5.250 from Equation 5.249 has been already given in Section 2.6 and Equations 2.97–2.109.

Very often, the capacitor manufacturers give the ESR at base frequency f_B, which is usually 100 or 120 Hz [28]. Let's rewrite Equation 5.250 in the following form

$$P_{\xi(CBUS)} = ESR(f_B)\sum_{k=1}^{+\infty} \frac{I_{C(RMS)(k)}^2}{\left(\frac{ESR(f_B)}{ESR(k f_0)}\right)} = ESR(f_B)\sum_{k=1}^{+\infty} \left(\frac{I_{C(RMS)(k)}}{k_{F(k f_0)}}\right)^2. \tag{5.251}$$

The frequency-dependent coefficients $k_{F(k)}$ are

$$k_{F(k f_0)} = \sqrt{\frac{ESR(f_B)}{ESR(k f_0)}}, \tag{5.252}$$

where $ESR(f_B)$ is the capacitor resistance at the base frequency f_B, (Ω).

The coefficients $k_{F(kf0)}$ are well known as the current scaling factors, usually given as a plot $k_{F(k)}$ versus frequency. An example of 450 V electrolytic capacitors is given in Figure 5.61.

The capacitor equivalent RMS current is computed from Equation 5.251 as

$$I_{CBUS(RMS)} = \sum_{k=1}^{+\infty} \left(\frac{I_{C(RMS)(k)}}{k_{F(kf0)}}\right)^2. \tag{5.253}$$

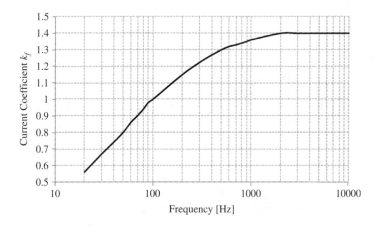

Figure 5.61   The current scaling factors K_f versus frequency [28]

Equation 5.253 is very important and will be used to calculate the capacitor load capability and the life expectancy. More details are given in Section 5.6.3.1.

Using a thermo-electrical analogy, Table 5.5, we can develop an equivalent circuit diagram of an electrolytic capacitor. A full lumped model is depicted in Figure 5.62a, where

$P_{\xi(BUS)}$ is the capacitor losses Equation 5.251, (W),
R_{co_T} is the core-to-terminal thermal resistance, (K/W),
R_{co_ca} is the core-to-case thermal resistance, (K/W),
R_{T_amb} is the terminal-to-ambient thermal resistance, (K/W),
R_{ca_amb} is the case-to-ambient thermal resistance, (K/W),
C_T is the terminal thermal capacitance, (J/K), and
C_{ca} is the case thermal capacitance, (J/K).

The thermal resistances R_{co_T}, R_{co_ca}, and R_{T_amb} are predominantly resistances of heat conduction, while the resistance R_{ca_amb} is the resistance of convection and radiation. The terminal-to-ambient conduction can be neglected in most cases, especially in the case of small snap-in capacitors. In the case of large screw terminal capacitors, the terminal to ambient conduction can be significant. For the sake of simplicity we will neglect it in the further analysis. Under this assumption we can develop a simplified model depicted in Figure 5.62b.

where $R_{in} \cong 1 \sim 3$ K/W is the capacitor internal resistance [34].

The case-to-ambient thermal resistance is

$$R_{ca\_amb} = \frac{1}{Ah_{total}}, \qquad (5.254)$$

Where
    A is the capacitor active surface (m²) and
    h_{total} is the total heat transfer coefficient (W/(m²K)).

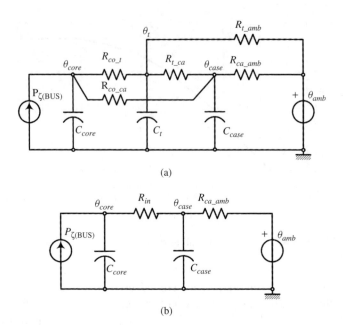

(a)

(b)

Figure 5.62 Thermal model of an electrolytic capacitor. (a) Full lumped parameters model and (b) simplified model

Since the heat is evacuated from the case via convection and radiation, the total heat transfer coefficient is

$$h_{total} = h_{conv} + h_{rad},$$ (5.255)

where
   h_{conv} is the convection heat transfer coefficient and
   h_{rad} is the radiation heat transfer coefficient.

The convection and therefore the total heat transfer coefficient strongly depend on the cooling air velocity. The total heat transfer coefficient can be approximated by

$$h_{total} \cong 5 + 17(v + 0.1)^{0.66},$$ (5.256)

where v is the cooling air velocity (m/s).
   Substituting Equation 5.256 into Equation 5.254 yields

$$R_{ca\_amb} = \frac{1}{A(5 + 17(v + 0.1)^{0.66})},$$ (5.257)

Now, finally having the capacitor thermal model, we can develop the equation of the capacitor temperature. The capacitor case and core temperatures are

$$\theta_{case}(s) = P_{\xi(CBUS)}(s)\frac{1}{1+sR_{co\_ca}C_{core}}\frac{R_{ca\_amb}}{1+sR_{ca\_amb}C_{case}} + \theta_{amb}(s),$$

$$\theta_{core}(s) = P_{\xi(CBUS)}(s)\frac{R_{co\_ca}}{1+sR_{co\_ca}C_{core}}\left(1+\frac{R_{ca\_amb}}{R_{co\_ca}}\frac{1}{1+sR_{ca\_amb}C_{case}}\right) + \theta_{amb}(s).$$

$$(5.258)$$

The core and case steady state temperatures are

$$\theta_{case} = P_{\xi(CBUS)}R_{ca\_amb} + \theta_{amb},$$

$$\theta_{core} = P_{\xi(CBUS)}(R_{co\_ca} + R_{ca\_amb}) + \theta_{amb}.$$

$$(5.259)$$

5.9 Summary

In this chapter, interface dc–dc converters for ultra-capacitor energy storage applications have been discussed. The background of dc–dc power conversion has been given and different converter concepts and topologies have been briefly compared.

- Full power rating versus fractional power rating converters.
- Non-isolated versus isolated converters.
- Two-level versus multi-level converters.
- Single-cell versus multi-cell interleaved converters.
- Cascade boost-buck versus buck-boost converters.

In power conversion and energy storage applications that do not strictly require isolation between the input and output, non-isolated converters are preferred. The main reasons are the better efficiency, higher power density, lower cost, and higher reliability than isolated type converters. If the input to output voltage is greater than 2, isolated type converters are preferred. A two-level multi-cell interleaved non-isolated dc–dc converter has been identified as the best candidate for low voltage (<1000 Vac or 1400 Vdc) power conversion applications.

The theoretical background of two-cell interleaved dc–dc converters has been given. The operating principle has been described in detail. The analysis was then extended to a general N-cell interleaved dc–dc interface converter. The converter's main parameters, such as the output current ripple and the dc bus current ripple, were analyzed and the solution was given as closed form equations.

To give practical value to this chapter, the converter design procedure was discussed in great detail. The design and selection of passive components, such as the output filter inductor and coupling transformer (ICT), the dc bus capacitor and output filter capacitor, have been described step by step. The selection process of power semiconductors was also addressed step by step. At the end of the chapter, thermal management of power converters was discussed in general. The origin of the conversion losses and the loss mechanisms in power semiconductors, transformers, inductors, and capacitors was fully addressed. Thermal modeling of power semiconductors as well as transformers, inductors, and capacitors was discussed in great detail. With the theoretical analysis, practical examples, and exercises presented, this chapter gives a clear overview of how to select and design an interface dc–dc converter for ultra-capacitor energy storage applications.

References

1. W. Li, Lv, X., Deng, Y. *et al.* (2009) A review of non-isolated high step-Up DC/DC converters in renewable energy applications. Applied Power Electronics Conference and Exposition 2009, February, 15–19, 2009, pp. 364–369.
2. Ni, L., D. J. Patterson and J. L. Hudgins, (2012) High power current sensor-less bidirectional 16-phase interleaved DC-DC converter for hybrid vehicle application, *IEEE Transaction Power Electronics*, **27**, 3, 1141–1151.
3. Matsui, M.A., F. Ued, K. Tsubo, and K. Iwata, (1993) A technique of parallel-connections of pulse width modulated NPC inverters by using current sharing reactors Proceedings of the 1993 IEEE-PES Conference, pp. 1246–1251.
4. Panov, Y. and M. M. Jovanovic, (2001) Design consideration for 12-V/1.5-V, 50-A voltage regulator modules, *IEEE Transactions on Power Electronic*, **16**, 6, 776–783.
5. Wong, P.-L., P. Xu, B. Yang, and F. C. Lee, Performance improvements of interleaving VRMs with coupling inductors, *IEEE Transactions on Power Electronics*, **16**, 4, 499–507, 2001.
6. Li, J., C.R. Sullivan, and A. Schultz, (2002) Coupled inductor design optimization for fast-response low-voltage DC-DC converters. Proceedings of the IEEE APEC Conference, 2002, Vol. 2, pp. 817–823.
7. Shen, J., K. Rigbers, and R. W. De Doncker, (2010) A Novel phase-interleaving algorithm for multi-terminal systems, *IEEE Transactions on Power Electronics*, **25**, 3, 741–750.
8. Hirakawa, M., Y. Watanabe, M. Nagano, *et al.* (2009) High power dc-dc converter using extreme close-coupled inductors aimed for electric vehicles. Energy Conversion Conference and Exposition, ECCE 2009, September 20–24, 2009, pp. 1760–1767.
9. Schroeder, J.C., B. Wittig, and F.W. Fuchs, (2010) High efficient battery backup system for lift trucks using interleaved-converter and increased EDLC voltage range. 36th Annual Conference on IEEE Industrial Electronics Society, IECON 2010, November 7–10, 2010, pp. 2334–2339.
10. Hiranuma, S., T. Takayanagi, N. Hoshi, *et al.* (2012) Experimental consideration on dc-dc converter circuits for fuel cell hybrid electric vehicle. Electric Vehicle Conference, IEVC 2012, March 8–12, 2012.
11. Hegazy, O., J. Van Mierlo, and P. Lataire, (2012) Analysis, modeling, and implementation of a multi-device interleaved dc-dc converter for fuel cell hybrid electric vehicles, *IEEE Transactions on Power Electronics*, **27**, 11, 4445–4458.
12. Zhu, J. and A. Pratt, Capacitor ripple current in an interleaved PFC converter, (2009) *IEEE Transactions on Power Electronics*, **24**, 6, 1506–1514.
13. Lee, W., B.-M. Han, and H. Cha, (2011) Battery ripple current reduction in a three-phase interleaved dc-dc converter for 5kW battery charger. Energy Conversion Conference and Exposition, ECCE 2011, September 17–22, 2011, pp. 3535–3540.
14. Zhang, M.T., Y. Jiang, F.C. Lee, and M.M. Jovanovic, (1995) Single-phase three-level boost power factor correction converter. Proceedings of the APEC'95, March 1995, Vol. 1, pp. 434–439.
15. Ruan, X., B. Li, Q. Chen, *et al.* (2008) Fundamental considerations of three-level dc-dc converters: topologies, analyses, and control. IEEE Transactions on Circuits and Systems-I: Regular Papers, December 2008, Vol. 55, No. 11, pp. 3733–3743.
16. Shen, M., F. Z. Peng, and L. M. Leon, (2008) Multilevel dc-dc power conversion system with multiple dc source, *IEEE Transactions on Power Electronics*, **23**, 1, 1190–1197.
17. Grbović, P. J., P. Delarue, P. Le Moigne and P. Bartholomeus, (2010) A bi-directional three-level dc-dc converter for the ultra-capacitor applications, *IEEE Transactions on Industrial Electronics*, **57**, 10, 3415–3430.
18. De Doncker, R. W. A. A., D. M. Divan, and M. H. Kheraluwala, (1991) A three-phase soft-switched high-power-density DC/DC converter for high-power applications, *IEEE Transactions on Industry Applications*, **27**, 1, 63–73.

19. Park, I.G. and S.I. Kim, (1997) Modelling and analysis of multi-interphase transformers for connecting power converters in parallel. Power Electronics Specialists Conference 1997.
20. Forest, F., T. A. Meynard, Member, E. L., V. Costan, E. Sarraute, A. Cunière, and T. Martiré, (2007) Optimization of the supply voltage system in interleaved converters using intercell transformers, *IEEE Transactions on Power Electronics*, **22**, 3, 934–942.
21. Labouré, E., A. Cunière and T. A. Meynard, (2008) A theoretical approach to inter-cell transformers, application to interleaved converters, *IEEE Transactions on Power Electronics*, **23**, 1, 464–474.
22. Tolstov, G. P. (1962) *Fourier Series*, Englewood Cliffs, Prentice-Hall.
23. Forest, F., E. Laboure and T. A. Meynard, (2009) Design and comparison of inductors and intercell transformers for filtering of PWM inverter output, *IEEE Transactions on Power Electronics*, **24**, 3, 812–821.
24. Colonel W. M., T. McLyman, (2011) *Transformer and Inductor Design Handbook*, Electrical and Computer Engineering, Boca Raton, CRC Press, 4th edn.
25. Hitachi Metals POWERLITE® C-Cores, Hitachi Metals www.hitachimetals.com (accessed 22 April 2013)
26. Ruan, X., B. Li, and Q. Chen, (2004) Three-level converters-a new approach for high voltage and high power DC-to-DC conversion. 33rd Annual Power Electronics Specialists Conference, PESC'02, Vol. 2, pp. 663–668.
27. TDK Epcos (2012) Film Capacitors for Industrial Applications, EPCOS Data Book 2012, www.epcos.com (accessed 22 April 2013).
28. TDK Epcos (2012) Aluminum Electrolytic Capacitors, General Technical Information, EPCOS Data Book 2013, www.epcos.com (accessed 22 April 2013).
29. Infineon http://www.infineon.com (accessed 22 April 2013)
30. Grbović, P. J. (2012) Art of control of advanced power semiconductors: from theory to practice. Full day tutorial, IPEMC (ECCE Asia) 2012, IEEE Energy Conversion Congress and Exposition, Harbin, Chine, June 2–6, 2012.
31. Linder, S. (2006) *Power Semiconductors*, Lausanne, EPFL Press, (Collection: Electrical Engineering).
32. Schnell, R. and N. Kaminski, Thermal Runaway During Blocking, Application Note 5SYA 2045-01, ABB Switzerland Ltd, Semiconductors, April 2005.
33. Caliskan, V.A, O. C. Verghese, and A. M. Stankovic, (1999) Multi-frequency averaging of dc-dc converters, *IEEE Transactions on Power Electronics*, **14**, 1, 124–133.
34. Albersten, A. *Electrolytic Capacitor Life Time Estimation*. Jinghai Europe GmbH, www.janghai-europe.com.
35. Karshenas, H. R., H. Daneshpajooh, A. Safaee, P. Jain and A. Bakhshai, (2011) *Bidirectional dc—dc converters for energy storage systems, energy storage in the emerging era of smart grids*, R. Carbone (Ed.)http://www.intechopen.com/books/energy-storage-in-the-emerging-era-of-smart-grids/bidirectional-dc-dcconverters-for-energy-storage-systems (accessed 29 April 2013).
36. Zhang, J., J.-S. Lai, R.-Y. Kim, and W. Yu, (2007) High-power density design of a soft-switching high-power bidirectional dc–dc converter, *IEEE Transaction Power Electronics*, **22**, 4, 1145–1153.
37. Zhao, C., S. D. Round, and J.W. Kolar, (2008) An isolated three-port bidirectional DC-DC converter with decoupled power flow management, *IEEE Transactions on Power Electronics*, **23**, 5, 2443–2453.
38. Inoue S. and H. Akagi, (2007) A bi-directional DC-DC converter for an energy storage system with galvanic isolation, *IEEE Power Electronics*, **22** 6, 2299–2306.
39. Ortiz, G., D. Bortis, J.W. Kolar, and O. Apeldoorn, (2012) Soft-switching techniques for medium-voltage isolated bidirectional DC/DC converters in solid state transformers. Proceedings of the 38th Annual Conference of the IEEE Industrial Electronics Society (IECON 2012), Montreal, Canada, October 25–28, 2012.
40. Van de Bossche, A. and V. C. Valchev, (2005) *Inductors and Transformers for Power Electronics*, Boca Raton, CRC Press.
41. Notaroš, B. M. (2010)*Electromagnetics*, Englewood Cliff, Prentice Hall, 1st edn.
42. Grbović, P.J. (2010) Ultra-capacitor based regenerative energy storage and power factor correction device for controlled electric drives. PhD Dissertation. Laboratoire d'Electrotechnique et d'Electronique de Puissance (L2EP) Ecole Doctorale SPI 072, Ecole Centrale de Lille, July 2010, Lille.
43. Parler, S.G. Jr., (1999) Thermal modeling of aluminum electrolytic capacitors. IEEE Industry Application Society Conference, October 1999.
44. Graovac D. and M. Pürschel, *IGBT Power Losses Calculation Using the Data-Sheet Parameters*, Application Note, V.1.1, Edition 2009-01-29, 2009, Infineon Technologies AG, Neubiberg.

Index

ABB, 315
air
    cooling, 193, 257, 258
    core inductor, 185
    forced-cooled, 207
alloy, 17
aluminum, 66, 315
amorphous, 258
amplifier, 208, 212
amplitude, 61–65
analytical
    determination, 35
    solution, 159, 193
angstroms, 6
angular
    frequency, 61, 66, 253, 307
    velocity, 13
aqueous electrolyte, 26
area-product, 256, 257
attenuation, 276, 291
automatic milling machines, 93
automotive, 17, 76, 77, 149, 155, 215,
    224, 231
autonomous
    power generation, 120
    electric supply, 85
    power generators, 113
    power source, 83, 120
    wind-diesel, 148
auxiliary
    energy system, 148
    power supply, 208, 209

power supplies, 226
switch mode power, 217
average
    losses, 69, 196, 197, 206, 207, 252,
        260, 306, 307, 310
    output, 235, 248
    power, 60, 61, 65, 97, 118, 141, 145,
        206
    value, 80, 242
    values, 235, 242
    variables, 299
    voltage, 156
    voltages, 164
averaging operator, 235, 242

back-to-back, 83, 113, 208, 226
back-up energy, 17
back-up system, 314
balancing
    capability, 180, 181
    circuits, 179
    inductor, 188, 189
    balancing resistance, 188
    resistor, 180
    resistors, 180
    switches, 188
bandwidth, 101, 121
base-plate, 301
bipolar junction transistor (BJT), 78
blade, 105, 108, 109, 111
blades, 111, 112
Boltzmann constant, 26

Ultra-Capacitors in Power Conversion Systems: Applications, Analysis and Design from Theory to Practice, First Edition. Petar J. Grbović.
© 2014 John Wiley & Sons, Ltd. Published 2014 by John Wiley & Sons, Ltd.